Land Surface Observation, Modeling and Data Assimilation

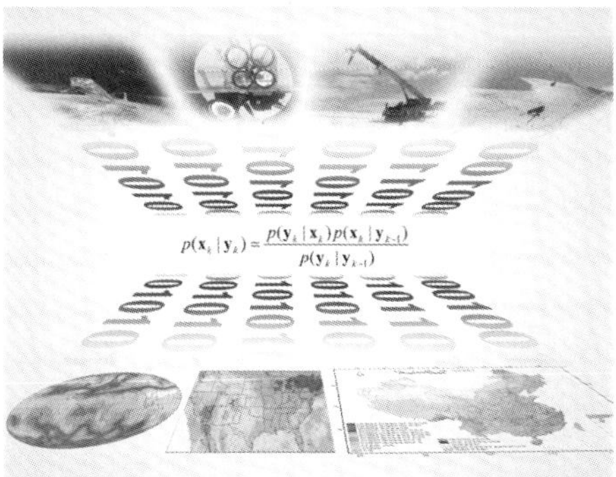

Land Surface Observation, Modeling and Data Assimilation

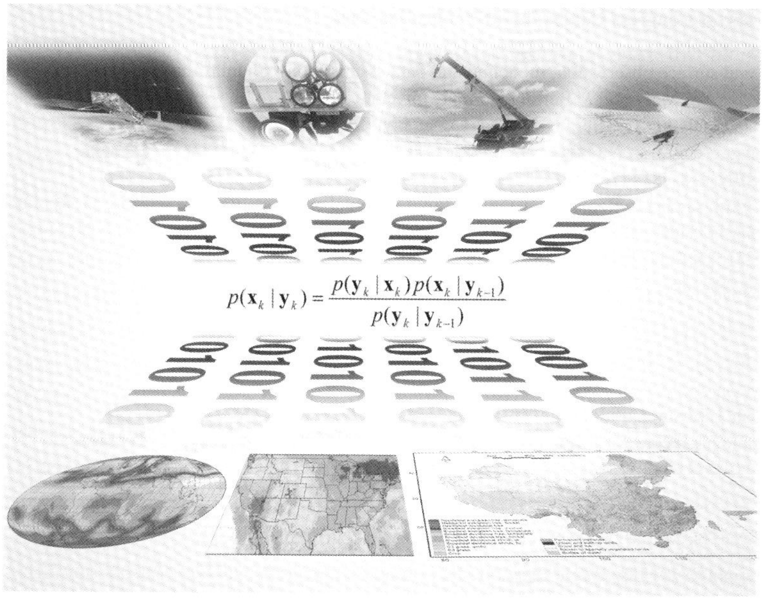

Editors

Shunlin Liang
University of Maryland, USA & Beijing Normal University, China

Xin Li
Cold and Arid Regions Environmental and Engineering Research Institute, Chinese Academy of Sciences, China

Xianhong Xie
Beijing Normal University, China

World Scientific

NEW JERSEY • LONDON • SINGAPORE • BEIJING • SHANGHAI • HONG KONG • TAIPEI • CHENNAI

Preface

Data assimilation is an advanced methodology that integrates various observations into model dynamics for improved model state estimation, model calibration, parameter estimation, uncertainty analysis, and data interpolation. Land data assimilation aims to reconstruct as accurately as possible the state of land surfaces using all available appropriate information, including observations and land models. This methodology plays an increasingly important role in better understanding hydrology and carbon cycles, and in many research activities and endeavors.

To help more researchers, especially young scientists, gain better access to the latest information and techniques in land surface studies and data assimilation, two summer schools were organized. Based on the success of the August 2007 Summer School and Workshop on Land Data Assimilation held in Lanzhou, the Second Summer School on Land Surface Observation, Modeling and Data Assimilation was held on July 13–15, 2010, at Beijing Normal University, China. A group of distinguished scientists was invited to present/deliver lecturers on four topics: remote sensing and other data products; land surface and distributed hydrological modeling; data assimilation methodology; and applications of data assimilation in hydrology, agriculture and ecology. More than 400 participants from dozens of countries attended. This book is based largely on the lecture notes from the summer school in 2010.

This book is organized with a focus on four components: (1) remote sensing observations and data products; (2) land surface modeling; (3) land data assimilation methodology; and (4) applications to climate prediction, agricultural monitoring and hydrology. All these components are connected logically, from data, modeling, data assimilation that integrates data and models, to applications.

The first three chapters deal mainly with satellite data products. Chapter 1 provides an overview of various remote sensing data products suitable for land data assimilation, including the forcing data products (e.g., radiation, air temperature, water vapor) of land surface models, and land surface high-level products (e.g., leaf area index, skin temperature, albedo) and their applications. It also introduces the land cover products for parameterization of the land surface dynamic models.

Since Chinese satellite data have not yet been widely introduced internationally, Chapter 2 describes the Chinese Fengyun 3 satellite series (FY-3), which is the second-generation of polar-orbiting meteorological satellites. The first two were launched in May 2008 and November 2010. Both satellites carry the same 11 sensors from the ultraviolet to microwave spectrum. Chapter 2 describes the sensors, ground segmentation, data service, and applications.

Chapter 3 introduces the data products derived mainly from the NASA Earth Observing System (EOS) program. The NASA satellite data are archived at twelve NASA data centers and can be downloaded from a center where the data are archived or through a centralized search and order system. This chapter briefly describes the land data products from NASA satellite missions as well as land surface assimilation models. The data search and order methods, data tools, as well as sample data applications, also are documented.

Land surface modeling is covered in Chapters 4 and 5. Chapter 4 introduces land surface modeling in drylands and high-elevation regions (mainly with reference to the Tibetan Plateau). The current issues in three aspects and the efforts to resolve them are discussed, including model uncertainty analysis, development of new model parameterization schemes, and their incorporation into land surface models.

Chapter 5 focuses on hydrologic modeling of the rainfall runoff processes. This chapter provides an overview of both the recent developments in models with various levels of sophistication and more advanced parameter estimation methods for model calibration. These two components, in conjunction with observations, are critical in a modeling system for predictions of future flow events.

Land data assimilation methodology is the primary focus of this book, which encompasses both theoretical and applied aspects with various novel approaches that integrate observations and models. Five chapters are devoted to this topic. Chapter 6 introduces the theories and methods of both continuous and sequential data assimilation and their applications in land surface studies. Both the recursive Bayesian filter of sequential data assimilation and the ensemble Kalman filter and their applications are presented.

Error characterization is critical for merging the models and observations effectively and the three chapters cover this topic. Chapter 7 provides an understanding of the impact of poor error assumptions on the performance of land data assimilation systems. This chapter describes recent research results that document the impact of incorrect model and observation error assumptions on the assimilation of surface soil moisture fields into a land surface model. It also reviews recent developments of adaptive filtering systems, which attempt to iteratively estimate modeling and observation error covariance information required by a sequential data assimilation filter.

Chapter 8 addresses the estimated error covariance matrix of both forecast and prior observation in the ensemble Kalman filter assimilation. It presents an adaptive inflation scheme for adjusting these two error covariance matrices by minimizing the -2log (likelihood) of innovation (observation-minus-forecast residuals) for a linear or tangent linear observation operator.

A more comprehensive review of the error related problems for land data assimilation systems is provided in Chapter 9. It exposes the basic components of practical estimation algorithms developed for different data assimilation systems, particularly on error descriptions and parameterizations.

Chapter 10 describes an extended Multiscale Kalman Smoother-based (MKS) framework to estimate the probability distribution of hydrological variables, given predictions from a model, observations and MKS parameters. The Expectation-Maximization (EM) algorithm is coupled to the MKS-based framework for improved estimations of the MKS parameters.

Chapter 11 presents the North American Land Data Assimilation System (NLDAS). It runs multiple land surface models (LSMs), included the Noah, Mosaic, Sacramento Soil Moisture Accounting (SAC-SMA), and Variable Infiltration Capacity (VIC) models, over the Continental United States (CONUS) to generate long-term hourly, 1/8th degree hydrological and meteorological products. NLDAS has also supported other modeling activities including high resolution 1km land surface modeling and the establishment of regional and global land data assimilation systems. The new version, NLDAS-2, operates in both a real-time monitoring mode and an ensemble seasonal hydrologic forecast mode for a variety of applications.

Chapters 12–14 presents three applications of land data assimilation methodology. Chapter 12 reviews recent progress on soil moisture data assimilation through a case study of the soil moisture initialization activities for NASA's seasonal and inter-annual climate prediction, which is one of the first to evaluate such potential using actual remote sensing data together with field observations to characterize model and observation errors.

Chapter 13 presents the agricultural applications of land data assimilation. This chapter starts with a general introduction of the procedure for assimilating remote sensing data into crop simulation models, followed by a brief description of the Decision Support System for Agrotechnology Transfer (DSSAT) model, and the description of several assimilation methods of various degrees of complexity, including direct input, sequential and variational assimilation techniques. Regional application examples also are provided to illustrate the procedure of sensitivity study, cost function construction, crop yield estimation, and hydrological simulation.

The final chapter presents another application of ensemble Kalman filter (EnKF) for state-parameter estimation of hydrologic models. A state-augmentation technique is demonstrated to estimate the states and parameters simultaneously with a simple two-parameter rainfall-runoff model and a distributed hydrological model, Soil and Water Assessment Tool (SWAT).

Acknowledgements

The "2nd Summer School on Land Surface Observing, Modeling and Data Assimilation" would not have been a success without the valuable contributions of many people and agencies. We would like to thank both Dr. Jiancheng Shi from the Institute of Remote Sensing and Digital Earth, Chinese Academy of Sciences (CAS), and Dr. Ailikun from IPO of Monsoon Asia Integrated Regional Study (MAIRS), who were also the organizers of the workshop. We would like to thank the sponsors for their generous support and the International Scientific Committee for its advice and guidance.

The sponsors include the College of Global Change and Earth System Science, Beijing Normal University; the United Nations Educational, Scientific and Cultural Organization (UNESCO) at Beijing; Monsoon Asia Integrated Regional Study (MAIRS); The Asia Pacific Network for Global Change Research (APN); Cold and Arid Regions Environmental and Engineering Research Institute, CAS; International Hydrology Program, Water and Development Information for Arid Lands-A Global Network (UNESCO IHP G-WADI); Institute of Tibetan Plateau Research, CAS; the Institute of Remote Sensing Digital Earth, CAS; and the State Key Laboratory of Remote sensing Sciences, Jointly Sponsored by Beijing Normal University and Institute of Remote Sensing and Digital Earth of Chinese Academy of Sciences.

This summer school was financially supported by the first six sponsors above and also by the research project "Generation and Application of Global Products of Essential Land Variables" (2009AA122100) under the "State Program for High-Tech Research and Development (863 program)". Special thanks to Dr. Bo Jiang, secretary of the summer school, for managing everything in perfect orders, and all the speakers for their outstanding lectures.

As editors of this book, we are deeply grateful for the work of all the contributing authors, three editors from the World Scientific

(Ms. Xiaohong Li, Ms. Lu Shan, and Mr. Steven Shi), and especially Ms. Liya Ma for her patience and supports.

We are greatly indebted to our families for their unwavering encouragement and full support.

Shunlin Liang, College of Global Change and Earth System Science, Beijing Normal University, China; Department of Geographical Sciences, University of Maryland, College Park, USA.

Xin Li, Cold and Arid Regions Environment and Engineering Research Institute, Chinese Academy of Sciences, China.

Xianhong Xie, College of Global Change and Earth System Science, Beijing Normal University, China.

Contents

Part 1

Observation

Chapter 1

Remote Sensing Data Products for Land Surface Data Assimilation System Application

Yunjun Yao*, Shunlin Liang*,† and Tongren Xu‡

*College of Global Change and Earth System Science, Beijing Normal University, Beijing, 100875

†Department of Geographical Sciences, University of Maryland, College Park, MD 20742, USA

‡School of Geography, Beijing Normal University, Beijing, 100875

This chapter introduces the main remote sensing data products and their applications in land surface data assimilation systems. Section 1.1 summarizes the basic concept of data assimilation and the importance of remotely sensed data for data assimilation; Sec. 1.2 describes the forcing data products (radiation, air temperature, water vapor, etc.) of land surface models (LSMs) in data assimilation systems; Sec. 1.3 introduces the land surface remote sensing data products and application for data assimilation; Sec. 1.4 describes the land cover types products for parameterization of the model and Sec. 1.5 is a summary of this chapter.

1.1 Introduction

For global change, remote sensing has provided an important data source. However, remote sensing only provides the instantaneous land surface information, and it is difficult to obtain the continuous spatio-temporal surface status variables, which limits its application in earth sciences. Within the dynamic framework of the surface process model, data assimilation can integrate other surface process models and various observation operators into the forecast system characterized by automatic adjustment of model track to reduce error based on direct and indirect observations from different

S. Liang, X. Li & X. Xie (eds.), *Land Surface Observation, Modeling and Data Assimilation*, 1–44.

resolutions datasets (Li *et al.*, 2007; Liang and Qin, 2008). Therefore, the combination of remote sensing data and process model can make up for the limitation of remotely sensed data, providing a novel approach for the remotely sensed applications of global change.

Data assimilation refers to the method of integrating numerical fitting results of physical process model and observation data by updating system status and variables to improve physical process simulation or forecast accuracy. Remote sensing data assimilation can effectively monitor environmental information related to climate, hydrology, and ecology on the regional and global scale.

The existing assimilation algorithms can be divided into variational assimilation and sequential assimilation. Variational assimilation is based on optimal control theory, which treats analysis values and observed values and their deviations from background fields as the objective function to search for the optimal solution to minimize throughout the entire assimilation period. Sequential assimilation only focuses on the solution of the optimal analysis values for a single observation period, constantly updating the forecast field of the model with newly observed data, to form the initial field of the next moment and to get a new forecast. Thus the forecast can proceed sequentially. Remote sensing data products used in data assimilation system include atmospheric forcing data (radiation, temperature, precipitation, water vapor, etc.), land surface remote sensing data (land surface temperature [LST], surface albedo, leaf area index [LAI], soil moisture, etc.), and data for parameterization of the model (mainly land cover types data). The following sections introduce these three types of data in detail.

1.2 Atmospheric Forcing Data

Atmospheric forcing data are mainly used as input data of land surface models (LSMs) for data assimilation. This section briefly introduces the common atmospheric data retrieved from remote sensing data, including radiation, air temperature, water vapor, precipitation, and other data.

1.2.1 *Radiation data*

Generally, surface radiation consists of shortwave radiation and longwave radiation. Shortwave radiation ($S_{\downarrow}$, insolation), i.e., surface downward shortwave radiation, refers to the total solar radiation reaching to the earth surface in spectral range of 0.15–5.00 μm, including direct solar radiation, scattered solar radiation, and solar radiation after multiple scattering. Surface shortwave radiation, as an important component of surface energy budget and net radiation, has played a vital role in surface radiation balance and energy exchange between earth and atmosphere. Radiation datasets can be divided into ground observation datasets, general circulation models (GCMs) radiation products, and satellite radiation product datasets.

1.2.1.1 *Surface radiation ground observation datasets and GCMs radiation products*

To obtain surface solar radiation, it is essential to carry out radiation measurement at certain meteorological or climatological observation stations. Global representative surface radiation observation networks are as follows: the Global Energy Balance Archive (GEBA) (Gilgen and Ohmura, 1999; Liang, 2004); the Baseline Surface Radiation network (BSRN) (Liang, 2004; Ohmura *et al.*, 1998); the Surface Radiation Budget Network (SURFRAD) (Augustine *et al.*, 2000; Augustine *et al.*, 2005; Liang, 2004); and FLUXNET (Baldocchi *et al.*, 2001; Liang, 2004). The ground measured data collected from these observation networks can be used to validate the accuracy of the empirical models or atmospheric radiative transfer model to forecast the radiation values from other stations.

Almost all of the GCMs include shortwave radiation; however, the spatial resolution of the existing GCMs is relatively low ($>1°$), with moderate temporal resolution (6 h) (Liang, 2004). Most of the GCMs tend to overestimate insolation (Liang, 2004; Wild, 2008). Liang (2004) compared surface downward shortwave radiation from different models and determined its considerable uncertainty.

1.2.1.2 *Satellite radiation datasets*

The estimation of insolation based on satellite remote sensing is the most practical and reliable method because remote sensing provides global cover compared to the ground point observations (Liang *et al.*, 2010), as it can cover nearly all global regions. At present, there are many algorithms available for calculating insolation, but the global insolation products are few. This is largely attributed to the applicability and limitation of the existing algorithms.

Since the 1960s, meteorological satellites have greatly improved our understanding of surface energy budget and radiation balance. There are many typical radiometers and sensors for surface solar radiation observations, such as the Earth Radiation Budget (ERB) sensor carried by Nimbus-7 (Jacobowitz and Tighe, 1984); Clouds and the Earth's Radiant Energy System (CERES) sensor carried by three satellites (Barkstrom and Smith, 1986); and Geostationary Earth Radiation Budget (GERB) sensor carried by Meteosat-8 and Meteosat-9 (Harries *et al.*, 2005). Multi-band sensors are also employed to generate surface solar radiation products, such as Spinning Enhanced Visible and Infrared Imager (SEVIRI) sensors provided by METEOSAT Second Generation (MSG) satellite and GOES-R ABI (Laszlo *et al.*, 2008), and Moderate Resolution Imaging Spectroradiometer (MODIS) sensors (Liang *et al.*, 2006). Three major methods are used to estimate insolation using satellite remote sensing: the statistical regression method, the physical model, and the look-up table, by simulating the interactions among solar radiation, surface, and atmosphere by virtue of the radiative transfer model. The popular satellite radiation products include International Satellite Cloud Climatology project (ISCCP) C1 data (Pinker and Laszlo, 1992); the Global Energy and Water Cycle Experiment (GEWEX) surface radiation data with the spatial resolution of 1°; the Clouds and the Earth's Radiant Energy System (CERES) radiation data with spatial resolution of 140 km; and Global Land Surface Radiation Products (GLASS) based on look-up table (Liang *et al.*, 2006).

By a series of validation and comparison tests, the results show that the values of both ISCCP-BR and ISCCP-PL are 12–16% and 8–12% larger than those of TOMAS-PAR, respectively. The validation at snowless stations in low-latitude regions illustrates that the RMSE of three data products are 28.1%, 13.7%, and 7.2%, respectively, with the biases of +25.9%, +12.0%, and +2.8% (Dye and Shibasaki, 1995; Schroeder *et al.*, 2009). Gui *et al.* (2010) compared the accuracy of the three shortwave radiation products and the estimated radiation based on the look-up table method in different regions, indicating that the accuracy of GWEXE products is higher than that of ISCCP and CERES products (Gui *et al.*, 2010a). Table 1.1 shows the representative global radiation products.

(1) ISCCP

The International Satellite Cloud Climatology Project (ISCCP), which began in 1982, is the first sub-project of World Climate Research Program (WCRP). It provides first-hand and systematic satellite weather nephograms on the global scale. Cloud and cloud properties and seasonal variations on the global scale can be calculated by analyzing satellite

Table 1.1. The representative global radiation products.

Radiation products	Name	Temporal coverage	Spatial resolution	Temporal resolution
PAR	ISCCP BR	1983–2007	280 km	3 h
	ISCCP PL	1983–2007	280 km	3 h
	TOMAS PAR	—	1°	Monthly
	GLASS	2008–2010	5 km	3 h
Insolation	ISCCP C1	1983–2007	280 km	3 h
	GEWEX	1983–2007	1°	3 h
	CERES	2000–	140 km	—
	GCMs	—	>1°	6 h
	GLASS	2008–2010	5 km	3 h

radiance. For ISCCP data products, the spatial resolution of photosynthetically active radiation (PAR) and insolation products is 280 km, with temporal resolution of 3 h and time span from 1983 to 2007. B3, BT are primary radiance products of single satellite in VIS (0.6 μm) and IR (11 μm) channels, respectively, generated by normalized calibration of randomly selected images on 4–7 km pixel scale, spatial resolution reduction to 30 km, and temporal sampling for 3 h. C1, D1, C2, D2 are the synthetic datasets of multiple satellites. Among these, the temporal resolution of CX, DX, C1, D1 are 3 h, and C2, D2 are monthly products. D-series products are the latest cloud parameter datasets based on C-series products by recalibration, modification of cloud detection threshold and radiation algorithm, and further gridding to improve the resolution.

(2) GEWEX

The Global Energy and Water Cycle Experiment (GEWEX) is an international research program initiated by the International Council for Science (ICSU) and the World Meteorological Organization (WMO) to study climate anomalies, long-term forecast, and disaster prevention and reduction for improving food production. The main products of GEWEX in the first stage include global cloud, precipitation, water vapor, surface radiation, and aerosol datasets during 1983–2007, with spatial resolution of 1° and temporal resolution of 3 h. These datasets, as the first global reference data for these variables, provide new understanding of seasonal, inter-annual, and regional climate changes. Currently, many major environmental satellite agencies participate in this program, and these datasets, with a time span of more than 20 years, have provided strong support for current climate assessment and validation of LSMs by combining field observation and remote sensing data.

(3) CERES

CERES is the acronym of the Clouds and the Earth's Radiant Energy System. The spatial resolution of radiation data provided by its radiometer is 140 km, with a time span from 2000 to now. These datasets derived from CERES are mainly used to analyze climate change and provide continuous experimental data for earth radiation budget at the top layer of the atmosphere such as radiation flux. These products also have supplied the long-term data for estimating atmospheric radiation flux and for cloud computing at the top layer of the atmosphere.

(4) GLASS

Liang *et al.* (2006) established the relation between radiation at the top layer of the atmosphere and atmospheric condition index using the look-up table method and then built the correlation between the atmospheric condition index and surface radiation to produce MODIS-PAR products with high resolution (4 km) daily from 2003 to 2005 in North America. The PAR products are released through the Oak Ridge National Lab (ORNL) DAAC (ftp://daac.ornl.gov/data/nacp/MODIS_PAR). Recently, Beijing Normal University produced Global Land Surface Radiation Products (GLASS) remote sensing radiation products based on the production and application of global land surface key variables supported by the key 863 High-Tech Research and Development Program of China. PAR and insolation products from 2008 to 2010 with a spatial resolution of 5 km and temporal resolution of 3 h are being produced (Zhang *et al.*, 2011).

1.2.2 *Air temperature*

Air temperature, also termed as the near-atmospheric temperature, is a physical variable that reflects the degree of hotness and coldness of air variation. Air temperature is one of the important model input parameters for data assimilation and also the key parameter for understanding earth energy, water, and carbon cycles. It has played a significant role for land process models, climate, and numerical weather forecast. Meteorological data are derived from point observation. Spatial interpolation methods are adopted to generate the regional air temperature due to the limited stations and uneven distribution with a poor spatial representative. However, the approaches for weighting optimization are different for different interpolation methods. For the sparse stations, different interpolation results will bring out (Hou *et al.*, 2010).

Remote sensing provides an important tool for near-air temperature retrieval on the regional scale. Extensive progress has been made on the retrieval of LST from thermal remote sensing. Although the retrieval error of LST for uniform land surface is within 1°C, the air temperature directly derived from remote sensing cannot meet the application requirement (Kerr *et al.*, 1992; Wan *et al.*, 2004; Hou *et al.*, 2010). Currently, there are a variety of algorithms for estimating air temperature from satellite remote sensing, and MODIS also provides the air temperature profiles products, but with lower accuracy. Yet many algorithms are still at the experimental stage. The common algorithms for near-air temperature estimation from remote sensing include the following three methods. The first method is to retrieve air temperature using temperature profiles from sounding data acquired from geostationary satellite systems and polar-orbiting sensors, such as AIRS. Based on the statistical relationship between the air profiles and the simulated radiation values, temperature profiles can be acquired from the radiation values of carbon dioxide, water vapor, and some other channels. This regression method is adopted to generate the initial guess field of physical algorithms. Smith and Woolf (1976) described this statistical regression algorithm in detail, which is the most popular method for estimating air temperature profiles from satellite at present. The second method is to directly estimate near-air temperature from the statistical relationship between air temperature and LST retrieved from remote sensing. The third method is based on the "trapezoid" or "triangular" spectral feature space between normalized difference vegetation index (NDVI) and LST.

Chen *et al.* (1983) found that the square of linear regression coefficient (R^2) between the near-air temperature estimated from thermal data of GOES stationary meteorological satellite and the observed air temperature at a height of 1.5 m is 0.76. Horiguchi *et al.* (1991) also pointed out that the biases between the LST derived from stationary meteorological satellites and the ground-measured air temperatures at the height of 1.5 m from many stations vary from 1 to 1.7 K. Green (2002) discovered a significant correlation between the surface temperature retrieved from AVHRR data and the monthly point mean temperature from meteorological stations over Africa and Europe. Recently, Hou *et al.* (2010) used LST derived from NOAA-AVHRR data and the ground-measured temperature to establish remotely sensed statistical method for air temperature estimation under sparse vegetation regions with different ranges of elevation. Although MODIS provides typical remote sensing air temperature profiles products, the point data from meteorological stations are used as inputs for data assimilation due to the lower accuracy of MODIS air temperature products.

1.2.3 *Water vapor*

Water vapor is one of the major meteorological parameters and its content varies dramatically with time and space. The spatio-temporal difference of water vapor content not only affects weather and climate changes but also is an important parameter to study the variation of land surface water and energy cycles. Almost all the current meteorological models can simulate water vapor with relatively low spatial resolution ($>1°$) and bring out the great uncertainty for some models. In many regions, especially secluded areas, where sounding data are not available for the estimation of the entire layer water vapor content, remote sensing has played an important role for acquiring water vapor content of the atmosphere.

Estimating atmospheric water vapor content usually requires thermal infrared (IR) and microwave remotely sensed data. Especially in recent years, the estimation of atmospheric water vapor content based on remote sensing is developing rapidly. With the improvement of the higher precision of satellite sensors, extensive progress has been made on the detection of atmospheric water vapor content using satellite data. Since the launch of meteorological satellites, many researchers are attempted to use satellite data to extract atmospheric water vapor, including the vertical and horizontal distributions of water vapor and their variations with the changes of time (Kaufman and Gao, 1992). For example, the Tropical Rain Measurement Mission (TRMM) provides water vapor profiles and precipitation data with low spatial resolution. The next section will introduce the TRMM precipitation datasets in detail. Multi-angle Imaging Spectro-Radiometer (MISR) and MODIS with high spatial resolution have provided water vapor products, of which MODIS water vapor products have both IR and near-IR products with a spatial resolution of 1 km. MODIS sensors consider various factors and the water-vapor-absorbed channel is sensitive to atmospheric water vapor to make MODIS near-IR products better suited for applications. As early as in 1992, Kaufman and Gao put forward the water vapor retrieval algorithm. Because IR water vapor products depend on the initially selected temperature and humidity profiles, the larger error will occur when LST is close to the atmospheric boundary temperature. Therefore, the near-IR water vapor products are more extensively applied in the retrieval of column water vapor content based on the solar radiation reflected by surface or cloud of near-IR channels.

The physical theory of the estimating atmospheric water vapor content from MODIS near-IR data is to detect the absorption of solar radiation reflected by water vapor. The water vapor content can be approximately calculated using the ratio of reflected solar radiations between atmospheric absorption bands and atmospheric window bands. The total water vapor content can be derived based on the difference of reflected solar radiation between water vapor absorption channel and the non-absorption channel. The center wavelengths of 0.865 μm, 0.905 μm, 0.936 μm, 0.940 μm, and 1.24 μm from MODIS data are used. The ratio between the water vapor absorption channel (0.905 μm, 0.936 μm, and 0.940 μm) and the atmospheric window channel (0.865 μm and 1.24 μm) is selected for water vapor retrieval. This ratio method can partially eliminate the influence of surface reflectance on the transmittance of water vapor as the wavelength changes. Water vapor column water content is estimated by establishing the look-up table between the simulation values based on radiative transfer equation and observed values (Meng *et al.*, 2007).

1.2.4 *Precipitation*

Precipitation refers to all solid and liquid water falling from clouds to the earth's surface (Michaelides *et al.*, 2009). It is an important component of the global water cycle and an important input variable for LSMs. Unlike other meteorological and hydrological variables, rainfall variability in time and space is large and often exhibits non-normal distribution; as a result, it is difficult to measure so far. Precipitation datasets can be divided into three categories: ground-measurements datasets by rain gauge, ground-based radar datasets, and remotely sensed precipitation datasets.

1.2.4.1 *Rain gauge measurement and ground-based radar precipitation datasets*

Precipitation is often acquired using rain gauge and ground-based radar, of which the former is the traditional instrument for ground-measurement of rainfall or snow. There are about 40 types rain gauges in use worldwide. The spatial resolution of the ground-based radar is about 1–2 km, and the revisit period is 15–30 min (Shelton, 2009). Although radar observation is also point observation, it can provide high frequent precipitation data with a wide spatial coverage, making it a very popular observation method.

At present, many countries have established radar observation networks. The US has established a new-generation radar system (NEXRAD) consisting of 159 Doppler weather radars with high resolution (Brown and Lewis, 2005). In Japan, the Japan Meteorological Agency (JMA) set up Automated Meteorological Data Acquisition System (AMeDAS), consisting of 20 ground-based radars for collecting regional weather data and validating weather forecast (Makihara *et al.*, 1996). In Canada, there is a weather radar system constituted by 31 weather radars, covering most regions in the country. So far, rain gauge and ground-based radar have been widely used in various fields, including short-term weather forecast, flood forecast, and hydrological modeling.

1.2.4.2 *Satellite precipitation datasets*

It is difficult to obtain global or regional precipitation data using only rain gauge or ground-based radar. However, satellites prove an effective means to obtain regional precipitation data. Over the past 40 years, precipitation retrieval is mainly based on passive visible (VIS), IR, and microwave remote sensing data from the sensors carried by geostationary (GEO) and low earth orbit (LEO) satellites. In November 1997, the Tropical Rain Measurement Mission (TRMM) carrying the first space-borne precipitation radar (PR) was successfully launched and is now still operating in orbit (Kummerow and Barnes, 1998). Currently, more than 60 precipitation retrieval algorithms based on empirical or physical models (Ebert and Manton, 1998; Kubota *et al.*, 2009) have been developed. Levizzani *et al.* (2007) reviewed the latest sensor technologies and precipitation retrieval algorithms. Table 1.2 shows four categories of retrieval algorithms: visible and infrared radiometer (VIS/IR), passive microwave (PMW), active microwave (AMW, radar), and multi-sensor combination (Multi-sensor). Among these algorithms, the VIS/IR precipitation retrieval algorithm is

Table 1.2. Satellite precipitation retrieval algorithms.

Sensors	Typical algorithms
VIS/IR	GPI; GMSRA; OPI; the Griffith–Woodley algorithm
PWM	The Wilheit algorithm; the SSM/I algorithm; GPROF
AWM	The TRMM standard PR algorithm
Multi-sensor	CMORPH; TMPA; GSMaP

based on the relationship between the air temperature of the cloud-top from IR images and the rainfall probability or intensity; PMW precipitation retrieval algorithm has been developed using microwave radiation equation; and radar precipitation retrieval algorithm is acquired using the estimation of corrected radar reflectance and the vertical profiles of precipitation rate. Later, Huffman *et al.* (2007) put forward the TRMM multi-satellite precipitation algorithm (TMPA). Okamoto *et al.* (2005) developed the GSMaP precipitation retrieval algorithm. TRMM, GSMaP, and GPCP are the main global satellite precipitation datasets, with extensive applications for global precipitation monitoring.

(1) TRMM

Tropical Rainfall Measurement Mission (TRMM) was jointly implemented by NASA and National Space Development Agency of Japan (NASDA) to observe and analyze the precipitation and energy exchange in tropical and sub-tropical regions (Kummerow and Barnes, 1998). TRMM satellite has been operating in low earth and circular orbit, covering 35°N–35°S with orbital inclination of about 35° and height of about 350 km, and can provide extensive data about tropical rainfall (Kummerow and Barnes, 1998). The TRMM scientific data information system (TSDIS) is involved in receiving, real-time processing, and post-processing of TRMM satellite data (http://tsdis.gsfc.nasa.gov/). TRMM data products from December 20, 1997, to now can be downloaded via Internet. There are two TRMM products: satellite standard products and ground validation (GV) products. Satellite standard products can be divided into (1) 1-level products, including calibrated VIRS reflectance, TMI brightness temperature, and other data; (2) 2-level products, including the earth physical parameters, TMI precipitation profiles (2A12), and other data; and (3) 3- and 4-level products, consisting of spatio-temporal averaged precipitation data.

(2) GSMaP

The Global Satellite Mapping of Precipitation (GSMaP) project from November 2002 to March 2008 has been sponsored by Japan Science and Technology Agency (JST). From 2007, the precipitation measurement mission science team of Japan Aerospace Exploration Agency (JAXA) undertook the project. GSMaP used multi-sensor datasets as the input to retrieve precipitation rate. Data sources include MWR datasets from LEO satellites and VIR/IR datasets from GEO satellites. Passive microwave sensors can be divided into both imagers and detectors. The former includes TRMM TMI, AMSR, AMSR-E, and SSM/I, while the latter only includes AMSU-B. IR datasets from GEO satellites are provided by Climate Prediction Center (CPC) and has been integrated into the current version of data processing system (Janowiak *et al.*, 2001). The spatial and temporal resolutions of data are 0.03635° (equal to 4 km in length at the equator) and 30 min, respectively. At present, standard GSMaP data products include GSMaP_TMI, GSMaP_MWR, GSMaP_MWR+, GSMaP_MVK, and GSMaP_MVK+, and these products are derived from different algorithms. GSMaP rainfall products tend to underestimate compared with the ground measurements, but the reason still remains unknown.

(3) GPCP

Global Precipitation Climatology Project (GPCP) was formally launched by the World Climate Research Programme (WCRP) in 1986 to obtain spatially and temporally weighted precipitation analysis products. By integrating observed precipitation from rain gauge and retrieved precipitation by satellite, GPCP has provided "quasi-standard" data for precipitation research. One of the major data sources of GPCP is IR radiation data from polar orbiting satellites, including GOES, GMS, Meteosat, and NOAA. Microwave data has been derived from SSM/I carried by DMSP satellites. GPCP includes three precipitation datasets, namely, the second version of GPCP monthly satellite and rain gauge (SG) precipitation products, SG precipitation products for every five months, and $1° \times 1°$ daily reanalysis precipitation products. All these products can be downloaded from the WMO World Data Center website (http://lwf.ncdc.noaa.gov/oa/wmo/wdcamet-NCDC.HTML). The resolution of 1 DD products is 1° with time coverage since January 1997. The resolution of precipitation products for five months is 2.5° with time coverage since 1979. CMAP rainfall products are the monthly precipitation products with 1° spatial resolution acquired by integrating reanalysis data from microwave sensors (SSMS/I, MSU, etc.), IR sensors (IR and GEO-IR, etc.), rain gauge, and numerical simulation.

1.3 Land Surface Remote Sensing Data Products

Land surface remote sensing products are mainly applied for data assimilation. This section introduces the common land surface remote sensing products, including LST, land surface emissivity, LAI, fraction of absorbed photosynthetically active radiation (FPAR), vegetation indices (VI), and soil moisture (SM).

1.3.1 *Land surface temperature*

Land surface temperature (LST) is the key parameter of climate, hydrological, ecological, and biogeochemical models and data assimilation, determining the longwave radiation of surface radiation budget (Cheng *et al.*, 2010; Li and Becker, 1993; Norman and Becker, 1995). Global or regional surface temperature is detected mainly by IR and passive microwave sensors. The spatial resolution of LST detected by IR sensors is relatively high, ranging from 90 m for ASTER to 1 km for MODIS. In contrast, the penetration ability of IR bands is limited and can only be used to obtain LST in clear sky. Little affected by atmosphere, microwave bands are able to acquire LST throughout the day. Compared to IR bands, the spatial resolution of LST derived by passive microwave is relatively lower, with poor accuracy. So far, the MODIS LST product is one of the most reliable LST products. For homogeneous surfaces, such as water bodies and sand, the retrieval accuracy of LST can be arrived at 1 K (Wan *et al.*, 2002, 2004); whereas the accuracy of LST retrieved from microwave data is far larger than 1 K. LST products can be divided into thermal IR LST and microwave LST.

1.3.1.1 *LST data of thermal IR remote sensing*

Table 1.3 shows the typical IR satellite data sources for LST retrieval. IR data for LST retrieval mainly includes NOAA/AVHRR, Landsat ETM+, MODIS, ASTER, AATSR, ABI/GOES-R, and SEVIRI/MSG. According

Table 1.3. The typical infrared satellite data sources for LST retrieval.

Remote sensing data	Bands	Spatial resolution (m)	Algorithms
AVHRR/NOAA	3, 4, 5	8000	Split window algorithm TISI algorithm
ETM+/Landsat 7	6	15	Single-channel algorithm
MODIS/EOS	20, 22, 23, 29, 31, 32, 33	1000	Split window algorithm day/night algorithm
ASTER/EOS	10,11,12,13,14	90	TES algorithm
AATSR/ENVISAT	6, 7	30	Split window algorithm
ABI/GOES-R	14, 15	1000	Split window algorithm
SEVIRI/MSG	9, 10	5000	Split window algorithm

1.3.4 *Fraction of absorbed photosynthetically active radiation*

The fraction of absorbed photosynthetically active radiation (FPAR) is the proportion of PAR absorbed by vegetation to incident solar radiation. As the primary biophysical parameter, FPAR can be used to estimate the net primary productivity and carbon dioxide absorbed from vegetation. FPAR is an important land surface parameter for crop models, net primary productivity models, climate model, ecological models, water cycle model, carbon cycle models, etc. (Liu *et al.*, 1997; Reich *et al.*, 1999; Cramer *et al.*, 1999; Scurlock *et al.*, 1999; Liang, 2004; Wu *et al.*, 2004).

Remote sensing can capture instantaneous land surface information and provide the data for FPAR and other parameters of broader region and time scope. Some sensors have provided global and regional LAI/FPAR data products, such as VEGETATION (VGT) sensors, advanced very-high-resolution radiometer (AVHRR), and moderate-resolution imaging spectroradiometer (MODIS). FPAR retrieval algorithms can be divided into two types: empirical retrieval algorithm and vegetation-spectral-model-based algorithm. Empirical retrieval algorithm includes LAI-based empirical algorithm and VI-based empirical algorithm; whereas vegetation-spectral-model-based algorithm can be divided into radiative-transfer-model-based algorithm and mixed-vegetation-spectral-model-based algorithm. Among the various FPAR products, radiative-transfer-model-based algorithm is the most widely used method for FPAR retrieval. Table 1.7 lists several typical FPAR products and two global FPAR products are described in the following sections.

1.3.4.1 *MODIS FPAR products*

MODIS provide the long-term earth observation data for dynamically monitoring global surface process and obtaining the information from the lower layer of the atmosphere. MODIS LAI/FPAR products are produced using 3D radiative transfer equations. The canopy structure is the most important variable of 3D radiative transfer model and varies among different types of vegetation. Therefore, three points should be considered when estimating canopy radiation: (1) structure of individual vegetation, trees, or community canopy; (2) vegetation elements (leaves, stems); and (3) optical properties of the land surface. The former relies on the physical properties of vegetation (water and pigment contents, etc.) and (3) atmospheric conditions, which have larger impacts on instantaneous

Table 1.7. Several typical FPAR products.

FPAR products	Spatial coverage	Temporal coverage	Spatial resolution	Temporal resolution	Sensors
MODIS FPAR	Global	2000–	1 km	8 days	TERRA-AQUA/MODIS
MISR	Global	2000–	1.1 km	9 days	—
JRC_FPAR	Europe/Global	—	2 km/10 km	—	—
CYCLOPES MERIS	Global	1997–2003, 2002–	1.2 km	Daily/10 days	SPOT/VEGETATION, NOAA/AVHRR, etc.
GLOBCARBON	Global	1998–2003	10 km	Daily/monthly	ATSR2, MERIS, etc.
SeaWiFs	Global	1997–2004	1.5 km	Daily/10-day/monthly	—

Table 1.8. Classification of land cover types.

Classification	IGBP (Type 1)	UMD (Type 2)	LAI/FPAR (Type 3)	NPP (Type 4)	FT (Type 5)
0	Water body	Water body	Water body	Water body	Water body
1	Evergreen needleleaf forest	Evergreen needleleaf forest	Evergreen needleleaf forest	Evergreen needleleaf forest	Evergreen needleleaf forest
2	Evergreen broadleaf forest	Evergreen broadleaf forest	Shrubland	Evergreen broadleaf forest	Evergreen broadleaf forest
3	Deciduous needleleaf forest	Deciduous needleleaf forest	Broadleaf crop	Deciduous needleleaf forest	Deciduous needleleaf forest
4	Deciduous broadleaf forest	Deciduous broadleaf forest	Savanna	Deciduous broadleaf forest	Deciduous broadleaf forest
5	Mixed forest	Mixed forest	Broadleaf forest	Annual broadleaf vegetation	Shrubland
6	Closed shrubland	Closed shrubland	Needleleaf forest	Annual grass vegetation	Grass
7	Open shrubland	Open shrubland	No vegetation	No vegetation	Cereal crop
8	Woody Savanna	Woody Savanna	Urban	Urban	Broadleaf crop
9	Savanna	Savanna			Urban/build up
10	Grassland	Grassland			Snow/ice
11	Permanent Wetland				Barren and sparse Vegetation
12	Crop land	Crop land			
13	Urban/build up	Urban/build up			
14	Crop land/natural vegetation mosaic				
15	Snow/ice				
16	Barren lands	Barren lands			
254	Unclassified	Unclassified	Unclassified	Unclassified	Unclassified
255	Filled value	Filled value	Filled value	Filled value	Filled value

Source: Referenced from http://datamirror.csdb.cn/modis/files/mod12q1.jsp.

solar radiation. For the initial MODIS LAI/FPAR algorithms, global land vegetation is divided into six vegetation types according to canopy structure. The latest Collection 5 LAI/FPAR products take advantage of the detailed algorithms to improve the quality of LAI/FPAR retrieval.

1.3.4.2 *JRC_FPAR products*

JRC_FPAR developed by European commission Joint Research Center is the FPAR product algorithm for vegetation conditions in Europe. The resolution of JRC_FPAR products on the global scale is 10 km, and 2 km on the European scale. JRC_FPAR as uses the physical model to retrieve FPAR. FPAR simulation is derived from the spectral range of PAR (400–700 nm). Therefore, continuous vegetation canopy model by combination with 6S model has been used to simulate land surface features and FPAR.

FPAR products are assimilated into ecological model and carbon cycle model for estimation of carbon flux. Veroustraete *et al.* (1996) suggested that the integration of FPAR products and ecological model would be a very promising area of scientific research. Stockli *et al.* (2008) assimilated MODIS FPAR and LAI products into phenological model and found that this could well constrain model parameters to obtain a more accurate net ecosystem exchange (NEE). Stockli *et al.* (2011) assimilated MODIS-FPAR and LAI products into the phenological model on the global scale to improve the prediction accuracy of global FPAR and LAI. Knorr *et al.* (2010) constructed the assimilation system of carbon cycle data based on the blade phenological model. By assimilation of FPAR products, this system synchronously optimized 14 phenological parameters and 20 vegetation photosynthetic parameters to reduce the uncertainty of these parameters and improve the accuracy of FPAR simulation.

1.3.5 *Vegetation indices*

Vegetation index (VI) is a simple, effective, and empirical measurement of surface vegetation conditions. Various VIs are established by combining visible and near-IR bands of satellites based on vegetation spectral features. At present, more than 40 VIs have been defined and widely used for global and regional vegetation coverage monitoring. VI based on remote sensing is an effective and convenient tool for monitoring global vegetation. Extensive observation data of land surface have been obtained via satellites, without the limitations of natural and social conditions, serving as an efficient means for monitoring land surface conditions and analyzing the influence of human activities and global or regional climate changes (Guo, 2003).

In 1969, Jordan put forward the first VI — ratio vegetation index (RVI) — using near-IR band and red band reflectances to characterize vegetation growth status. To eliminate the influence of soil, NDVI was designed and put forward to reflect vegetation changes. NDVI can remove the irradiance changes caused by the instrument calibration, solar angle, topography, cloud shadows, and atmospheric conditions, thus enhancing the ability for response to vegetation variation. For this reason, NDVI is the most widely used VI so far. Various satellites have provided NDVI products (Tucker, 1979). To reduce the influence of atmosphere for NDVI, Kaufman and Tanre (1992) developed the atmospheric-resistant vegetation index (ARVI) based on the fact that the red channel was much more affected by atmosphere than the near-IR channel, by correcting the radiation of red channel when defining NDVI. By doing this, the dependence of VIs on atmosphere was greatly reduced. To reduce the interference of soil and vegetation canopy, Huete (1988) proposed the soil-adjusted vegetation index (SAVI) and later developed the modified soil-adjusted vegetation index (MSAVI). Considering the mutual influences between atmosphere and soil, Liu (1995) introduced a feedback component, i.e., enhanced vegetation index (EVI), to modify them simultaneously. EVI reduces the influence of soil and atmosphere using soil adjustment parameter L and atmospheric correction parameters C_1 and C_2.

Both AVHRR-NDVI and MODIS-NDVI are still the most widely used VIs at present. In 1997, the International Geosphere-Biosphere Program (IGBP) completed VI with spatial resolution of 1 km and temporal resolution of 10 days on the global scale. These data and the previously published data formed a complete data sequence: the time coverage was from 1992 to 1996; the spatial resolutions were 1 km, 2 km, 4 km, 8 km, and 16 km, respectively; and the data contained 10 bands (Teillet *et al.*, 2000; Wang and Liu, 2003). Almost all of the remote sensing data can be used to retrieve VIs at present. MODIS VIs, as a group of products, are the representative VIs. The primary inputs for MODIS are NIR and Red bands, with spatial resolution of 250 m and Blue band with spatial resolution of 500 m. The primary products are two VIs with spatial resolution of both 250 m and 1000 m: NDVI (normalized difference VI) and EVI.

Remotely sensed albedo and VIs products are assimilated into crop models to improve the yield prediction. Mo *et al.* (2005) performed linear interpolation of NOAA 10-day composite VIs products over time at first and then retrieved daily LAI data using empirical formula. Directly inputting LAI data into SVAT crop model, they successfully simulated the spatial distribution of per unit area yield for winter wheat and summer maize in North China Plain. Remote sensing data (Dente *et al.*, 2008) or products (Fang *et al.*, 2008) could be assimilated, as well as spectral reflectances (Guerif and Duke 1998; Wang, 2008; Xiao *et al.*, 2009; Xiao *et al.*, 2011a), VIs (Launay and Guerif, 2005; Ma *et al.*, 2008), and back-scattering coefficients (Shen *et al.*, 2009). The radiative transfer model of atmosphere-crop-soil will have to combine with crop model when assimilating spectral reflectances, VIs, or back-scattering coefficients (Liang, 2009).

1.3.6 *Soil moisture*

Soil moisture is a component of energy and water exchange between the hydrosphere, atmosphere, and biosphere and is also an important parameter for characterizing land surface drought. Soil moisture is one of the important parameters for LSM parameterization. The spatial distribution and variations of soil moisture can obviously affect the energy budget of the earth–atmosphere system, soil temperature, and agricultural moisture content.

Remotely sensed data integrates spatial heterogeneity of land surfaces and can be used to retrieve global or regional soil moisture. The widely used optical remote sensing method for soil moisture retrieval is developed from the correlation between visible-near-IR-based band VIs (such as NDVI) and soil moisture. The soil moisture data can also be derived from the relationship among surface temperature, thermal inertia, and soil moisture (Watson, 1971). In addition, the triangular or trapezoid methods from spectral features based on both VIs and LST are also widely used to monitor soil moisture (Sandholt *et al.*, 2002). However, the accuracy and timeliness of optical remote sensing for monitoring soil moisture is limited. Microwave remote sensing can penetrate surface soil and has distinctive superiority in monitoring soil moisture because it is all-weather and all-time. The effective records of passive microwave observation start from the scanning multi-channel microwave radiometer (SMMR) lown on the Nimbus-7 earth satellites launched in 1978. From 1979 to 1987, it observed the entire globe every six days. Currently, typical microwave soil moisture products include SMM, SSM, and AMSR-E. Table 1.4 shows the three products. SMMR was carried by Nimbus-7 satellite, with the spatial resolution of 150 km and the lowest frequency of 6.6 GHz. Many research shows that 6.6 GHz and 10.7 GHz channels under low vegetation are sensitive to soil moisture. SSM/I was launched in 1987 with the lowest frequency of 19.3 GHz, mainly for monitoring vegetation. In terms of its spatial resolution, both SMM and SSM are not suitable for land surface monitoring.

AMSR-E was developed based on both SMM and SSM, with greatly improved spatial resolution. The two sensors have the same instrument parameters, and the greatest difference is that AMSR crosses the equator at about 10:30 a.m., while AMSR-E at about 1:30 p.m. Except the polar regions, AMSR is able to cover the whole globe once in ascending or descending orbit within two days. The spatial and temporal resolutions of AMSR-E soil moisture product are 0.25° and one day, respectively. Owe and van de Griend (1990) developed a large-scale soil moisture retrieval model by establishing a LSM using SMMR-6.6 GHz microwave data. The derived soil moisture based on this model relies on whether the data at sparsely distributed meteorological stations are consistent with surface soil conditions at the time the satellite overpasses this region. However, studies indicate that spatial microwave data has good correlation with soil moisture (Mao, 2007).

Currently, assimilation of soil moisture has been made by assimilating microwave brightness temperature data to improve the forecast accuracy of soil moisture profiles. Houser *et al.* (1998) summarized the available soil moisture data assimilation methods (direct insertion, statistical correction, relaxation method, statistical interpolation, etc.), and constructed soil moisture assimilation scheme using distributed hydrological model TOPLATS (Topmodel-based Land Atmosphere Transfer Scheme) as the model operator. They also assimilated soil moisture retrieved from L-band microwave data (PBMR, push broom microwave radiometer). By comparison, they found that the performance of Newton's method was the best among these methods. Galantowiez *et al.* (1999) designed soil moisture and temperature assimilation scheme by extending Kalman filter, assimilating L-band (1.4 GHz) brightness temperature into the discrete water and heat expansion model. Walker *et al.* (2001) developed the soil moisture profiles assimilation scheme from water and heat transfer model, using ideal data to assimilate surface soil moisture and temperature with direct insertion and extended Kalman filter (EKF), respectively, so as to improve the soil moisture profiles. The results showed that EKF was superior over direct insertion, as the latter failed to improve soil moisture profiles. When there were accurate driving data available, the influence of observation interval on the assimilation was not obvious.

Reichle *et al.* (2001, 2002) constructed a soil moisture data assimilation scheme using both the variational method and the land surface process model, and improved soil moisture estimation through the assimilation of microwave brightness temperature data. The meteorological data land types and soil texture with high resolution were used to drive the land surface process model and assimilate brightness temperature data with low resolution. The results showed that the scheme could effectively improve the accuracy of soil moisture estimation. This provides a new method for scale conversion. Margulis (2002) designed the soil moisture assimilation scheme using Ensemble Kalman filter (EnKF) and NOAH model and performed the L-band brightness temperature assimilation experiment combined with SGP97 experiment data. Compared with land surface measurements, the assimilation results of soil moisture and latent heat flux were obviously better than the results from model simulation, confirming the reliability of EnKF for analyzing nonlinear model operators. Crow and Wood (2003) assimilated microwave brightness temperature data into hydrological model using EnKF, improving the simulation accuracy of the hydrological model. Yang *et al.* (2007) assimilated AMSR-E microwave bright temperature data into SiB2 using double-scanning technology. They not only successfully estimated a variety of model parameters but also improved the accuracy of the model in simulating soil moisture profiles.

1.4 Data for Parameterization of Models

1.4.1 *Land cover types*

Land cover refers to the formed covering of earth's surface under the influence of natural process and human activities, such as surface vegetation, soil, glaciers, lakes, marshes, and roads (Zhao, 2003). Remote sensing technology has witnessed rapid development in recent decades, with extensive applications in land cover classification. Many researchers have attempted to use NDVI to design regional land cover database. But NDVI cannot directly provide land cover types, and it can only quantify land surface biological and physical activities. Townshend (1994) accomplished the supervised classification of South America using global NDVI composite datasets. Although they did not use the ground-measurements to validate their results, they found that the accuracy of trained samples could be effectively improved by adding images for time series. In addition, Koomanoff produced global vegetation map using the annual NDVI composite products of NOAA. The results showed that there were nine vegetation types independent of the phenology of NDVI (Wei and Wang, 2010).

MODIS land cover product is one of the famous land cover datasets. Since its launch in July 1999, the Terra satellite can provide global land cover product with spatial resolution of 1 km for each quarter. Recently, land cover change product with the spatial resolution of 1 km has been developed. Both these products depend on raster database with the spatial resolution of 1 km, generated by MODIS level-2 and -3 level products (Cihlar, 2000). MODIS land cover type products include data classification products extracted from Terra satellite data each year using different classification scheme of land cover characteristics. The basic land cover can be divided into 17 types defined by IGBP (International Geosphere-Biosphere Program), including 11 natural vegetation types, 3 developed lands types, and 3 non-vegetation land types (see Table 1.10). The annual composite raster datasets of MODIS Terra land cover types with spatial resolution of 1 km include 5 different land cover classification systems, and data classification is based on decision tree supervised classification algorithm. The first system is global vegetation classification scheme of IGBP; the second is vegetation classification scheme provided by the University of Maryland (UMD); the third is MODIS LAI/fraction of PAR component (LAI/FPAR) extraction scheme; the fourth is MODIS net primary productivity (NPP) extraction scheme; and the fifth is plant functional type (PFT) classification scheme.

Considering that model parameters vary for different vegetation types, land cover type data is usually taken as the input datasets of models parameterization for remote sensing data assimilation. It has played a significant role in crop growth simulation, land surface water cycle, and land surface process simulation.

1.5 Summary

This chapter reviews the remote sensing products for data assimilation system, mainly including atmospheric forcing data, land surface remote sensing products, and model parameterization data. Among these, atmospheric forcing data mainly act as model input data for the data assimilation system, including radiation data, atmosphere temperature, water vapor, and precipitation, and land surface remote sensing products are the main data sources for data assimilation, including LST, albedo, LAI, FPAR, VI, soil moisture, etc. Data for parameterization of the models only serves as inputs to adjust model parameters, which is mainly land cover type data. These remote sensing products provide abundant data sources for the extensive application of data assimilation in hydrology, meteorology, ecology, and other fields.

Currently, many remote sensing products have been widely used for data assimilation (LST, leaf area index, etc.), but some remote sensing products (such as land surface emissivity) are seldom applied in data assimilation. Moreover, the influence of clouds on various satellite products is great, limiting their application in the data assimilation. To improve the retrieval accuracy of remote sensing products and to provide temporal seamless remote sensing products with certain spatial resolution on a global scale, much work remains to be done.

References

Augustine, J. A., Deluisi, J. and Long, C. N. (2000). SURFRAD-A national surface radiation budget network for atmospheric research. *Bulletin of the American Meteorological Society*, **81**, 2341–2357, (see Table 1.8).

Augustine, J. A., Hodges, G. B., Cornwall, C. R., Michalsky, J. J. and Medina, C. I. (2005). An update on SURFRAD — The GCOS surface radiation budget network for the continental United States. *Journal of Atmospheric and Oceanic Technology*, **22**, 1460–1472.

Bacour, C. and Breon, F. (2005). Variability of biome reflectance directional signatures as seen by POLDER. *Remote Sensing of Environment*, **98**, 80–95.

Baldocchi, D., Falge, E., Gu, L., Olson, R., Hollinger, D., Running, S., Anthoni, P., Bernhofer, C., Davis, K. and Evans, R. (2001). FLUXNET: A new tool to study the temporal and spatial variability of ecosystem-scale carbon dioxide, water vapor, and energy flux densities. *Bulletin of the American Meteorological Society*, **82**, 2415–2434.

Barkstrom, B. R. and Smith, G. L. (1986). The earth radiation budget experiment: Science and implementation. *Reviews of Geophysics*, **24**, 379–390.

Bicheron, P. and Leroy, M. (2000). BRDF signatures of major biomes observed from space. *Journal of Geophysical Research*, **105**, 26669–26681.

Boni, G., Entekhabi, D. and Castelli, F. (2001). Land data assimilation with satellite measurements for the estimation of surface energy balance components and surface control on evaporation. *Water Resources Research*, **37**, 1713–1722.

Bouman B. A. M. (1992). Linking physical remote sensing models with crop growth simulation models, applied for sugar beet. *International Journal of Remote Sensing*, **13**, 2565–2581.

Brown, R. A. and Lewis, J. M. (2005). Path to NEXRAD: Doppler radar development at the National Severe Storms Laboratory. *Bulletin of the American Meteorological Society*, **86**, 1459–1470.

Caparrini, F., Castelli, F. and Entekhabi, D. (2004). Estimation of surface turbulent fluxes through assimilation of radiometric surface temperature sequences. *Journal of Hydrometeorology*, **5**, 145–159.

Chen, E., Allen, L. H., Bartholic, J. F. *et al.* (1983). Comparsion of winter-noctual geostationary satellite infrared-surface temperature with shelter-height temperature in Florida. *Remote Sensing of Environment*, **13**, 313–327.

Chen, J. M., Pavlic, G., Brown, L. *et al.* (2001). Derivation and validation of Canada-wide coarse-resolution leaf area index maps using high-resolution satellite imagery and ground measurements. *Remote Sensing of Environment*, **80**, 165–184.

Cheng, J., Liang, S., Weng, F., Wang, J. and Li, X. (2010). Comparison of radiative transfer models for simulating snow surface thermal infrared emissivity. *IEEE Journal of Selected Topics in Earth Observations and Remote Sensing*, **3**, 323–336.

Cheng, J., Ren, H. Z. and Liang, S. L. (2011). GLASS-global land surface broad-band emissivity product: Algorithm theoretical basis document. Version 1.0., pp. 1–56.

Cihlar, J. (2000). Land cover mapping of large areas from satellites: Status and research priorities. *International Journal of Remote Sensing*, **21**, 1093–1114.

Clevers, J. G. P. W. and van Leeuwen, H. J. C. (1996). Combined use of optical and microwave remote sensing data for crop growth monitoring. *Remote Sensing of Environment*, **56**, 42–51.

Cramer, W., Kicklighter, D. W., Bondeau, A. *et al.* (1999). Comparing global models of terrestrial net primary productivity (NPP): Overview and key results. *Global Change Biology*, **5**, 1–15.

Crow, W. T. and Wood, E. F. (2003). The assimilation of remotely sensed soil brightness temperature imagery into a land surface model using ensemble Kalman filtering: A case study based on ESTAR measurements during SGP97. *Advances in Water Resources*, **26**, 137–149.

Dente, L., Satalino, G., Mattia, F. *et al.* (2008). Assimilation of leaf area index derived from ASAR and MERIS data into CERES-wheat model to map wheat yield. *Remote Sensing of Environment*, **112**, 1395–1407.

Dickinson, R. (1995). Land processes in climate models. *Remote Sensing of Environment*, **51**, 27–38.

Dye, D. G. and Shibasaki, R. (1995). Intercomparison of global PAR data sets. *Geophysical Research Letters*, **22**, 2013–2016.

Ebert, E. E. and Manton, M. J. (1998). Performance of satellite rainfall estimation algorithms during TOGA COARE. *Journal of the Atmospheric Sciences*, **55**, 1537–1557.

Fang, H., Liang, S., Hoogenboom, G. *et al.* (2008) Corn-yield estimation through assimilation of remotely sensed data into the CSM-CERES-maize model. *International Journal of Remote Sensing*, **29**(10), 3011–3032.

Fang, H., Liang, S. and Hoogenboom, G. (2011). Integration of MODIS LAI and vegetation index products with the CSM-CERES-maize model for corn yield estimation. *International Journal of Remote Sensing*, **32**(4), 1039–1065.

Galantowiez, J. F., Entekhabi, D. and Njoku, E. G. (1999). Test of sequential data assimilation for retrieving profile soil moisture and temperature from observed L-band radio brightness. *IEEE Transactions on Geoscience and Remote Sensing*, **37**, 1860–1870.
near IR from EOS/MODIS, *IEEE Transactions on Geoscience and Remote Sensing*, **30**, 871–884.
Gao, F., Schaaf, C., Strahler, A., Roesch, A., Lucht, W. and Dickinson, R. (2005). MODIS bidirectional reflectance distribution function and albedo climate modeling grid products and the variability of albedo for major global vegetation types. *Journal of Geophysical Research*, **110**, D01104.
Geiger, B., Roujean, J., Carrer, D. and Meurey, C. (2005). Product User Manual (PUM) land surface albedo. LSA SAF Internal Documents, 41.
Gilgen, H. and Ohmura, A. (1999). *The Global Energy Balance Archive*. Boston, MA, ETATS-UNIS American Meteorological Society.
Govaerts, Y., Pinty, B., Taberner, M. and Lattanzio, A. (2006). Spectral conversion of surface albedo derived from Meteosat first generation observations. *IEEE Geoscience and Remote Sensing Letters*, **3**, 23–27.
Govaerts, Y. and Lattanzio, A. (2007). Retrieval error estimation of surface albedo derived from geostationary large band satellite observations: Application to Meteosat-2 and Meteosat-7 data. *Journal of Geophysical Research*, **112**, D05102.
Govaerts, Y., Lattanzio, A., Taberner, M. and Pinty, B. (2008). Generating global surface albedo products from multiple geostationary satellites. *Remote Sensing of Environment*, **112**, 2804–2816.
Green, R. (2002). The potential of Pathfinder AVHRR data for providing surrogate climatic variables across Africa and Europe for epidemiological applications. *Remote Sensing of Environment*, **79**(2–3), 166–175.
Products. WCRP Series Report No. 128 and WMO TD-No. 1430, 1–55.
Guerif, M. and Duke, C. (1998). Calibration of the SUCROS emergence and early growth module for sugar beet using optical remote sensing data assimilation. *European Journal of Agronomy*, **9**, 127–136.
Gui, S., Liang, S. and Li, L. (2010a). Evaluation of satellite-estimated surface longwave radiation using ground-based observations. *Journal of Geophysical Research*, **115**(D18), D18214.
Gui, S., Liang, S., Wang, K. and Li, L. (2010b). Validation of three satellite-estimated land surface downward shortwave radiation datasets. *IEEE Geoscience and Remote Sensing Letters*, **7**(4), 776–780.
Guo, N. (2003). Vegetation index and its advances. *Arid Meteorology*, **21**(4), 71–75 (in Chinese).
Harries, J. E., Russell, J. E., Hanafin, J. A. *et al.* (2005). The geostationary earth radiation budget project. *American Meteorological Society*, **86**, 945–960.
Horiguchi, I., Tani, H. and Motoki, T. (1991). Accurate estimation of 1.5 m-height air temperature by GMS IR data. *Proceedings of the 24th International Symposium on Remote Sensing of the Environment*, Rio de Janeiro, pp. 301–307.

Houser, P. R., Shuttleworth, W. J. and Famiglietti, J. S. (1998). Integration of soil moisture remote sensing and hydrologic modeling using data assimilation. *Water Resources Research*, **34**, 3405–3420.

Hou, Y. Y., Zhang, J. H. and Yan, H. (2010). Air temperature retrieval from remote sensing data at regional level. *Meteorological Monthly*, **36**(4), 75–79 (in Chinese).

Huang, C. L., Li, X. and Lu, L. (2008). Retrieving soil temperature profile by assimilating MODIS LST products with ensemble Kalman filter. *Remote Sensing of Environment*, **112**, 1320–1336.

Huete, A. R. (1988). A soil adjusted vegetation index (SAVI) [J]. *Remote Sensing of Environment*, 25, 295–309.

Huffman, G. J., Alder, R., Bolvin, D. T., Gu, G., Nelkin, E. J., Bowman, K. P., Hong, Y., Stocker, E. F. and Wolff, D. B. (2007). The TRMM multisatellite precipitation analysis (TMPA): Quasi-global, multiyear, combined-sensor precipitation estimates at fine scales. *Journal of Hydrometeorology*, **8**, 38–55.

Jacobowitz, H. and Tighe, R. (1984). The earth radiation budget derived from the NIMBUS 7 ERB experiment. *Journal of Geophysical Research*, **89**, 4997–5010.

Janowiak, J., Joyce, R. J. and Yahosh, Y. (2001). A real-time global half-hourly pixel-resolution IR dataset and its applications. *Bulletin of American Meteorological Society*, **82**, 205–217.

Jimenez-Munoz, J. C. and Sobrino, J. A. (2003). A generalized single-channel method for retrieving land surface temperature from remote sensing data. *Journal of Geophysical Research*, **108**(D22), 4688.

Kaufman, Y. J. and Gao, B. C. (1992). Remote sensing of water vapor in the near IR from EOS/MODIS. *IEEE Transactions on Geoscience and Remote Sensing*, **30**(5), 871–884.

Kaufman, Y. J. and Tanre, D. (1992). Atmospherically resistant vegetation index (ARVI) for EOS-MODIS. *IEEE Transactions on Geoscience and Remote Sensing*, **30**, 261–270.

Kerr, Y. H., Lagouarde, J. P. and Imberton, J. (1992). Accurate land surface temperature retrieval from AVHRR data with use of an improved split window algorithm. *Remote Sensing of Environment*, **41**, 197–209.

Kimes, D. S. (1991). Radiative transfer in homogeneous and heterogeneous vegetation canopies. *Photon-Vegetation Interactions: Applications in Optical Remote Sensing and Plant Physiology*, J. R. R. B. Myneni (ed.), pp. 339–388, Springer-Verlag.

Knorr, W., Kaminski, T., Scholze, M., Gobron, N., Pinty, B., Giering, R. and Mathieu, P. (2010). Carbon cycle data assimilation with a generic phenology model. *Journal of Geophysical Research*, **115**.

Kubota, T., Ushio, T., Shige, S., Kida, S., Kachi, M. and Okamoto, K. (2009). Verification of high-resolution satellite-based rainfall estimates around Japan using a gauge-calibrated ground-radar dataset. *Journal of the Meteorological Society of Japan*, **87A**, 203–222.

Kumar, P. and Kaleita, A. L. (2003). Assimilation of near-surface temperature using extended Kalman filter. *Advances in Water Resources*, **26**, 79–93.

Kummerow, C. and Barnes, W. (1998). The Tropical Rainfall Measuring Mission (TRMM) sensor package. *Journal of Atmospheric and Oceanic Technology*, **15**, 809–817.

Laszlo, I., Ciren, P., Liu, H., Kondragunta, S., Tarpley, J. D. and Goldberg, M. D. (2008). Remote sensing of aerosol and radiation from geostationary satellites. *Advances in Space Research*, **41**, 1882–1893.

Launay, M. and Guerif, M. (2005). Assimilating remote sensing data into a crop model to improve predictive performance for spatial applications. *Agriculture, Ecosystems and Environment*, **111**, 321–339.

Leroy, M., Deuzé, J., Bréon, F., Hautecoeur, O., Herman, M., Buriez, J., Tanré, D., Bouffies, S., Chazette, P. and Roujean, J. (1997). Retrieval of atmospheric properties and surface bidirectional reflectances over land from POLDER/ADEOS. *Journal of Geophysical Research*, **102**, 17023.

Levizzani, V., Bauer, P. and Turk, F. J. (2007). *Measuring Precipitation from Space: EURAINSAT and the Future*. Dordrecht: Springer.

Li, X., Huang, C. L., Che, T., Jin, R., Wang, S. G., Wang, J. K., Gao, F., Zhang, S. W., Qiu, C. J., and Wang, C. H., 2007. Development of a Chinese land data assimilation system: Its progress and prospects. *Progress in Nature Science*, 17(8): 881–892.

Liang, S. (2004). *Quantitative Remote Sensing of Land Surfaces*. Hoboken, New Jersey: Wiley & Sons.

Liang, S. (2009). *Quantitative Remote Sensing of Land Surfaces*. China: Science Press (in Chinese).

Liang, S., Fang, H., Chen, M., Shuey, C., Walthall, C., Daughtry, C., Morisette, J., Schaaf, C. and Strahler, A. (2002). Validating MODIS land surface reflectance and albedo products: Methods and preliminary results. *Remote Sensing of Environment*, **83**, 149–162.

Liang, S., Shuey, C., Russ, A., Fang, H., Chen, M., Walthall, C., Daughtry, C. and Hunt, R. (2003). Narrowband to broadband conversions of land surface albedo: II. Validation. *Remote Sensing of Environment*, **84**, 25–41.

Liang, S. L., Zheng, T., Liu, R. G., Fang, H. L., Tsay, S. C. and Running, S. (2006). Estimation of incident photosynthetically active radiation from moderate resolution imaging spectrometer data. *Journal of Geophysical Research — Atmospheres*, **111**, 13.

Liang, S. and Qin, J. (2008). Data assimilation methods for land surface variable estimation. In Liang, S. (ed.), *Advances in Land Remote Sensing: System, Modeling, Inversion and Applications*, pp. 319–339. Dordrecht: Springer.

Liang, S., Wang, K., Zhang, X. and Wild, M. (2010). Review of estimation of land surface radiation and energy budgets from ground measurements, remote sensing and model simulation. *IEEE Journal of Special Topics in Applied Earth Observations and Remote Sensing*, **3**, 225–240.

Li, Z.-L. and Becker, F. (1993). Feasibility of land surface temperature and emissivity determination from AVHRR data. *Remote Sensing of Environment*, **43**, 67–85.

Liu, H. Q. (1995) A feedback based modification of the NDVI to minimize canopy background and atmospheric noise. *IEEE Transactions on Geoscience and Remote Sensing*, **33**: 457–465.

Liu, J., Chen, J. M., Cihlar, J. *et al.* (1997). A process based Boreal ecosystem productivity simulator using remote sensing inputs. *Remote Sensing of Environment*, **62**, 158–175.

Liu, Q., Qu, Y., Wang, L. Z., Liu, N. F., Liu, S. H. and Liang, S. L. (2011). GLASS-global land surface broadband albedo product: Algorithm theoretical basis document. Version 1.0, pp. 1–50.

Lucht, W., Schaaf, C. and Strahler, A. (2002). An algorithm for the retrieval of albedo from space using semiempirical BRDF models. *IEEE Transactions on Geoscience and Remote Sensing*, **38**, 977–998.

Ma, X. L., Wan, Z., Moeller, C. C., Menzel, W. P. and Gumley, L. E. (2002). Simultaneous retrieval of atmospheric profiles, land-surface temperature, and surface emissivity from Moderate-Resolution Imaging Spectrometer themal infrared data: Extension of a two-step physical algorithm. *Applied Optics*, **41**, 909–924.

Ma, Y., Wang, S., Zhang, L. *et al.* (2008). Monitoring winter wheat growth in North China by combining a crop model and remote sensing data. *International Journal of Applied Earth Observation and Geoinformation*, **10**, 426–437.

Maas, S. J. (1988). Using satellite data to improve model estimates of crop yield. *Agronomy Journal*, **80**, 655–662.

Maas, S. J. (1991). Use of remotely-sensed information in plant growth simulation models. *Advances in Agronomy*, **1**, 17–26.

Maignan, F., Bréon, F. and Lacaze, R. (2004). Bidirectional reflectance of earth targets: Evaluation of analytical models using a large set of spaceborne measurements with emphasis on the hot spot. *Remote Sensing of Environment*, **90**, 210–220.

Makihara, Y., Uekiyo, N., Tabata, A. and Abe, Y. (1996). Accuracy of radar-AMeDAS precipitation. *IEICE Transactions on Communications*, **E79-B**, 751–762.

Mao, K. B. (2007). The study of algorithm for retrieving land surface temperature and soil moisture from thermal and microwave data. Ph.D. Dissertation, Chinese Academy of Sciences, pp. 35–46 (in Chinese).

Margulis, S. A., McLaughlin, D., Entekhabi, D. and Dunne, S. (2002). Land data assimilation and estimation of soil moisture using measurements from the Southern Great Plains 1997 field experiment. *Water Resources Research*, **38**, 1229.

Martonchik, J., Diner, D., Kahn, R., Ackerman, T., Verstraete, M., Pinty, B. and Gordon, H. (2002a). Techniques for the retrieval of aerosol properties over land and ocean using multiangle imaging. *IEEE Transactions on Geoscience and Remote Sensing*, **36**, 1212–1227.

Martonchik, J., Diner, D., Pinty, B., Verstraete, M., Myneni, R., Knyazikhin, Y. and Gordon, H. (2002b). Determination of land and ocean reflective, radiative, and biophysical properties using multiangle imaging. *IEEE Transactions on Geoscience and Remote Sensing*, **36**, 1266–1281.

Martonchik, J., Pinty, B. and Verstraete, M. (2002c). Note on an improved model of surface BRDF-atmospheric coupled radiation. *IEEE Transactions on Geoscience and Remote Sensing*, **40**, 1637–1639.

McMillin, L. M. (1975). Estimation of sea surface temperature from two infrared window measurements with different absorption. *Journal of Geophysical Research*, **20**, 11587–11601.

Meng, X. H., Lu, S. H. and Zhang, T. T. (2007). Testing, improvement and application of MODIS near infrared water vapor products — Taking Jinta oasis in Heihe River Basin as a case study. *Journal of Infrared and Millimeter Waves*, **26**(2), 107–111 (in Chinese).

Michaelides, S., Levizzani, V., Anagnostou, E., Bauer, P., Kasparis, T. and Lane, J. E. (2009). Precipitation: Measurement, remote sensing, climatology and modeling. *Atmospheric Research*, **94**, 512–533.

Mo, X., Liu, S., Lin, Z. *et al.* (2005). Prediction of crop yield, water consumption and water use efficiency with a SVAT-crop growth model using remotely sensed data on the North China plain. *Ecological Modelling*, **183**, 301–322.

Muller, J.-P. (2008). ATBD: BRDF correction and albedo retrieval. Available at: www.brockmann-consult.de/albedomap/documentation.html.

Myneni, R. B., Asrar, G. and Gerstl, S. A. W. (1990). Radiative transfer in three-dimensional leaf canopies. *Transport Theory and Statistical Physics*, **19**, 205–250.

Norman, J. M. and Becker, F. (1995). Terminology in thermal infrared remote sensing of natural surfaces. *Remote Sensing Reviews*, **12**, 159–173.

NASA AMSR-E. Available at: http://wwwghcc.msfc.nasa.gov/AMSR/.

Ohmura, A., Dutton, E. G., Forgan, B., Fröhlich, C., Gilgen, H., Hegner, H., Heimo, A., König-Langlo, G., McArthur, B., Müller, G., Philipona, R., Pinker, R., Whitlock, C. H., Dehne, K. and Wild, M. (1998). Baseline surface radiation network (BSRN/WCRP): New precision radiometry for climate research. *Bulletin of the American Meteorological Society*, **79**, 2115–2136.

Okamoto, K., Iguchi, T., Takahashi, N., Iwanami, K. and Ushio, T. (2005). The global satellite mapping of precipitation (GSMaP) project. *25th IGARSS Proceedings*, pp. 3414–3416.

Owe, M. and van de Griend, A. A. (1990). Daily surface soil moisture model for large area semi-arid land application with limited climate data. *Journal of Hydrology*, **121**, 119–132.

Prince, S. D. and Coward, S. N. (1995). Global primary production: A remote sensing approach. *Journal of Biogeography*, **22**, 2829–2849.

Pinker, R. T. and Laszlo, I. (1992). Global distribution of photosynthetically active radiation as observed from satellites. *Journal of Climate*, **5**, 56–65.

Pinty, B., Roveda, F., Verstraete, M., Gobron, N., Govaerts, Y., Martonchik, J., Diner, D. and Kahn, R. (2000a). Surface albedo retrieval from Meteosat 1: Theory. *Journal of Geophysical Research*, **105**, 18099.

Pinty, B., Roveda, F., Verstraete, M., Gobron, N., Govaerts, Y., Martonchik, J., Diner, D. and Kahn, R. (2000b). Surface albedo retrieval from Meteosat 2: Applications. *Journal of Geophysical Research*, **105**, 18113.

Prince, S. D. and Coward, S. N. (1995). Global primary production: A remote sensing approach. *Journal of Biogeography*, **22**, 2829–2849.

Qin J., Liang, S., Liu, R., Zhang, H. and Hu, B. (2007). A weak-constraint-based data assimilation scheme for estimating surface turbulent fluxes. *IEEE Geoscience and Remote Sensing Letters*, **4**(4), 649–653.

Qin, Z. and Karnieli, A. (2001). Mono-window algorithm for retrieving land surface temperature from Landsat TM data and its application to the Israel–Egype border region. *International Journal of Remote Sensing*, **22**, 3719–3746.

Reich, P. B., Turner, D. P. and Bolstad, P. (1999). An approach to spatially distributed modeling of net primary production (NPP) at the landscape scale and its application in validation of EOS NPP products. *Remote Sensing of Environment*, **70**, 69–81.

Reichle, R. H., McLaughlin, D. B. and Entekhabi, D. (2001). Variational data assimilation of microwave radio brightness observations for land surface hydrology applications. *IEEE Transactions on Geoscience and Remote Sensing*, **39**, 1708–1718.

Reichle, R. H., Mclaughlin, D. B. and Entekhabi, D. (2002). Hydrologic data assimilation with the ensemble Kalman filter. *Monthly Weather Review*, **130**, 103–114.

Russell, M., Nunez, M., Chladil, M., Valiente, J. and Lopez-Baeza, E. (1997). Conversion of nadir, narrowband reflectance in red and near-infrared channels to hemispherical surface albedo. *Remote Sensing of Environment*, **61**, 16–23.

Rutan, D., Charlock, T., Rose, F., Kato, S., Zentz, S. and Coleman, L. (2006). Global surface albedo from CERES/TERRA surface and atmospheric radiation budget (SARB) data product. The proceedings 12th conference on Atmosphere radiation, Madison, WI, July 10–14.

Sandholt, I., Rasmussen, K. and Andersen, J. (2002). A simple interpretation of the surface temperature vegetation index space for assessment of surface moisture status. *Remote Sensing of Environment*, **79**(2–3), 213–224.

Schaaf, C., Gao, F., Strahler, A., Lucht, W., Li, X., Tsang, T., Strugnell, N., Zhang, X., Jin, Y. and Muller, J. (2002). First operational BRDF, albedo nadir reflectance products from MODIS. *Remote Sensing of Environment*, **83**, 135–148.

Schaaf, C., Martonchik, J., Pinty, B., Govaerts, Y., Gao, F., Lattanzio, A., Liu, J., Strahler, A. and Taberner, M. (2008). Retrieval of surface albedo from satellite sensors. *Advances in Land Remote Sensing*, 219–243.

Schroeder, T. A., Hember, R., Coops, N. C. and Liang, S. (2009). Validation of solar radiation surfaces from MODIS and reanalysis data over topographically complex terrain. *Journal of Applied Meteorology and Climatology*, **48**, 2441–2458.

Scurlock, J. M. O., Cramer, W., Olson, R. J. *et al.* (1999). Terrestrial NPP: Toward a consistent data set for global model evaluation. *Ecological Applications*, **9**(3), 913–919.

Shelton, M. L. (2009). *Hydroclimatology*. Cambridge: Cambridge University Press.

Shen, S. H. H., Yang, S. B., Li, B. B. *et al.* (2009). A scheme for regional rice yield estimation using ENVISAT ASAR data. *Science in China Series D: Earth Sciences*, **39**(6), 763–773.

Smith, W. L. and Woolf, H. M. (1976). The use of eigen-vectors of statistical covariance matrices for interpreting satellite sounding radiometer observations. *Journal of Atmosphereic Sciences*, **33**, 1127–1140.

Stockli, R., Rutishauser, T., Dragoni, D., O'Keefe, J., Thornton, P. E., Jolly, M., Lu, L. and Denning, A. S. (2008). Remote sensing data assimilation for a prognostic phenology model. *Journal of Geophysical Research*, **113**, G04021.

Stockli, R., Rutishauser, T., Baker, I., Liniger, M. and Denning, A. (2011). A global reanalysis of vegetation phenology. *Journal of Geophysical Research*, **116**.

Teillet, P. M., Saleous, N. E. L., Hansen, M. C. *et al.* (2000). A valuation of the global 12 km AVHRR land dataset. *International Journal of Remote Sensing*, **21**(10), 1987–2021.

Townshend, J. R. G. (1994). Global data sets for land applications from the advanced very high resolution radiometer: An introduction. *International Journal of Remote Sensing*, **15**, 3319–3332.

Tucker, C. J. (1979). Red and photographic infrared linear combinations for monitoring. *Remote Sensing of Environment*, **8**, 127–150.

van Leeuwen, W. and Roujean, J. (2002). Land surface albedo from the synergistic use of polar (EPS) and geo-stationary (MSG) observing systems: An assessment of physical uncertainties. *Remote Sensing of Environment*, **81**, 273–289.

Veroustraete, F., Patyn, J. and Myneni, R. B. (1996). Estimating net ecosystem exchange of carbon using the normalized difference vegetation index and an ecosystem model. *Remote Sensing of Environment*, **58**, 115–130.

Walker, J. P., Willgoose, G. R. and Kalma, J. D. (2001). One-dimensional soil moisture profile retrieval by assimilation of near-surface observations: A comparison of retrieval algorithms. *Advances in Water Resources*, **24**, 631–650.

Wan, Z. and Li, Z.-L. (1997). A physics-based algorithm for retrieving land-surface emissivity and temperature from EOS/MODIS data. *IEEE Transactions on Geoscience and Remote Sensing*, **35**, 980–996.

Wan, Z., Zhang, Y. L., Zhang, Q. C. and Li, Z.-L. (2002). Validation of the land surface temperature products retrieved from terra moderate resolution imaging sepctrometer data. *Remote Sensing of Environment*, **83**, 163–180.

Wan, Z., Zhang, Y. L., Zhang, Q. C. and Li, Z.-L. (2004). Quality assessment and validation of the MODIS global land surface temperature. *International Journal of Remote Sensing*, **25**, 261–274.

Wan, Z. M. (1999). MODIS land surface temperature algorithm theoretical basis document (LST ATBD OL). Version 3.3. Available at: http://eospso.gsfc.nasa.gov.

Wan, Z. M., Feng, Y. Z., Zhang, Y. L. *et al.* (1998). Land-surface temperature and emissivity retrieval from MODIS Airborne Simulator (MAS) data. *Summaries of the Seventh JPL Airborne Earth Science Workshop*, **3**, 57–66.

Wang, D. W. (2008). Methods and application of assimilating remote sensing data and crop growth model, Ph.D. Dissertation, Beijing Normal University (in Chinese).

Wang, J. R. (1985). Effect of vegetation on soil moisture sensing observed from orbiting microwave radiometers. *Remote Sensing of Environment*, **17**, 41–151.

Wang, J. and Li, X. (2008). Knowledge database and inversion. In Liang, S. (ed.), *Advances in Land Remote Sensing: System, Modeling, Inversion and Application*, Chapter 8, pp. 203–217. New York: Springer.

Wang, Z. X. and Liu, C. (2003). From AVHRR-NDVI to MODIS-EVI: Advances in vegetation index research. *Acta Ecologica Sinica*, **23**(5), 979–987 (in Chinese).

Wei, Y. X. and Wang, L. W. (2010). Progress in research on land cover products of MODIS. *Spectroscopy and Spectral Analysis*, **30**(7), 1848–1852 (in Chinese).

Weng, F. and Grody, N. C. (1998). Physical retrieval of land surface temperature using the special sensor microwave imager. *Journal of Geophysical Research*, **103**, 8839–8848.

Wild, M. (2008). Short-wave and long-wave surface radiation budgets in GCMs: A review based on the IPCC-AR4/CMIP3 models. *Tellus A*, **60**, 932–945.

Wu, B. F., Zeng, Y. and Huang, J. L. (2004). Overview of LAI/FPAR retrieval from remotely sensed data. *Advances in Earth Sciences*, **19**(4), 585–590 (in Chinese).

Xiao, Z., Liang, S., Wang, J. *et al.* (2009). A temporally integrated inversion method for estimating leaf area index from MODIS data. *IEEE Transactions on Geoscience and Remote Sensing*, **47**(8), 2536–2545.

Xiao, Z., Liang, S., Wang, J. *et al.* (2011a). Real-time inversion of leaf area index from MODIS time series data. *Remote Sensing of Environment*, **115**(1), 97–106.

Xiao, Z., Liang, S., Wang, J. *et al.* (2011b). GLASS-global land surface leaf area index product: Algorithm theoretical basis document. Version 1.0. 1–27.

Xu, T., Liu, S., Liang, S. and Qin, J. (2011a). Improving predictions of water and heat fluxes by assimilating MODIS land surface temperature products into the common land model. *Journal of Hydrometeorology*, **12**(2), 227–244.

Xu T., Liang, S. and Liu, S. (2011b). Estimating turbulent fluxes through assimilation of geostationary operational environmental satellites data using ensemble Kalman filter. *Journal of Geophysical Research*, **116**, D09109, doi:10.1029/2010JD015150.

Yang, K., Takahiro, W. and Toshio, K. (2007). An auto-calibration system to assimilate AMSR-E data into a land surface model for estimating soil moisture and surface energy budge. *Journal of the Meteorological Society of Japan*, **85**, 229–242.

Zhang, X. T., Zhou, G. Q., Wu, H. R., *et al.*, (2011). GLASS-global and surface insolation and PAR product: algorithm theoretical basis document version 1.0.1–70.

Zhao, Y. S. (2003). *Theory and Methods of Analysis of Remote Sensing Applications*. Beijing: Science Press (in Chinese).

Chapter 2

Second-Generation Polar-Orbiting Meteorological Satellites of China: The Fengyun 3 Series and Its Applications in Global Monitoring

Peng Zhang

National Satellite Meteorological Center,
China Meteorological Administration, 100081, Beijing

The Fengyun 3 series (hereafter abbreviated as FY-3) are the second-generation polar-orbiting meteorological satellites of China. The first two satellites of the FY-3 series, FY-3A and FY-3B, were launched on May 27, 2008, and November 5, 2010, respectively. FY-3A is a monitoring orbit satellite, with the descending node at 10:00, and FY-3B is an afternoon orbit satellite, with the ascending node at 13:30. These two satellites together make up a global observation constellation with an enhanced temporal interval. FY-3A and FY-3B carry the same instruments. A total of 11 onboard instruments cover the spectrum from ultraviolet to microwave in order to measure the characteristics of the atmosphere, land, ocean, cryosphere, and biosphere. In this chapter, the instruments, ground segment designs, and the data service are introduced in detail. The data applications in environmental monitoring, weather prediction, and climate change research are presented. Improvements in the FY-3 series regarding Earth observations are summarized in comparison with the FY-1 series.

2.1 Historical Review of Chinese Meteorological Satellites

The Chinese involvement with meteorological satellites began in 1969. Since then, two parallel lines of development have been followed: receiving, processing, and utilizing data from foreign satellites; and programming, developing, and deploying Chinese meteorological satellites (Fang *et al.*, 2004).

Chinese meteorological satellites fall into two categories (Li, 2001; Meng, 2004): the polar orbit series and the geostationary orbit series. Each

S. Liang, X. Li & X. Xie (eds.), *Land Surface Observation, Modeling and Data Assimilation*, 45–66.

Table 2.1. Chinese meteorological satellites.

Satellite name	Type	Launch date	Function
FY-1A	Polar	Sept. 7, 1988	R&D
FY-1B	Polar	Sept. 3, 1990	R&D
FY-1C	Polar	May 10, 1999	Operation
FY-1D	Polar	May 15, 2002	Operation
FY-3A	Polar	May 27, 2008	R&D
FY-3B	Polar	Nov. 5, 2010	R&D
FY-2A	Geostationary	June 10, 1997	R&D
FY-2B	Geostationary	June 25, 2000	R&D
FY-2C	Geostationary	Oct. 18, 2004	Retired
FY-2D	Geostationary	Dec. 8, 2006	Operation
FY-2E	Geostationary	Dec. 23, 2008	Operation

satellite is named with an Arabic numeral and a letter. The Arabic numerals represent the satellite series, odd numbers being the polar orbit satellites and even numbers representing the geostationary satellites. The letter represents the sequence number within the series. So far, six polar orbit satellites and five geostationary satellites have been launched successfully (Table 2.1).

2.2 Mission of the Fengyun 3 Series

The Fengyun 3 series is the second-generation polar-orbiting meteorological satellites of China. To meet new and higher requirements in modern meteorological services, especially in numerical weather predictions (NWPs), these satellites are designed to perform global, three-dimensional, quantitative, and multi-spectral observations under all weather conditions (i.e., cloud-free and cloudy conditions) with multiple sensors onboard (Fan, 2000; Zhang, 2001).

FY-3A was the first satellite of the Fengyun 3 series. It was launched on May 27, 2008, from the Taiyuan launch center. FY-3B was the second, launched on November 5, 2010, from the same launch station. The morning satellite FY-3A and the afternoon satellite FY-3B together make up a global observation system with an enhanced temporal interval. Compared with the single payload of the Fengyun 1 series, the number of onboard instruments in the FY-3 series has increased to 11. The FY-3 series thus represent a new chapter in the history of Chinese meteorological satellites and satellite meteorology (Yang, 2008).

The first two satellites in the FY-3 series, i.e., FY-3A and FY-3B, are research and development (R&D) satellites whose projected lifetime is three years. The FY-3 satellites describe a sun-synchronous, near-polar orbit at about 830 km altitude. The attitude control uses three-axis stabilization techniques. The details of the FY-3A and FY-3B platforms can be found in the referenced technical reports (National Satellite Meteorological Center, 2004; National Satellite Meteorological Center, 2008a).

2.3 The Payloads on FY-3A and FY-3B

FY-3A and FY-3B have identical instruments onboard. There are 11 types of instruments that cover the spectrum from ultraviolet to microwave. The instruments include the Visible and Infrared Radiometer (VIRR), Infrared Atmospheric Sounder (IRAS), Microwave Temperature Sounder (MWTS), Microwave Humidity Sounder (MWHS), Medium Resolution Spectral Imager (MERSI), Microwave Radiation Imager (MWRI), Solar Backscatter Ultraviolet Sounder (SBUS), Total Ozone Unit (TOU), Earth Radiation Measurement (ERM), Solar Irradiance Monitor (SIM), and the Space Environment Monitor (SEM). Of these, the IRAS, MWTS, and MWHS make up the Vertical Atmospheric Sounding System (VASS). The VIRR is the only instrument inherited from the formal FY-1 series platform. The other instruments are all in orbit for the first time.

- The VIRR is a 10-channel VIS/IR radiometer for multi-purpose imagery with 1.1 km resolution at nadir. The swath of the VIRR is 2800 km.
- The MERSI is a 20-channel VIS/IR radiometer. There are 19 channels in the VIS/NIR/SWIR bands and one in the broad TIR band at 10.0–12.5 μm. The spatial resolution at nadir is twofold: 250 m (for four VIS/NIR channels and one TIR channel) and 1 km (for the other channels). The swath width is similar to VIRR at 2800 km.
- The MWRI is a 10-channel microwave radiometer with conical scanning at five frequencies. All frequencies are in dual polarization. Spatial resolution is 9.5 km $\times$ 15 km at 90 GHz and 30 km $\times$ 50 km at 19 GHz. The swath of the MWRI is 1400 km.
- The IRAS is a 26-channel IR radiometer for temperature and humidity sounding. The spatial resolution is 17 km and the swath width is 2250 km.
- The MWTS is a four-channel microwave radiometer for nearly-all-weather temperature sounding with a spatial resolution of 70 km at

54 GHz. The MWTS performs on cross-track scanning mode with a swath of 2200 km.

- The MWHS is a five-channel microwave radiometer with four frequencies (one frequency in dual polarization) for nearly-all-weather humidity sounding. The spatial resolution is 15 km in the 183 GHz band. The swath width is 2700 km with cross-track scanning.
- The TOU and SBUS make up a suite of two UV spectro-radiometers. The TOU measures total ozone amount with six channels in the 308–360 nm range, with a spatial resolution of 50 km and a 3000 km swath. The SBUS measures ozone profile with channels in the range of 252–340 nm. The spatial resolution is 200 km at nadir viewing without side scanning.
- The ERM is a two-broadband channel radiometer for earth-reflected solar flux and earth-emitted thermal flux over short (0.2–4.3 μm) and total (0.2–50 μm) wavebands. The ERM has two working modes. One is the cross-track scanning mode, with 28 km spatial resolution at a 2° narrow field of view (NFOV); the swath width is 2300 km. The other is the nadir viewing mode with a 120° wide field of view.
- The SIM is a three-channel radiometer over 0.2–50 μm waveband for the total incident solar flux. It views the sun near the north polar area.
- The SEM is the only *in situ* instrument to measure charged particles in the solar wind.

From the viewpoint of data application, the payload instruments on the FY-3 can be considered as analogous to sensors that are widely used elsewhere in the world. In fact, the VIRR is an expanded AVHRR instrument, and the MERSI is a MODIS-similar sensor. Both these optical imagers can provide surface characteristics (including cloud surface, land surface, and ocean surface) and aerosol information. The VASS set is made up of ATOVS-similar instruments for atmospheric sounding. The MWRI is an AMSR-similar instrument, except that the low frequency on the 6.9 GHz channel is not included. The SBUS and TOU are SBUV-similar and TOMS-similar sensors, respectively, which can provide ozone profile and total ozone amount separately. The ERM is a CERES-similar instrument inherited from ERBE. The detailed specifications of these 11 payload instruments and their data applications are listed in Table 2.2 (National Satellite Meteorological Center, 2004; National Satellite Meteorological Center, 2008a; Zhang *et al.*, 2009).

Table 2.2. The specification of 11 onboard payloads of FY-3A and FY-3B.

Name	Specification		Purpose
VIRR	Spectral range	0.43–12.5 μm	Cloud image, cirrus and cloud phase, vegetation, sediment, snow and ice, land surface temperature, sea surface temperature, water vapor content
	Number of channels	10	
	Scan range	±55.4°	
	Spatial resolution at nadir	1.1 km	
	Cal accuracy in VNIR	5–10%	
	Cal accuracy in IR	1 K (270 K)	
	Quantization	10 bits	
VASS IRAS	Spectral range	0.69–15.0 μm	Atmospheric temperature profile, atmospheric moisture profile, total ozone amount, outgoing long wave radiation
	Number of channels	26	
	Scan range	±49.5°	
	Spatial resolution at nadir	17 km	
	Cal accuracy in VNIR	5–9%	
	Cal accuracy in IR	1 K (270 K)	
	Quantization	13 bits	
MWTS	Spectral range	50–57 GHz	
	Number of channels	4	
	Scan range	±48.3°	
	Spatial resolution at nadir	50–75 km	
	Cal accuracy	1.2 k	
	Sensitivity (NEΔN)	0.4–0.55 k	
	Quantization	13 bits	
MWHS	Spectral range	150–183 GHz	
	Number of channels	5	
	Scan range	±53.35°	
	Spatial resolution at nadir	15 km	
	Cal accuracy	1.5 k	
	Sensitivity (NEΔN)	1.1–1.2 k	
	Quantization	14 bits	
MERSI	Spectral range	0.40–12.5 μm	Ocean color, aerosol, water vapor content, cloud properties, vegetation, surface properties, surface temperature, snow and ice
	Number of channels	20	
	Scan range	±55.4°	
	Spatial resolution at nadir	0.25–1 km	
	Cal accuracy in VNIR	5–10%	
	Cal accuracy in IR	1 K (270 K)	
	Quantization	12 bits	
MWRI	Spectral range	10–89 GHz	Precipitation rate, liquid water content, water vapor content, soil moisture, sea ice, sea surface temperature, snow cover, ice cover
	Number of channels	10	
	Scan range	±55.4°	
	Spatial resolution at nadir	15–85 km	
	Cal accuracy	1–2.8 K	
	Quantization	12 bits	

(*Continued*)

Table 2.2. (*Continued*)

Name	Specification			Purpose
SBUS	Spectral range		0.16–0.4 μm	Ozone profile
	Number of channels		12	
	Spatial resolution at nadir		200 km	
	Quantization		16 bits	
	Stray light		10^{-6}	
	Cal accuracy		3% (160–250 nm)	
	Cal accuracy in diffuse		2% (250–400 nm)	
	Reflection board		3%	
TOU	Spectral range		0.3–0.36 μm	Total ozone amount
	Number of channels		6	
	Scan range		±54°	
	Spatial resolution at nadir		50 km	
	Quantization		12 bits	
	Stray light		10^{-3}	
	Cal accuracy		2%	
ERM	Spectral range	Total band	0.2–50 μm	Terrestrial radiation
		Solar band	0.2–3.8 μm	
	Number of channels	Narrow FOV	2	
		Broad FOV	2	
	Scan range		±50° (narrow)	
	Sensitivity (NEΔN)		$0.4\,\mathrm{Wm^{-2} \cdot sr^{-1}}$	
	Cal accuracy	Total band	0.8%	
		Solar band	1%	
	Stability within 2 years		<1%	
SIM	Spectral range		0.2–50 μm	Solar irradiation
	Sensitivity (NEΔN)		$0.2\,\mathrm{Wm^{-2}}$	
	Cal accuracy		0.5%	
	Quantization		16 bits	
	Stability within 2 years		<0.02%	
SEM	Heavy ions, high energy proton, high energy electron, radiation dose, satellite surface charging monitoring, single event upset			Space environment

2.4 Ground Segment Designs for FY-3A and FY-3B

The Chinese meteorological satellite engineering system comprises five parts: satellite segment, launch pad segment, launch vehicle segment, measurement and control segment, and ground segment (Li and Dong, 2008). The ground segment responds to data receiving, data pre-processing, and data processing. It is the key to sustaining and promoting satellite data applications in the meteorological services (Xu *et al.*, 2006).

The ground segment of FY 3 series includes the data processing center, operation and control center, ground receiving center, and data archiving and service center. It is divided into 10 technical parts from the point of view of engineering and construction (National Satellite Meteorological Center, 2006). These technical parts are the Data Acquisition System (DAS), Computer and Network System (CNS), Operation Control System (OCS), Data Pre-Processing System (DPPS), Products Generation System (PGS), Quality Control System (QCS), Utilization Demonstration System (UDS), Archive and Service System (ARSS), Monitoring and Analysis System (MAS), and Simulation and Technical Support System (STSS). The purpose of the airborne-based field experiment is to test the engineering model of the payload instruments. The ground segment framework of FY-3A and FY-3B is shown in Fig. 2.1.

The ground segment of FY-3A and FY-3B has five ground stations to receive the satellite direct-broadcasting data. Four of them are inside China and the other is located near the northern polar region. Such a distribution, especially the ground station in the polar region, guarantees that global

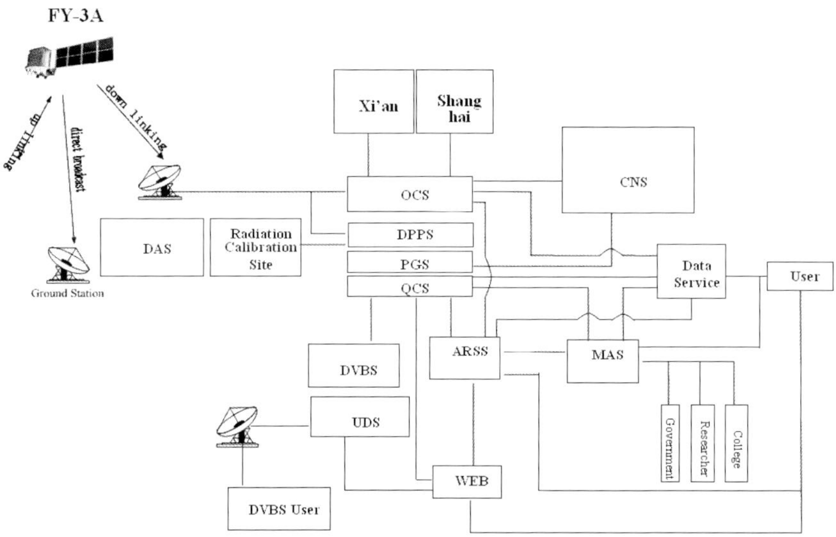

Fig. 2.1. The ground segment framework for FY-3A and FY-3B.

Table 2.3. Ground station distribution.

Station name	Longitude	Latitude
Beijing Station	116° 16′ 36″ E	40° 03′ 06″ N
Guangzhou Station	113° 20′ 20″ E	23° 09′ 52″ N
Wulumuqi Station	87° 34′ 08″ E	43° 52′ 17″ N
Jiamusi Station	130° 22′ 48″ E	46° 45′ 20″ N
Kiruna Station	21° 02′ E	67° 32′ N

data can be received and collected within about 3 h. The ground stations receive all of the L-band HRPT data, X-band MPT data, and DPT data. Table 2.3 gives information about the current ground stations constructed for the FY-3A and FY-3B.

2.5 Standard Product in Levels 1 and 2

In the FY-3 ground segment, the level-1 product is the radiance data measured directly by the instruments. The level-1 product is generated by the DPPS system and the geolocation and calibration information are included. The level-2 product is the geophysical and geochemical parameters, such as atmospheric temperature profile, column water vapor content, etc. The level-2 product is the indirectly measured information retrieved through inversion algorithms. The level-2 product is produced by the PGS system. The data received from all of the instruments are processed in the data processing center into archived level-1 and -2 products in near real time. Table 2.4 lists the level-2 products generated from the PGS currently covering the atmosphere, land, ocean, cryosphere, radiation, and space environment (National Satellite Meteorological Center, 2008b; Yang, *et al.*, 2010).

Both level-1 and -2 datas are recorded in HDF format. To facilitate data utilization, level-1 data are defined as instrument-dependent radiance data with information for geolocation and radiance calibration. Level-1 data are stored without projection in 5-min granule files for VIRR and MERSI, and in one-orbit files for the other instruments. The data in level 2 and above are geophysical-chemical-related parameters derived from inversion algorithms. Level-2 data are stored with projection in 648 block files for VIRR and MERSI, each covering a $10° \times 10°$ area of the earth, and in one file for the other instruments, covering globally. This design for satellite data storage represents a compromise between data volume and data utilization efficiency and is illustrated in Fig. 2.2.

Table 2.4. Level-2 products generated from PGS.

Discipline	Parameters	FY-3A sensors
Atmosphere	Cloud/fog properties	MERSI/VIRR/IRAS/MWRI/MWHS
	Total water vapor	VIRR/MERSI/MWRI
	Precipitation	MWHS/MWRI
	Aerosol properties	MERSI/VIRR
	Atmospheric temperature and humidity	IRAS/MWTS/MWHS
	Total ozone and ozone profile	TOU/SBUS/IRAS
Land	Land cover	VIRR/MERSI
	Surface temperature	VIRR
	Vegetation dynamics	VIRR/MERSI
	Fire and flood monitoring	VIRR/MERSI
	Soil moisture	MWRI
Ocean	Sea surface temperature	VIRR/MWRI
	Sea surface color	MERSI
Cryosphere	Sea ice	MERSI/VIRR/MWRI
	Snow cover	VIRR/MWHS/MERSI/MWRI
Radiation	earth's radiation and solar irradiance	ERM/SIM
Space Environment	High energy particles	SEM
	Radiation dose	

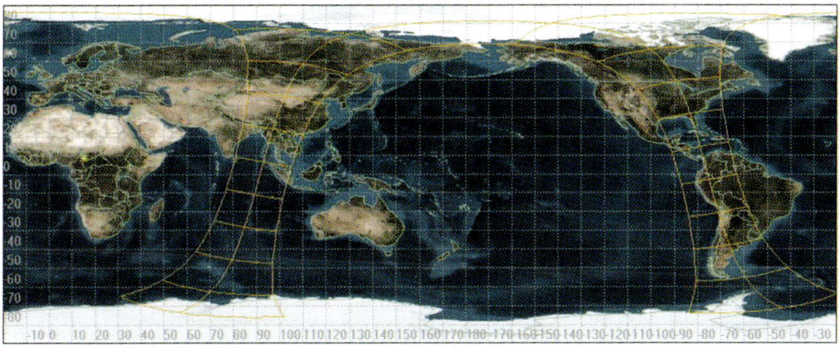

Fig. 2.2. Data storage concept. The yellow solid line corresponds to level-1 data storage and the white dashed line to level-2 data storage, with the world map as background.

2.6 Data Archives and Service

As shown in Fig. 2.3, after level-1 and -2 products are generated by the DPPS and PGS in the FY-3 ground segment, these products are archived in the data archive center through the ARSS in either on-line, near-line, or

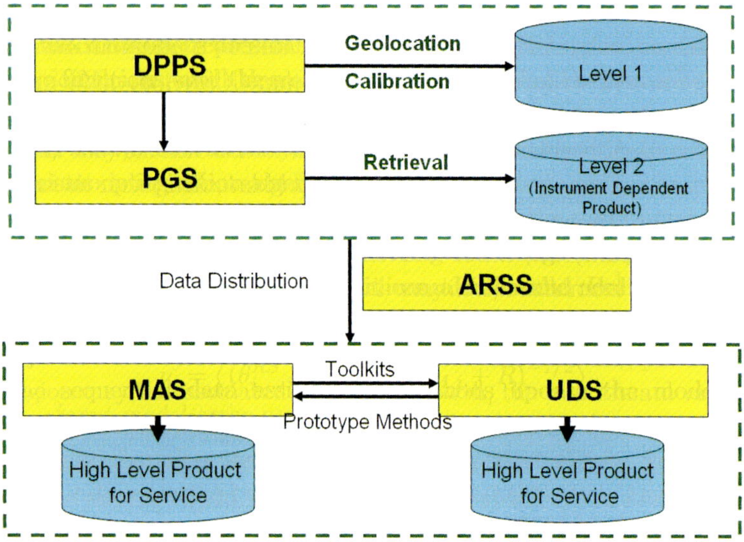

Fig. 2.3. FY-3A ground segment related with data application.

off-line mode. Data from the last one month are stored on an operational magnetic disk in the on-line mode; the last three months' data are stored on cache magnetic disk in the near-line mode; and historical data are stored in a tape library in the off-line mode.

To facilitate the use of the data, a special toolkit is provided for applying FY-3 satellites data to meteorology and environmental services. This software package has temporarily been named MAS. MAS is a user-oriented application software system designed for FY-3 satellite data visualization, processing, analysis, and service in a unified platform. Currently, seven functions are included: multi-source data reading and display, image processing, remote sensing processing tools, product monitoring analysis, thematic map of product monitoring, operations management, and product distribution monitoring. Levels-1 and -2 data can be delivered to UDS end-users together with the MAS software package to promote the utilization and incorporation of FY-3 data in meteorological services.

For FY-3A or FY-3B, each satellite will collect about 380 gigabytes of data in level-1 per day and 850 gigabytes of data in level-2 and above per day. It is a difficult challenge to deliver such a huge quantity of data to the users as soon as possible. Five means of access have been developed for acquisition of the FY-3 data, depending on the user's service rank. They are the web-based service, DVBs broadcasting service, FTP push service, FTP pull service, and manual service, the last four of which are the operational means of distributing data to authorized international, regional and national users in near real time. The manual service responds to emergency requirements.

Unauthorized users can acquire FY-3 data through the web-based service after free registration (http://FY3.satellite.cma.gov.cn). FY-3 data can be searched, categorized, and downloaded by instrument, temporal coverage, and spatial coverage, as shown in Fig. 2.4.

2.7 Demonstration of Multi-disciplinary Data Utilization

The three basic scientific goals of FY-3A and FY-3B are to improve the accuracy of NWP, to enhance the capacity for environmental monitoring, and to promote the utilization of data in climate change research. Since the launch of FY-3A and FY-3B, their data have been utilized in global synoptic weather monitoring, typhoon monitoring, NWP by data assimilation, sea and inland water body monitoring, fire monitoring, aerosol and air quality monitoring, monitoring of the hot-island effect in cities, global sea shelf monitoring, global ozone monitoring, etc.

2.7.1 *Synoptic weather monitoring*

Figure 2.5 shows the global mosaic image in MERSI channels 3, 2, 1 with 250 m resolution. MERSI can acquire such images daily. The inter-tropical convergence zone (ITCZ), tropical depression, and sub-tropical high pressure can be seen clearly.

2.7.2 *Typhoon monitoring*

Figure 2.6 shows an image of typhoon Fung-wong, as monitored by MERSI on 27 July, 2008, in which the eye of the typhoon and the surrounding spiral cloud can be clearly distinguished. This provides very useful information for typhoon positioning and classification as a strong typhoon.

In order to evaluate the image interpretation, a small area centered near the eye of the typhoon was depicted in three different spatial resolutions, corresponding to the observation from FY-2 (5 km), FY-1 (1.1 km), and

Fig. 2.4. FY-3 web-based data service.

Fig. 2.5. Global mosaic image in MERSI channels 3, 2, 1 with 250 m resolution. (*Courtesy*: Qian Huang, NSMC.)

Fig. 2.6. Typhoon Fung-wong monitored by MERSI on 27 July, 2008.

FY-3 (0.25 km) series. As Fig. 2.7 illustrates, the image is hard to distinguish in 5 km spatial resolution. The subtle structure of the clear sky in the eye of the typhoon and the convection in the spiral cloud band can be seen more clearly at 250 m spatial resolution.

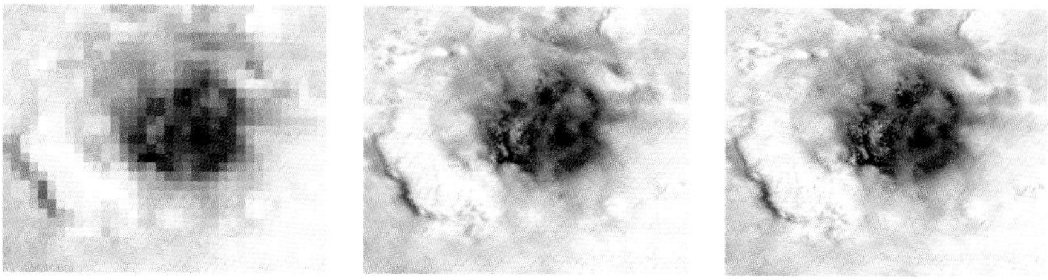

(a) (b) (c)

Fig. 2.7. Part of typhoon observed in three spatial resolutions: (a) 5 km; (b) 1.1 km; (c) 0.25 km.

2.7.3 *NWP by data assimilation*

The principal purpose of the FY-3 series is to improve the accuracy of NWP through the assimilation of satellite data. The first stage of satellite data assimilation is to simulate the satellite observation with the forward model based on a calculation of atmospheric radiation. Figure 2.8 shows an example.

Lu (2010, 2011) introduced FY-3A sounding data into the ECMWF integrated forecasting system (IFS) to compare the VASS with ATOVS. The data quality from the FY-3A instruments were analyzed through a careful assessment of the first guess departures, comparisons with analogous ATOVS instruments, and a comprehensive series of observing system experiments (OSEs). The results show that the quality of data from the FY-3A instruments is good, as summarized in Fig. 2.9, which shows a comparison of first guess departures (standard deviations) from key channels of the FY-3A instruments and equivalent ATOVS instruments.

A satellite data assimilation experiment was designed by Lu (2010) to validate the use of FY-3 sounding data in NWP forecasting. A control experiment was set as baseline, based on the ECMWF operational system (OSE), and the assimilation of data from FY-3 instruments (MWTS, MWHS, and IRAS) into OSE was used as a verification experiment. The results are shown in Fig. 2.10. The curve above the zero line indicates a positive benefit from FY-3A data for the 00Z forecast of the 500 hPa and 200 hPa geopotential height in the northern hemisphere (NH, left in Fig. 2.10) and the southern hemisphere (SH, right in Fig. 2.10). The experiments were run from July 20 to November 1, 2008, and 93 cases were selected for statistical analysis. The results are encouraging and indicate that FY-3A atmospheric sounding data can have a positive impact on

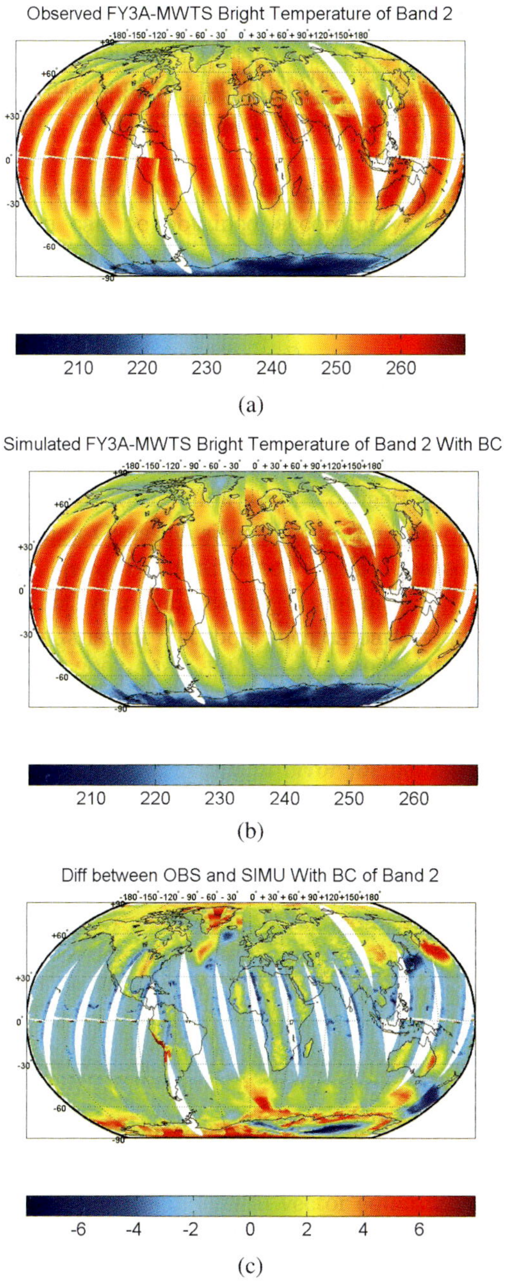

Fig. 2.8. FY-3 satellite data simulated from the forward model and NWP outputs: (a) observed MWTS channel 2 image; (b) simulated MWTS channel 2 image; and (c) difference between observation and simulation. (*Courtesy*: Qifeng Lu, NSMC.)

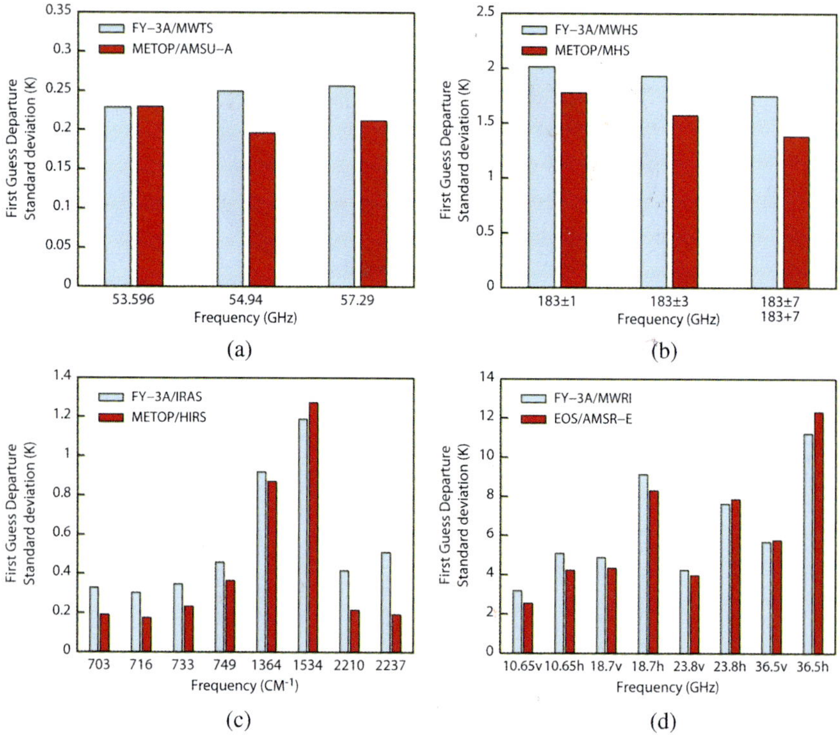

Fig. 2.9. Standard deviations of first guess departures for FY-3A and MetOp-A/Aqua equivalent instruments and channels. (*Courtesy*: Qifeng Lu, NSMC.) Statistics are derived from data for the period from August 28 to September 28, 2008: (a) MWTS and AMSU-A; (b) MWHS and AMSU-B; (c) IRAS and HIRS; and (d) MWRI and AMSR-E.

NWP forecasting. As a package, the FY-3A instruments show an impact equivalent to a significant fraction (e.g., 30% at day 5 in SH 500 hPa geopotential height) of that of the MetOp-A ATOVS suite.

2.7.4 *Ozone monitoring*

The total column ozone retrieval algorithm of TOU is similar to the algorithm used by TOMS/NIMBUS-7 and OMI/AURA. The total column ozone is derived from a pair of measurements of backscattered solar ultraviolet radiation at two wavelengths where ozone absorption has an apparent difference (Wang *et al.*, 2010). Figure 2.11 gives the total ozone amount retrieved from TOU/FY-3, GOME-2/Metop, and OMI/Aura, showing global consistency with a difference within 3%.

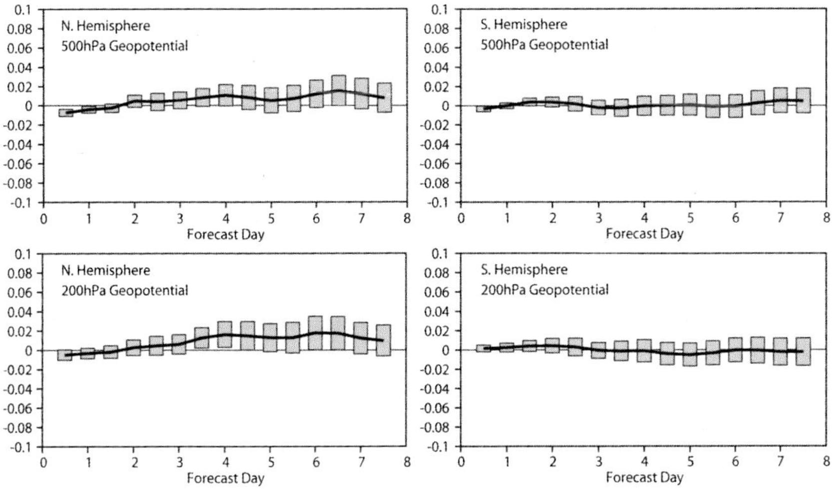

Fig. 2.10. Normalized differences in the RMS forecast error between the Full + FY3A experiment and the Full system experiment for the 00Z forecast of 500 hPa and 200 hPa geopotential height for the NH (left) and the SH (right). (*Courtesy*: Qifeng Lu, NSMC.)

The ozone profile retrieval algorithm of SBUS is similar to the algorithm used by BUV/NIMBUS-4 and SBUV in the NOAA series. The ozone profile is deduced from 12 wavelengths in the ozone ultraviolet absorption spectrum, using an iterative method of calculation of the radiation transfer equation (Huang *et al.*, 2010).

Figure 2.12 shows the ozone hole monitored from TOU on November 1, 2008. The region on the cold color side corresponds to the ozone hole in the image. The ozone valley over the Tibetan Plateau also can be monitored by the TOU on board FY-3A and FY-3B.

2.7.5 *Air quality monitoring*

The aerosol loading in the atmosphere indicates the air quality. The aerosol product can be deduced from MERSI measurements. The aerosol retrieval algorithm of MERSI is similar to the algorithm used by MODIS/EOS. Information from the 0.47 μm, 0.65 μm, and 2.1 μm bands is used to retrieve aerosol loads over ocean and land, respectively. Figure 2.13 shows a haze layer monitored by MERSI/FY-3A during the 2008 Olympic Games in the southwestern region of Beijing. The retrieved aerosol optical thickness at 550 nm shows heavy aerosol loading in this area.

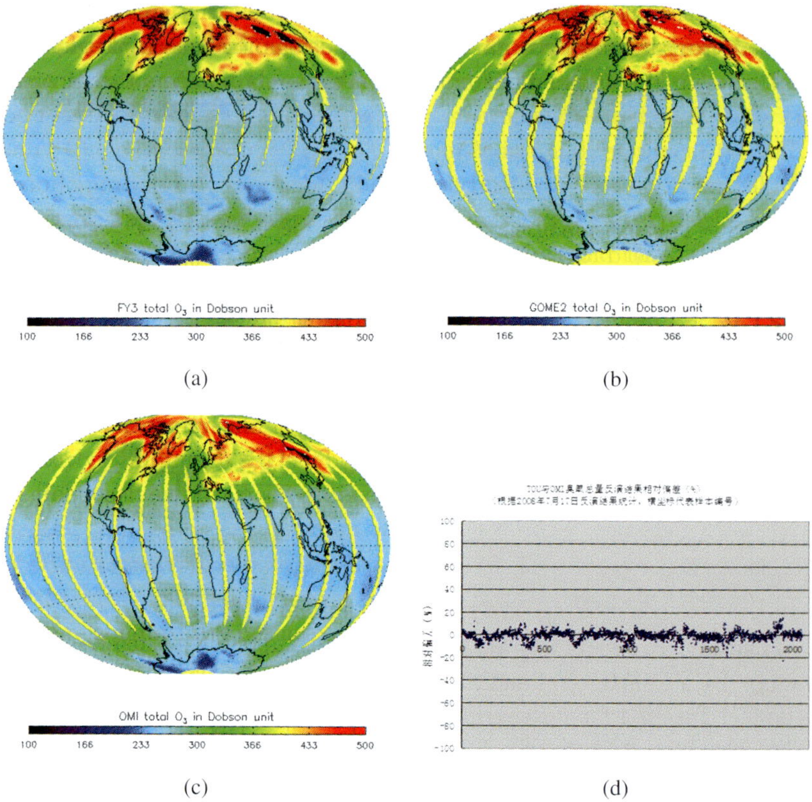

Fig. 2.11. Global total column ozone distribution monitored by (a) TOU/FY-3A, (b) GOME 2/Metop, (c) OMI/Aura, and (d) the consistency between TOU and OMI. (*Courtesy*: Weihe Wang, NSMC.)

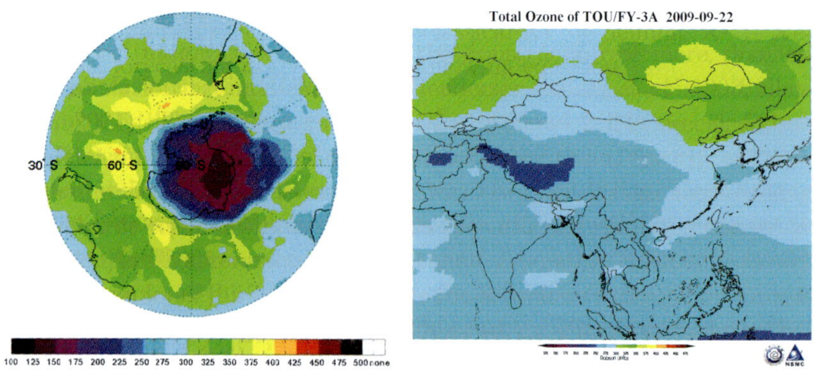

Fig. 2.12. Total ozone in DU monitored from TOU. (*Courtesy*: Weihe Wang, NSMC.)

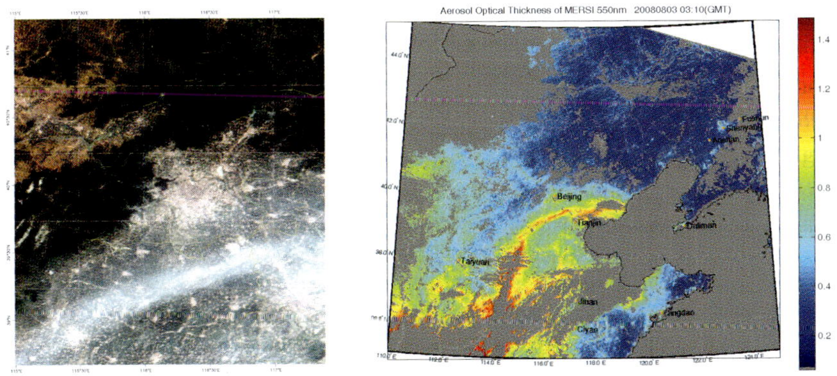

Fig. 2.13. Air quality monitoring by MERSI during the Beijing 2008 Olympic Games. (*Courtesy*: Xiaojing Li, NSMC.)

2.8 Conclusions

China has become one of the few countries to operationally maintain sun-synchronous and geostationary meteorological satellites. Chinese meteorological satellites are becoming an important component of the space-based global observation system.

In comparison with the first-generation polar-orbiting meteorological satellites in China, the current FY-3 satellites are involved in microwave imaging, sounding, atmospheric chemistry, and Earth radiation budget monitoring, using the 11 onboard instrument systems. Optical imaging resolution has been improved from 1 km to 250 m, making FY-3A and FY-3B not only meteorological satellites but also Earth observation satellites.

FY-3A and FY-3B constitute a sun-synchronous constellation that provides global observation of the Earth four times per day. These observation data from the FY series have been widely utilized in weather analysis, numerical weather forecasting, climate prediction, and environmental and disaster monitoring. Analysis of the consistency between observation and simulation from the ECMWF Integrated Forecasting System has shown the FY-3 data to be of good quality overall.

Current FY-3 data delivery services include the web-based services, FTP push service, FTP pull service, manual service, and CMACAST service. Users can access the data through the webpage (http://FY3.satellite.cma.gov.cn) after a quick and free registration. It is believed that the FY-3 series satellites not only benefit the nation of China but also make a valuable contribution to the international meteorological, hydrological, and environmental community.

Acknowledgments

The authors would like to thank their colleagues working in the National Satellite Meteorological Center for their contributions to the FY-3A engineering project.

References

Fan, T. X. (2000). The roles of Fengyun 3 meteorological satellites. Technologic Note of Satellite Meteorology, National Satellite Meteorological Center, 162 (in Chinese).

Fang, Z. Y., Xu, J. M. and Zhao, F. S. (2004). Progress and prospect of China meteorological satellite and satellite meteorology. *ACTA Meteorological Sinica*, **62**(5), 550–560 (in Chinese, with English abstract).

Huang, F. X., Liu, N. Q., Zhang, P. *et al.* (2010). Comparison and evaluation of retrieved ozone profiles from solar backscatter ultraviolet sounder on FY-3A satellite. *Optics and Precision Engineering*, **18**(7), 1568–1576 (in Chinese).

Li, Q. (2001). Development of Chinese geostationary meteorological satellite. *Spacecraft Recovery and Remote Sensing*, **22**(1), 13–19 (in Chinese, with English abstract).

Li, Q. and Dong, Y. H. (2008). Achievement and forecast of meteorological satellite technology in China. *Aerospace Shanghai*, **1**, 1–10 (in Chinese, with English abstract).

Lu, Q. F., Bell, W., Bauer, P., Bormann, N. and Peubey, C. (2010). An evaluation of FY-3A data, ECMWF Technical Memorandum, 631.

Lu, Q. F., Bell, W., Bauer, P., Bormann, N. and Peubey, C. (2011). An evaluation of FY-3A satellite data for numerical weather prediction. *Quarterly Journal of the Royal Meteorological Society*, **137**, 1298–1311.

Meng, Z. H. (2004). The polar orbit meteorological satellite in China. *Engineering Science*, **6**(10), 1–5 (in Chinese, with English abstract).

National Satellite Meteorological Center (2004). User requirement of Fengyun 3 series (in Chinese).

National Satellite Meteorological Center (2006). Initiative design on the first engineering stage of Fengyun 3 series meteorological satellite application system (in Chinese).

National Satellite Meteorological Center (2008a). FY-3: A new era of Chinese meteorological satellites. FY-3A introduction brochure (in Chinese).

National Satellite Meteorological Center (2008b). Brief catechism of FY-3A polar orbital meteorological satellite (in Chinese).

Wang, W. H., Zhang, X. Y., An, X. Q. *et al.* (2010). Analysis for retrieval and validation results of FY-3 Total Ozone Unit (TOU). *Chinese Science Bulletin*, **55**(26), 3073–3043.

Xu, J. M., Niu, Y. S., Dong, C. H., Zhang, W. J. and Yang, J. (2006). Ground segments for FY meteorological satellites. *Engineering Science*, **8**(11), 13–18 (in Chinese, with English abstract).

Yang, J. (2008). Development and applications of China's Fengyun (FY) meteorological satellite. *Spacecraft Engineering*, **17**(3), 23–28 (in Chinese, with English abstract).

Yang, J., Dong, C. H. *et al.* (2010). *Operational Products and Applications on New Generation Polar Orbit Fengyun Meteorological Satellite*. China: Science Press (in Chinese).

Zhang, W. J. (2001). Status and development of FY series of meteorological satellites. *Aerospace Shanghai*, **2**, 8–13 (in Chinese, with English abstract).

Zhang, P., Yang, J., Dong, C. H. *et al.* (2009). General introduction on payloads, ground segment and data application of Fengyun 3A. *Frontiers of Earth Science in China*, **3**(3), 367–373.

Chapter 3

NASA Satellite and Model Land Data Services: Data Access Tutorial*

Suhung Shen[†,‡,¶] Gregory Leptoukh[†] and Hongliang Fang[§]

[†]*NASA Goddard Earth Sciences Data and Information Services Center, Greenbelt, MD, USA*

[‡]*George Mason University, Fairfax, VA, USA*

[§]*Institute of Geographic Sciences and Natural Resources Research (IGSNRR), Chinese Academy of Sciences, Beijing, China*

The NASA Earth Observing System (EOS) program began data acquisition in the early 1990s, focusing on observation of measurements for climate studies, such as temperature, precipitation, wind, greenhouse gases, aerosols, vegetation index, soil moisture, snow/ice, and ocean color. The huge amount of data acquired from satellites are archived and distributed at 12 NASA data centers. Data can be downloaded from a center where the data are archived or through a centralized search and order system. This document provides brief information about the land data products from NASA satellite measurement missions as well as land surface assimilation models. The data search and order methods, data tools, as well as sample data applications are documented.

3.1 Introduction of NASA Land Products

3.1.1 *NASA satellite missions on land observations*

The satellite observations on land surface processes began with the launch of the first in a series of Landsat satellites developed by National Aeronautics and Space Administration (NASA) and operated by US Geological Survey (USGS) (http://landsat.gsfc.nasa.gov/). Landsat-1 was launched on July 23, 1972, known as the Earth Resources Technology Satellite

---

*This chapter is based on training class notes for *The 2nd Summer School on Land Surface Observing, Modeling and Data Assimilation*, July 13–16, 2011, Beijing, China.

¶Corresponding author. Email: suhung.shen@nasa.gov

S. Liang, X. Li & X. Xie (eds.), *Land Surface Observation, Modeling and Data Assimilation*, 67–88.

(ERTS). Landsat-1 operated until January 1978, outliving its design life by five years. Six Landsat satellites were launched following Landsat-1. Currently, Landsat-5 and Landsat-7 are in operation. About 38 years of land surface data were collected from Landsat satellites. The next-generation land observation satellites, Landsat Data Continuity Mission (LDCM), was launched in February 2013.

The NASA Earth Observing System (EOS) program began data acquisition in the early 1990s, focusing on climate science areas: radiation, clouds, water vapor, and precipitation; the oceans; greenhouse gases; land surface hydrology and ecosystem processes; glaciers, sea ice, and ice sheets; ozone and stratospheric chemistry; and natural and anthropogenic aerosols. Detailed information about the NASA EOS program is documented in the *Earth Science Reference Handbook* by Parkinson *et al.* (2006) and can be found at the NASA EOS project science office website (http://eospso.gsfc.nasa.gov/). More than 20 satellites have been launched since 1991. Land surface products are obtained from a number of instruments, such as ETM on Landsat 7 (1999), MODIS on Terra (2000), and Aqua (2002); ASTER on Terra (2000); and SeaWiFS on OrbView-2 (1997). Other relevant measurements for land surface study in EOS program include precipitation from TRMM (1997), soil moisture and snow/ice water from AMSR-E on Aqua (2002), etc.

3.1.2 *NASA satellite land products, processing levels, resolutions, and data format*

Most NASA satellite land measurements are summarized in the NASA Land Measurement Portal (http://landportal.gsfc.nasa.gov/). The portal lists land products in four categories: Surface Radiation Budget, Vegetation Parameters, Land-Cover/Land-Use Changes, and Land Hydrosphere. The portal provides product metadata, including instrument name, spatial and temporal resolution, data archiving location, data access policy, etc. The following are some MODIS land products listed in the portal. Information about Land Cover Types of 30-m resolution from Landsat are also available in the portal:

- Land Surface temperature/Emissivity from MODIS at 1 km, 6 km, or 0.05° for daily, 8-day duration, and monthly.
- Vegetation index from MODIS at 250 m, 500 m, 1 km, or 0.05° for 16-day duration, and monthly.

Table 3.1. Typical NASA satellite data processing levels.

Levels	Description	Data type
Level-0	Source data, the raw radiance counts at full resolution	swath
Level-1A	The raw radiance counts with time and geolocation referred	swath
Level-1B	Calibrated radiance, or processed to sensor units, such as brightness temperature at full resolution	swath
Level-2	Derived geophysical variables (e.g., land surface temperature, vegetation index, and land cover type, etc.) after performing atmospheric correction at the same resolution and location of Level-1 data.	swath
Level-3	Geophysical variables mapped on uniform space-time grid scales at reduce resolution.	grid
Level-4	Geophysical variables derived from lower level data (i.e., variables are not measured by the instrument, but instead are derived from multiple measurements by an analysis model)	grid

- Land cover types from MODIS at 250 m or 500 m for yearly; Thermal Anomalies/Fire from MODIS at 500 m or 1 km for daily, 8-day duration, and monthly.
- Snow cover from MODIS at 500 m and 0.05° for daily and 8-day duration.
- Evapotranspiration from MODIS at 0.05° for 8-day duration and annually.

Most standard NASA satellite data have three processing levels, as listed in Table 3.1. Some data have been processed to Level-4. Level-0 and Level-1A products are the raw radiance counts at full (or instrument) resolution. For example, the spatial resolutions of the MODIS L0 products are 250 m, 500 m, or 1 km at nadir for spectral bands 1–2, 3–7, and 8–36, respectively. Level-1B products are calibrated radiances or brightness temperatures at full resolution. Level-2 products are derived geophysical variables (e.g., land surface temperature, vegetation index, cover type, etc.) after performing atmospheric correction. Level-3 products are geophysical variables mapped on uniform space-time grid scales at reduced resolution. For example, the global daily (MOD11C1) and monthly (MOD11C3) data are land surface temperatures at 0.05° resolution from MODIS-Terra. Level-3 data are used most popularly because their regular spatial and temporal structure allows easy analysis. Level-4 products are geophysical variables that are not retrieved directly from satellite measurements, but derived from these measurements by an analysis model. Example of a Level-4 product is a merged field from multiple instruments.

The common data formats of NASA EOS satellite data are HDF (Hierarchical Data Format, http://www.hdfgroup.org/), HDF-EOS (Hierarchical Data Format — Earth Observing System, http://hdfeos.org/, http://www.hdfgroup.org/hdfeos.html), or NetCDF (Network Common Data Form, http://www.unidata.ucar.edu/software/netcdf/). All these data formats are self-describing, enable an application to interpret the data structure and contents without any outside information, and are platform-transparent. Commercial and free tools are available to support the above-mentioned data formats.

3.1.3 *NASA land assimilation model products*

The Global Land Data Assimilation System (GLDAS) is a land surface data assimilation system developed by the Hydrological Science Branch at NASA Goddard. The system integrates satellite and ground-based observational data products, using advanced land surface modeling and data assimilation techniques, for generating optimal fields of land surface states (e.g., soil moisture and surface temperature) and fluxes (e.g., evaporation and sensible heat flux) parameters (Rodell *et al.*, 2004). The current GLDAS includes data from four land surface models (LSMs): Mosaic, Noah, the Community Land Model (CLM), and the Variable Infiltration Capacity (VIC). More details about GLDAS models can be found at GLDAS Readme (ftp://agdisc.gsfc.nasa.gov/data/s4pa/GLDAS/README.GLDAS.pdf) and NASA Land Data Assimilation System website (http://ldas.gsfc.nasa.gov/).

The temporal resolution for the GLDAS products is 3-hourly. Monthly products are also generated through temporal averaging of the 3-hourly products. Data is available from January 1, 1979 to the present. The spatial resolutions are $0.25^\circ \times 0.25^\circ$ or $1^\circ \times 1^\circ$. The regional higher resolution (0.125°, hourly) land assimilation data (NLDAS) are available over North America. The data are archived and distributed by the NASA Goddard Earth Science Data and Information Services Center (GES DISC) (http://disc.sci.gsfc.nasa.gov/) through direct ftp service, the search and ordering system, Mirador (http://mirador.gsfc.nasa.gov/), and via GrADS Data Server (GDS) http://agdisc.gsfc.nasa.gov/dods/. The GDS service allows a user to access the data, perform subsetting and analysis operations without first downloading them. A more advanced online visualization and analysis tool (Giovanni) (http://disc.sci.gsfc.nasa.gov/giovanni/) is also available to GLDAS and NLDAS data, which provides a simple and intuitive way to visualize, analyze, and access vast amounts of data without resorting to data download.

3.2 Search and Order NASA Earth Science Data Products

3.2.1 *NASA earth science data centers*

The Earth Observing System Data and Information System (EOSDIS) manages and distributes NASA Earth sciences data through 12 Distributed Active Archive Centers (DAACs). Each data center serves a specific Earth system science discipline and provides users with data products, services, and data-handling tools unique to the center's specialty (http://earthdata.nasa.gov/data/data-centers). The data product can be found through the centralized searching and ordering system or from a specific data center where data are archived. A brief summary of land product data centers is in Sec. 3.2.3.

3.2.2 *Find and access data from the centralized systems*

3.2.2.1 *Search and access data from Reverb*

Reverb (http://reverb.echo.nasa.gov) is a web-based metadata and service discovery tool. Reverb, originally Warehouse Inventory Search Tool (WIST), is a primary access point of more than 2000 NASA EOS and other Earth science data sets. You may browse data products without having to login or by login as a guest user. However, registered users may access to some restricted data. Here is an example of steps to order MODIS-Aqua monthly land surface temperature from Reverb:

(a) In "Search Options" on the left column, select "Platform & Instrument" as Aqua and MODIS, select "Processing Level" as 3, and enter the "Search terms" as land surface temperature. It will list all Level-3 land surface temperature products from MODIS-Aqua.
(b) Select one or more data set, say "MODIS/Aqua Land Surface Temperature/Emissivity Monthly L3 Global 0.05Deg CMG V005".
(c) Refine the list in (b) by defining "Spatial Search" area: click, hold, and drag on the map, or enter longitude/latitude values of four corner points. The default area is global. In this case, the default option is selected.
(d) Further refine the list by giving temporal range as "2009-01-01" to "2009-12-31".
(e) Click on "Search for Granules".

(f) Twelve granules (files) are found and displayed on the searched result page. In addition to granule ID, the result page lists other information about the searched granules, such as metadata, data archive center, image quick look, and source page about data quality etc. Now, select items or click "All on page" to add granules into the shopping cart, then click on "View Items in Cart" at the bottom of the page.
(g) Select data granules that you wish to order, then click on "Order Selected".
(h) Fill the "Order Form", then click on "Proceed".
(i) Review items and specify ordering option at this time if not defined, then click on "Proceed".
(j) At this time, you will see a page that contains your contact/shipping information, and ordered items. If satisfied, click on "Submit Order".
(h) You will receive an e-mail containing data downloading instruction.

3.2.2.2 *Find data from ECHO*

NASA EOS Clearinghouse (ECHO) (https://www.echo.nasa.gov/) is a warehouse database to archive EOS product metadata from a variety of science disciplines and domains, including Climate Variability and Change, Carbon Cycle and Ecosystems, Earth Surface and Interior, Atmospheric Composition, Weather, and Water and Energy Cycle. All products in Reverb system are in ECHO database. Products in ECHO can be navigated by topics, instrument, campaign, data center, etc. On the search or navigation result page, metadata, such as short description of the product, temporal coverage, and the link to data ordering page, are provided. ECHO has been working with other organizations to provide their Earth science metadata alongside NASA's for users to search and access.

3.2.2.3 *Find data from GCMD*

The Global Change Master Directory (GCMD) (http://gcmd.gsfc.nasa.gov) is a comprehensive source of satellite and *in situ* Earth science data that is collected by not only NASA but also other international data agency and Universities, with broad coverage of the atmosphere, hydrosphere, oceans, solid Earth, and biosphere. The system provides keyword search and navigation to find the metadata of a data product. For example to find the same MODIS-Aqua monthly land surface temperature product:

- Click on LAND SURFACE → LAND TEMPERATURE → LAND SURFACE. More than 250 items are listed; refine by entering "MODIS" in Refine by Full Text, item number is reduced to 63; refine again by entering "monthly", the item number is reduced further. (Note, the searched item numbers subject to change since the system is updated actively.)

- Click "MODIS/Aqua Land Surface Temperature/Emissivity Monthly L3 Global CMG V005". It displays a page with metadata information of this product, including product description, geographic coverage, spatial resolution, instrument/platform, data source link, etc.
- Under "Related URL", a link was provided to the FTP service (Data Pool) of the product, where you may download data through FTP.

3.2.3 *Find land data from data archive centers*

The NASA data center archives and serves a specific Earth science discipline. For example, physical oceanography data are archived at PO.DAAC (http://podaac.jpl.nasa.gov/) and socioeconomic data are archived at SEDAC (http://sedac.ciesin.columbia.edu/). The land products are archived and distributed at several data centers described below. Additional information about MODIS land products is available at the NASA MODIS Land project (http://modis-land.gsfc.nasa.gov/).

3.2.3.1 *LP DAAC (https://lpdaac.usgs.gov/)*

Land Processes Distributed Active Archive Center (LP DAAC) processes, archives, and distributes many land data and products derived from the EOS sensors. LP DAAC holds land data from MODIS and ASTER on Terra, and MODIS on Aqua. Archived MODIS data are Level-2 (swath) and Level-3 (grid) in HDF-EOS format. The spatial resolution range of the MODIS Level-3 products vary from 250 m to 6000 m. The ASTER data are Level-1 (swath), Level-2 (swath), and Level-3 (grid) in HDF-EOS or GeoTIFF format. The Level-3 data at resolution of 1 km or higher are stored in $10^\circ \times 10^\circ$ tiled files.

3.2.3.2 *GES DISC (http://disc.gsfc.nasa.gov/)*

Goddard Earth Sciences Data and Information Services Center (GES DISC) archives and distributes satellite and model data of multiple disciplines, including hydrology, atmospheric composition and dynamic, as well as land surface process. The archived satellite land products at GES DISC are global products at $1^\circ \times 1^\circ$ resolution and regional products at higher resolution (1 km and 500 m) for Monsoon Asia region. Global and North American regional LSM data are also available at GES DISC.

3.2.3.3 *NSIDC (http://nsidc.org/daac/)*

The National Snow and Ice Data Center (NSIDC) is part of the Cooperative Institute for Research in Environmental Sciences at the University of Colorado at Boulder. NSIDC provides cryosphere-related data and information, including snow, ice, glaciers, frozen ground, and climate interactions from field and satellite observation. Example satellite land data are soil moisture, and snow water equivalent from AMSR-E.

3.2.3.4 *OBPG (http://oceancolor.gsfc.nasa.gov/)*

The Ocean Biology Processing Group (OBPG) at NASA GSFC process and archive ocean biology data from sensors, such as SeaWiFS, MODIS, OCTS, and MERIS etc. Land vegetation index (NDVI) at 4 km and 9 km resolution from SeaWiFS are produced and distributed by OBPG as the additional products.

3.2.4 *Access Landsat data*

All Landsat data in the USGS archive are accessible for free since 2008. The data are available from Landsat-1 through Landsat-7 since 1972, with the resolution as high as 30 m. Data and information are available from following services:

- USGS GioVis, The Global Visualization Viewer (http://glovis.usgs.gov/): A web-based tool to search and order data by clicking on an interactive map or by entering the geographic coordinates (latitude and longitude) of the site you are searching for.
- USGS Earth Explorer (http://edcsns17.cr.usgs.gov/EarthExplorer/): A web-based tool that allows you to customize and tailor your search parameters for Landsat data.
- GLCF — Global Land Cover Facility at University of Maryland (http://glcf.umiacs.umd.edu/index.shtml): This site provides satellite imagery of land cover and other land products from ASTER, IKONOS, Landsat, MODIS, QickBird, Orbview, and SRTM, as well as products from AVHRR, GOES, Landsat, and MODIS, etc.
- LEDAPS — Landsat Ecosystem Disturbance Adaptive Processing System (http://ledaps.nascom.nasa.gov/): This site is a NASA-funded project to map North American forest disturbance since 1975 from the Landsat and ASTER satellite data.

- Land Measurement Portal (http://landportal.gsfc.nasa.gov/): This portal provides comprehensive metadata of land measurement from NASA EOS, NOAA and other agencies and institutions.

3.2.5 *Access data from GES DISC*

Through the centralized data search and ordering system, such as Reverb, a user can order data archived at multiple data centers. However, this process is relatively slow. If the center where the data archived is known, the data can be accessed from that center directly. In general, the data center provides more tools to serve the data. For example, GES DISC is one of the 12 NASA Earth Science data centers, holding data and information of precipitation, atmospheric chemistry and dynamics, and land surface process from satellite observation and assimilation models. All data from GES DISC are open to the public. No registration is required to access the data. This section describes several data accessing methods at GES DISC, including Mirador, OPeNDAP, GDS, OGC Web Map Service, and Giovanni.

3.2.5.1 *Mirador (http://mirador.gsfc.nasa.gov/)*

Mirador is a major data search and access system at GES DISC, which has a very simple and easy-to-use interface that can perform Google-like keyword search for Earth science data (Lynnes *et al.*, 2009). Mirador features include data file hit estimator, an interactive shopping cart, and spatial and parameter subsetting (this feature is available for selected products). In addition, the system enables search through navigation of projects and science area by applying semantic web technology. A list of products available in the Mirador system can be obtained by clicking "Data Holding" from the Mirador page. The following is an example for accessing soil moisture from GLDAS LSM. From the Mirador page (http://mirador.gsfc.nasa.gov/):

- **Using Keyword Search**
  (a) Enter Keyword as "soil moisture"; location [(minLat, minLon), (maxLat, maxLon)]: (15, 5), (55, 135) over China; and Time Span: 2000-01-01, 2000-01-31.
  The system returns a list of products found that contains soil moisture parameter. The list has product information, file

numbers, parameters, and spatial and temporal resolution. If the list is too long, and narrowing it down is preferable, go back to "keyword" search page, and add more keyword. For example, add model name in Keyword as "GLDAS CLM soil moisture".

(b) Select the product, say "GLDAS CLM Land Surface Model L4 Monthly 1.0 degree × 1.0 degree (GLDAS_CLM10_M)", then click on "Add selected files to cart".

(c) Select service by clicking available services, say "convert to NetCDF". This is optional. If no service is selected, the original data format and structure will be provided.

(d) Click on "Continue to shopping cart".

(e) Click on "check out" (the order cart can be modified at this time).

(f) Follow instructions to download data. Multiple download methods are available. You may select the one that works best for you.

- **Using Projects Navigation**

  There are two types of projects, one is satellite missions, such as AIRS, OMI, and TRMM, and the other is research projects, such as A-Train, NEESPI, and GLDAS.

  In this case, navigate in the following path to get the above same soil moisture product.

  Projects → GLDAS → Common Land Model (CLM) → GLDAS_CLM10_M.001 → 2000.

  Then do the steps (c) to (f) in Keyword Search.

- **Using Science Area Navigation**

  This is for navigation parameters starting with science areas, for example: Science Areas → Water and Energy Cycles → Soil → Average Layer Soil Moisture → GLDAS_CLM10_M.001→2000.

  Then do steps (c) to (f) in Keyword Search.

3.2.5.2 *GDS and OPeNDAP*

Some data at GES DISC are available through OPeNDAP and GrADS Data Service (GDS) services, which are software framework that allows one to work on remote data without downloading the data. Data in OPeNDAP service can be accessed directly by using tools, such as IDV, Panoply, Ferret, and GrADS, etc. GDS provides subsetting and analysis services across the internet for any GrADS-readable dataset. For example, GLDAS datasets are available through GDS. The following are examples to access GLDAS data in GDS service:

- Perform global averaging analysis for soil moisture content

```
$ grads
ga-> sdfopen
http://agdisc.gsfc.nasa.gov:80/dods/_expr_{GLDAS_MOS10SUBP_
3H}{aave(soilm3.1,lon=-180,lon=180,lat=-60,lat=90)}{0:360,-60:90,
1:1,00z22jan1980:21z22jan1980}
ga-> d
Result value = 443.626 mm
```

- Sample GrADS script to access GLDAS data via GDS
  Save the following script into a file named: *map_gldas.gs*

```
'reinit'
'sdfopen http://agdisc.gsfc.nasa.gov:80/dods/GLDAS_NOAH025_M'
'set lon -180 180'
'set lat -60 90'
'set gxout grfill'
'set grads off'
'set t 92'
'd avgsurft'
'set rbcols'
'run cbarn'
'draw title GLDAS Noah Monthly 0.25 degree Average Surface
Temperature\in October, 2007 [k]'
'printim GLDAS_NOAH025_M_avgsurft.A200710.001.gif white'
```

  Run the script as following and generate the global image of average surface temperature for October 2007 as shown in Fig. 3.1.

```
$grads
ga->map_gldas
```

3.2.5.3 *Online visualization tools*

GES DISC provides quick and easy tools to allow users exploring data easily online without downloading data to the local system. Open Geospatial Consortium (OGC) Web Map Service (WMS) is an interface that allows the use of data and enables clients to build customized maps with data coming from a different network. Currently, data in OGC WMS at GES DISC are these from AIRS, TRMM, and OMI. Many satellite and model data are available through Giovanni, which is a GES DISC developed online

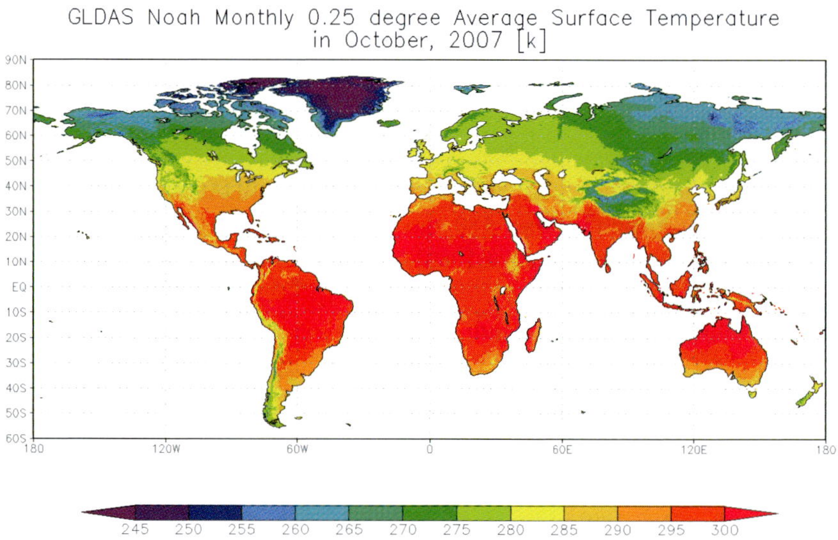

Fig. 3.1. Sample image generated from GLDAS data through GDS service at GES DISC.

visualization and analysis tool. More about Giovanni will be described in the Sec. 3.3.1.

3.3 NASA Online Visualization Services

Online visualization tools are provided by different NASA groups to allow quick exploration of Earth science data. Each tool was designed for specific purpose and has its unique features. Five popularly used online visualization services are described briefly in the following.

3.3.1 *Giovanni (http://disc.sci.gsfc.nasa.gov/giovanni/)*

Giovanni (GES-DISC Interactive Online Visualization ANd aNalysis Infrastructure) is a web-based application developed by GES DISC, that can visualize and analysis data online by a few clicks without downloading the data (Acker and Leptoukh, 2007; Berrick *et al.*, 2009). The Giovanni database holds more than 1000 geophysical parameters. Visualization and statistical analysis features include maps, time series, cross-section, profile, scatter plots, correlation maps, etc. The Giovanni system generates images in formats as png, and KMZ for Google Earth. The processed data can be downloaded in formats as ASCII, hdf, or NetCDF. To use Giovanni, only internet and a browser are needed; there is no need to install data processing software or downloading data. The current Giovanni system consists of more than 30 portals (interfaces) for different collection of data. Sample Giovanni portals that have land products are: (a) global land assimilation models, http://gdata1.sci.gsfc.nasa.gov/daac-bin/G3/gui.cgi?instance_id=GLDAS10_M; (b) global atmospheric assimilation model, http://gdata1.sci.gsfc.nasa.gov/daac-bin/G3/gui.cgi?instance_id=MERRA_MONTH_2D; and (c) land data collections for MAIRS project: http://gdata1.sci.gsfc.nasa.gov/daac-bin/G3/gui.cgi?instance_id=mairs_monthly. It is requested that the GES DISC be specifically and clearly acknowledged if Giovanni (or data downloaded from Giovanni) is used for data analyses and visualizations in publications, posters, oral presentations, reports, web pages, and other types of scientific media. Examples of acknowledge statements can be found on the Giovanni page.

3.3.2 *MODIS Rapid Response System (http://lance.nasa.gov/imagery/rapid-response/)*

MODIS Rapid Response System is an image gallery that provides high-quality, near real-time and past-time true-color and false-color imagery to support monitoring and study natural hazards, such as fire, dust storm, and hurricane etc. Images are at resolutions of 1 km, 500 m, or 250 m generated from Level-2 products and in formats of gif, KMZ for Google Earth, and Worldfile for GIS. Links are provided to data used to generate MODIS images.

3.3.3 *NASA Earth Observations (NEO) (http://neo.sci.gsfc.nasa.gov)*

This tool is designed for outreach and education purpose. A user can explore remote sensing products from an easy use web interface by generating images on-the-fly. The generated image can be mapped onto Google Earth as well.

3.3.4 *NASA Earth Observatory (http://earthobservatory.nasa.gov)*

Earth Observatory is another outreach and education site of NASA. In addition to pre-generated high-quality images, a story/article is written about the image. Images and stories about natural hazards (fires, dust storms, floods, droughts, volcanoes, and earthquakes) are documented. You may sign up to receive daily or weekly electronic newsletters to keep track of recent natural hazards stories. Global maps and articles of precipitation, surface temperature anomaly, snow cover, vegetation, etc., are also available.

3.3.5 *NASA visible earth (http://visibleearth.nasa.gov)*

This is a huge archive of high-quality images and animations from many instruments of NASA, NOAA, DOD, and non-US agencies. Images can be searched by key words, satellite, sensor, country, collection, and GCMD topics, etc., and be downloaded freely.

3.4 Support Research Projects and Sample Usage of Data and Services

This section introduces two projects at GES DISC to support researches on land cover and land use changes and climate variations by providing data and services. Samples to plot satellite and model data by using Giovanni are presented.

3.4.1 *NASA data to support research projects: NEESPI and MAIRS*

The NASA NEESPI and MAIRS data portals at GES DISC are funded by the NASA land cover and land use change program to support international programs, Northern Eurasia Earth Science Partnership Initiative (NEESPI) (http://neespi.org/), and Monsoon Asia Integrated Regional Study (MAIRS) (http://www.mairs-essp.org/), by providing satellite remote sensing and model data and information. The NEESPI and MAIRS programs focus on studies of land processing and climate variations at Northern Eurasia and Monsoon Asian regions, respectively.

The NEESPI data support project (http://disc.sci.gsfc.nasa.gov/neespi/) was started in 2006 (Leptoukh *et al.*, 2007). The infrastructure of an automated data management system was established to support the data end-to-end, including data transfer, data format convert, and data ingest into archive and distribute system. Tools are provided to read and analysis data. Products collected and processed are land cover types, land surface temperature, soil moisture, vegetation index, active fire, and snow/ice, etc.

The spatial resolution of products is $1^\circ \times 1^\circ$ for global coverage. Customized Giovanni interfaces for monthly, 8-day, and daily data have been created to allow exploring data from multiple instruments of multiple disciplines.

Following successful support of the NEESPI program by NASA satellite remote sensing data, the MAIRS data support project (http://disc. sci. gsfc.nasa.gov/mairs/) began in 2009. It uses and leverages the established data management and service infrastructure for NEESPI. Customized Giovanni portals have been created for MAIRS that contains satellite observations from multiple sensors and model output from GLDAS and MERRA. Higher-resolution satellite data, such as 1-km land surface temperature, vegetation index, etc., are integrated into the system to support MAIRS regional studies. Due to large overlap of geographic coverage and many similar scientific interests of NEESPI and MAIRS, the collected data and developed tools will serve for both projects.

3.4.2 *Sample plots by using Giovanni*

3.4.2.1 *Basic visualization features*

Longitude–latitude map and time series are basic visualization features of Giovanni for analyzing a single parameter. The plotted image or time series can be modified to have the best results. For example, the map can be viewed with different projections, such as Equidistant Cylindrical, North/South polar stereographic, and Robinson projection; the color palette and size of images are adjustable; and the axis of time series can be modified as well. Other features for analyzing a single parameter include cross-section plots (time-longitude/latitude, vertical profile, etc.), zonal mean, animation, trend of a time series, as well as climatology and anomaly analysis for some products.

The Giovanni database contains more than 1000 geophysical parameters observed by many satellite sensors and output from numerical models. In addition to the basic visualization functions for studying a single parameter, the system is able to compare similar parameters from different sources, such as difference or overlay of two parameters and scatter plots. An analysis of the relationship between any two parameters is possible by calculating the correlation coefficient.

3.4.2.2 *Monitoring natural hazard events*

Natural hazard events, such as dust storms, hurricanes, and forest fires, occur frequently around the world. The NASA satellite data and

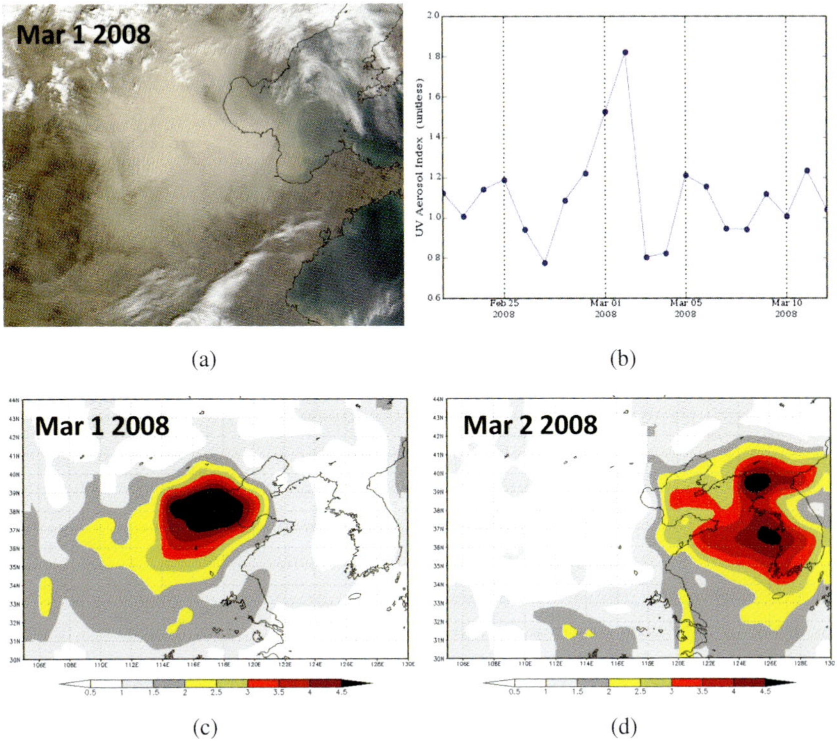

Fig. 3.2. (a) True color images from MODIS Terra on March 1 03:20Z and (from NASA MODIS Rapid Response System) (upper left) time series of daily UV aerosol index from February 22 to March 12 over (110°E–130°E, 30°N–40°N) (upper right); (b) daily UV aerosol index from OMI of March 1 and March 2 (lower left and middle); time series of daily UV aerosol index from OMI of March 1 and March 2 (lower).

visualization tools allow one to view and study events easily. Figure 3.2 illustrates a dust storm over East China during March 1–2, 2008. The true color images captured by MODIS Terra (from NASA MODIS Rapid Response System) on March 1, 03:20Z, 2008 show a heavy yellow dust plume over East China near Beijing. It moved eastward and appeared over Yellow Sea on March 2. The storm was observed by a different instrument, OMI on AURA as well. The lower left and middle images of Fig. 3.2 show daily UV aerosol index from OMI for March 1 and 2, respectively, generated by using Giovanni. The maximum value of the elevated UV aerosol index values exceed 4.5 over the location where the dust plume appears in the true color image. The time series of daily OMI UV aerosol index from February 22

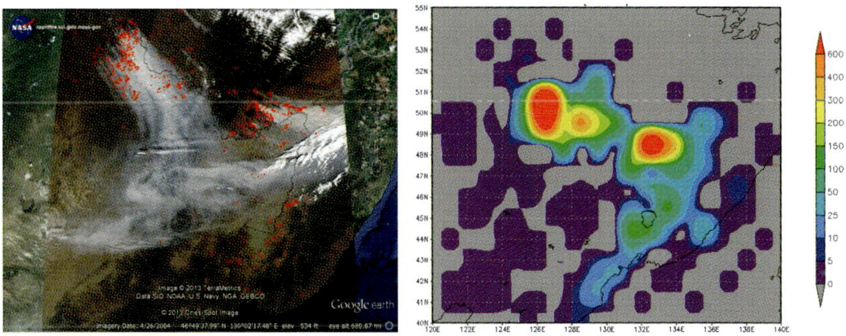

Fig. 3.3. The forest fire over Northeast China captured by MODIS-Aqua on October 15, 04:20 UTC (image from NASA MODIS Rapid Response System displayed on Google Earth) (left panel) and monthly active fire counts from MODIS-Terra in October 2004 (right panel).

to March 12 over East China and Yellow Sea (110°E–130°E, 30°N–44°N), plotted by using Giovanni, gives a different view of the significant aerosol increases on March 1 and 2.

The next case is the forest fire over Northeast China in October 2004. The fire broke out in the afternoon of October 14, 2004, in Heihe, Heilongjiang, China. Due to dry and windy weather, the fire lasted for about 6 days and destroyed ~1000 km^2 of forest land (State Forestry Administration in China 1995–2006). The fire was observed by a number of NASA satellites. Figure 3.3 shows the true color image of fire captured by MODIS-Aqua on October 15, 04:20 UTC (from NASA MODIS Rapid Response System). The image is displayed on Google Earth, indicating clearly the geolocation of the fire. The heavy smoke and soot caused by the fire are observed even in the monthly mean aerosol optical depth from MODIS-Terra (Figure 3.3, right panel) generated from Giovanni. The smoke and gases from the fire were observed clearly from space. Figure 3.4 shows images (from Giovanni) for UV aerosol index and column NO_2 from the Ozone Monitoring Instrument (OMI), and CO volume mixing ratio at 407 hPa from AIRS averaged from October 13–16, 2004. A study of burned area, annual emissions of polycyclic aromatic hydrocarbons (PAHs) by Yuan *et al.* (2008) indicated that approximately 30 Mg of PAHs were released within a week. The article stated that the smoke from the forest fires, together with conflagrations spread over the Sea of Japan, caused considerable haze over Hokkaido, Japan.

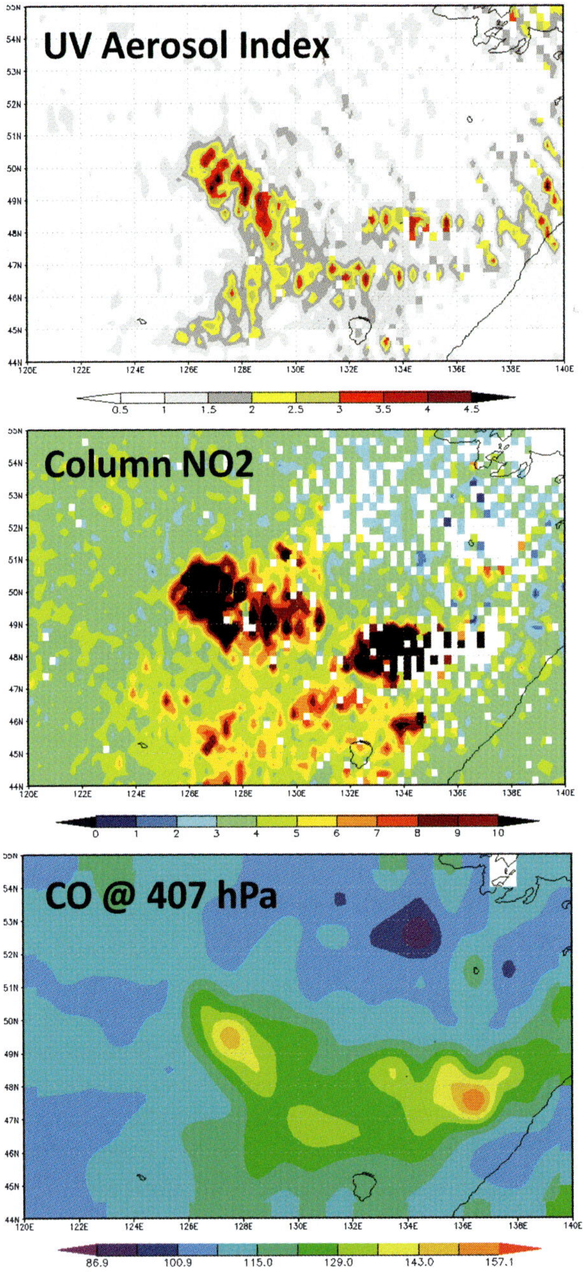

Fig. 3.4. UV aerosol index from OMI (upper), column NO_2 from OMI (center), and CO volume mixing ratio at 407 hPa from AIRS (lower) averaged from October 13–16, 2004.

3.4.2.3 *Studying seasonal and interannual variations*

This example is the extension of Shen *et al.* (2009)'s work by using newer version and including more data to study seasonal and interannual variations of monthly active fires over temperate Europe. Satellite observations show that in temperate Europe, fires or hot spots occur mainly over croplands as shown in Fig. 3.5 (upper image). Time series of fire pixel count

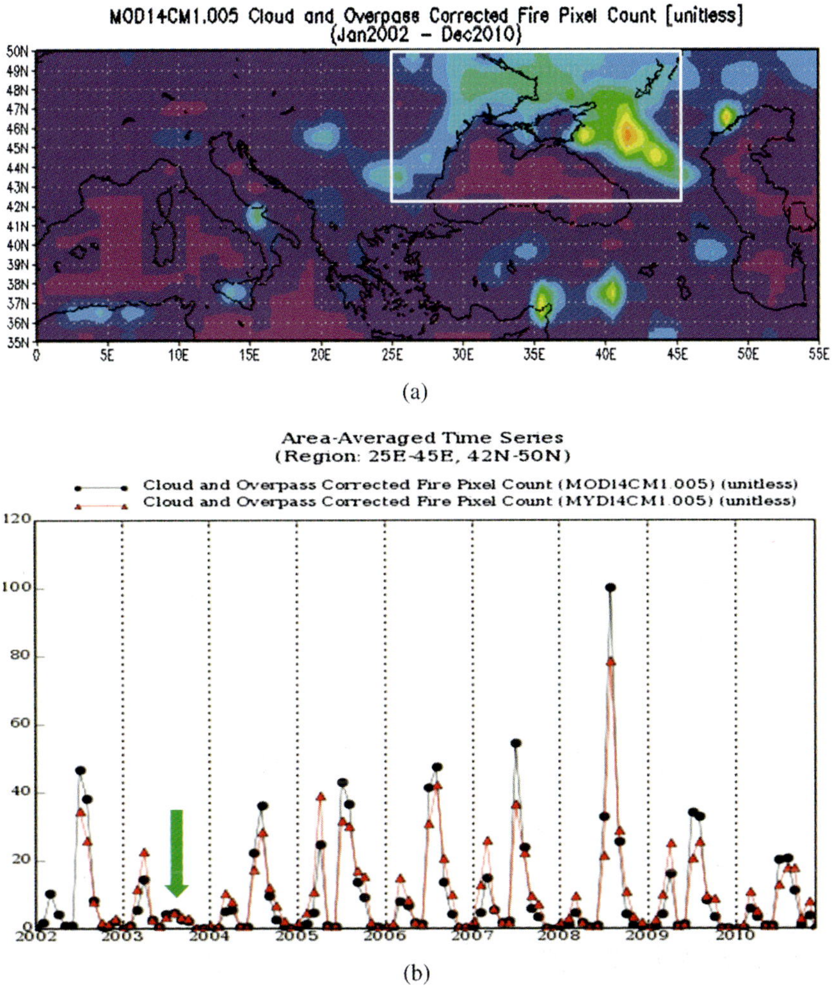

Fig. 3.5. (a) Averages of MODIS fire pixel count from 2002.01 to 2010.12 (upper); (b) time series of monthly fire pixel count over the boxed area in (a) (lower).

over the major fire region (boxed area, 25°E–45°E, 42°N–50°N) shows a clear seasonal cycle of maximum in summer (July–August), second peak in spring (March–April), and minimum during winter. The results from MODIS-Terra and MODIS-Aqua are consistent. The active fires over the croplands may be associated with agricultural burning of crop residue for inexpensive and quick stubble removal while adding nutrients to the soil and killing weeds and pests at the same time.

The interannual variation of the fire count value is significant. In particular, the value of 2003 summer is very low, even lower than the value in the spring. The averaged summer fire count value is reduced from 2001 to 2003 and then increased until 2008. Taking into consideration that the fire in this region may be associated with crop growth, the interannaul variations of precipitation and vegetation index are studied. For doing this, monthly time series of the same area are generated from Giovanni for precipitation, NDVI, and daytime land surface temperature (LST). It is found that the precipitation in the spring to early summer is anomalous low in 2003 which has caused anomalous low crops (NDVI) in this region (not shown).

3.4.2.4 *Urbanization and local climate*

NASA, GES DISC, MAIRS project has processed 8-day 1-km MODIS LST for the Monsoon Asia region (60°E–150°E, 0°–60°N), which is feasible to study regional climate variations, such as that associated with urbanization. Images in Fig. 3.6 are daytime LST of the summer (June–August) for

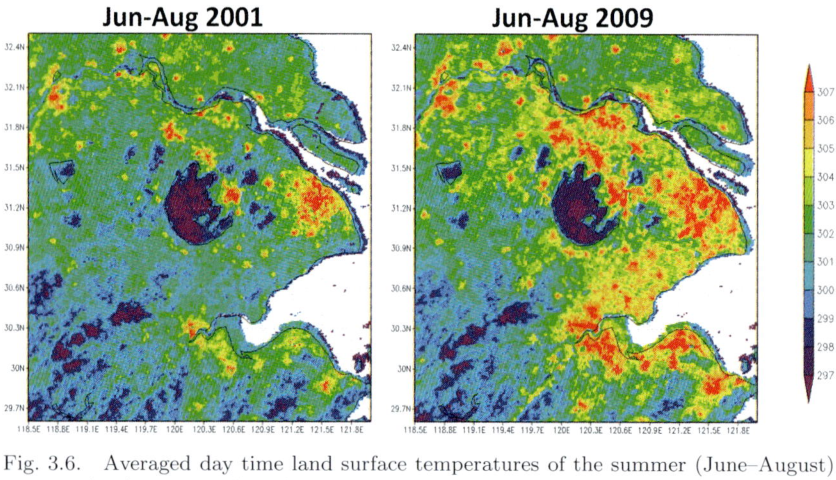

Fig. 3.6. Averaged day time land surface temperatures of the summer (June–August) for 2001 (left) and 2009 (right) from MODIS-Terra over the Yangtze River Delta region.

2001 (left) and 2009 (right) from MODIS-Terra over the Yangtze River Delta region. The urban areas are significantly warmer than the rural by about 5°K due to urban heating island effects. More interestingly, comparing 2001, the warm areas of 2009 are clearly expanded around many cities, such as Shanghai, Hangzhou, Suzhou, etc., indicating the fast urbanization in the Yangtze River Delta region during last decade.

More sample studies by using Giovanni system can be found in the list of publication and science focus articles on the GES DISC web page.

3.5 Summary

The NASA EOS program has collections of atmospheric, land, oceanic, and socioeconomic data from satellite observation and model output, managed by EOSDIS project. Due to extreme large volumes, the data are archived in 12 data centers. The majority land data archived at four data centers: LP DAAC, NSIDC, GES DISC, and OBPG. The data can be accessed either from the centralized system, such as Reverb, GCMD, and ECHO, or from the center where the data are archived. Each data center provides a number of tools to service the data. For example, data at GES DISC are serviced through one or more methods: Mirador, OPeNDAP, GDS, OGC Web Map Service, and Giovanni.

In addition to online data downloading services, NASA provides a number of visualization services, such as Giovanni, NEO, MODIS Rapid Response System, Earth Observatory, Visible Earth, etc., each possessing its own unique features. In particular, Giovanni, an interactive online visualization and analysis system, is extremely useful to quickly explore data. Sample studies by using Giovanni are presented, such as monitoring events, seasonal and interannual variation, as well as urbanization.

Moreover, NASA EOSDIS publishes bulletin annually since 1994, named *Sensing Our Planet*. The bulletin contains featured research articles that illustrate the use of Earth-observing remote sensing data from NASA earth science data centers. The articles in *Sensing Our Planet* are available online at http://nasadaacs.eos.nasa.gov/articles/index.html. Hard copies can be ordered by sending email to nasadaacs@eos.nasa.gov.

Most NASA data are open to the public with no charge. We recommend users to acknowledge data producer and data center/service when publishing paper or in any type of use of the data. Questions regarding data accessing can be sent to help-desk at each data center.

Acknowledgments

This project is supported by NASA ROSES 2008 (NNH08ZDA001N-LCLUC) program.

Relevant Links

Global Change Master Director (GCMD): http://gcmd.gsfc.nasa.gov
NASA EOS program: http://eospso.gsfc.nasa.gov/
NASA EOS data and documents searching system, ECHO: http://www.echo.nasa.gov/
NASA EOS data order primary entry point, Reverb: http://reverb.echo.nasa.gov/
NASA Earth System Science Data and Services: http://nasadaacs.eos.nasa.gov/
NASA Land Measurement Portal: http://landportal.gsfc.nasa.gov/
NASA MAIRS data support project: http://disc.gsfc.nasa.gov/mairs/
US Landsat program: http://landsat.gsfc.nasa.gov/
Web-based visualization tool, Giovanni: http://daac.gsfc.nasa.gov/giovanni

References

Acker, J. and Leptoukh, G. (2007). Online analysis enhances use of NASA Earth science data. *EOS Transactions American Geophysical Union*, **88**, 14–17.

Berrick, S. W., Leptoukh, G., Farley, J. D. and Rui, H. (2009). Giovanni: A Web service workflow-based data visualization and analysis system. *IEEE Transactions on Geoscience and Remote Sensing*, **47**(1), 106–113.

Leptoukh, G., Csiszar, I., Romanov, P., Shen, S., Loboda, T. and Gerasimov, I. (2007). NASA NEESPI data center for satellite remote sensing data and services. *Environmental Research Letters*, **2**, 045009.

Lynnes, C., Strub, R., Seiler, E., Joshi, T. and MacHarrie, P. (2009). Mirador: A simple, fast search interface for global remote sensing data sets. *IEEE Transactions on Geoscience and Remote Sensing*, **47**(1), 92–96.

Parkinson, C., Ward, A. and King, M. D. (eds.) (2006). *Earth Science Reference Handbook — A Guide to NASA's Earth Science Program and Earth Observing Satellite Missions*. Washington, DC: National Aeronautics and Space Administration.

Rodell, M., Houser, P. R., Jambor, U., Gottschalck, J., Mitchell, K., Meng, C.-J., Arsenault, K., Cosgrove, B., Radakovich, J., Bosilovich, M., Entin, J. K., Walker, J. P., Lohmann, D. and Toll, D. (2004). The global land data assimilation system. *Bulletin of the American Meteorological Society*, **85**(3), 381–394.

Shen, S., Leptoukh, G., Loboda, T., Csiszar, I., Romanov, P. and Gerasimov, I. (2009). The NASA NEESPI data portal to support studies of climate and environmental changes in non-boreal Europe. In *Regional Aspects of Climate-Terrestrial-Hydrologic Interactions in Non-boreal Eastern Europe*, Groisman, P. Y. and Ivanov, S. (eds.), pp. 9–16, p. 255. Dordrecht: Springer.
State Forestry Administration in China. (1995–2006). *China Forestry Yearbook*. Beijing, China: China Forestry Press.
Yuan, H., Tao, S., Li, B., Lang, C., Cao, J. and Coveney, R. M. (2008). Emission and outflow of polycyclic aromatic hydrocarbons from wildfires in China. *Atmospheric Environment*, **42**(28) 6828–6835.

Part 2

Modeling

Chapter 4

Land Surface Process Study and Modeling in Drylands and High-Elevation Regions

Yingying Chen and Kun Yang

Institute of Tibetan Plateau Research, Chinese Academy of Sciences (CAS)

Drylands and high-elevation regions have sensitive responses to global changes. Land–atmosphere interactions in these regions also play an essential role in atmospheric circulation. However, some key processes of land–atmosphere interactions in these regions remain to be understood. In this chapter, we introduce existing problems of land surface modeling in drylands and high-elevation regions (mainly with reference to the Tibetan Plateau) and the efforts to resolve these issues, including the following three aspects. First, to identify land surface modeling uncertainties, we performed a climate model intercomparison and evaluation against observations in these regions. Second, to address the modeling deficiencies, a parameterization scheme for thermal roughness length was developed for drylands; a single-source model and an inverse estimation system for soil properties were developed to account for the impacts of the soil vertical heterogeneity; and a soil resistance scheme was developed for evaporation. Third, the parameterizations were incorporated into land surface models (LSMs) and their effectiveness was evaluated.

4.1 Brief Review of Land Surface Models

There is increasing evidence that the land surface significantly influences weather and climate on local, regional, and global scales, on a timescale ranging from seconds to millions of years, by exchanging momentum, energy, and mass with the overlying atmosphere (e.g., Jarvis, 1976; Dickinson, 1992; Sellers, 1992; Foley *et al.*, 1996; Pielke *et al.*, 1998; Chen *et al.*, 2001; Pitman, 2003; Koster *et al.*, 2004; Los *et al.*, 2006). As the interface between the solid earth and the atmosphere, the land surface provides the link between several scientific disciplines, and land

S. Liang, X. Li & X. Xie (eds.), *Land Surface Observation, Modeling and Data Assimilation*, 89–124.

surface modeling has been subject to intense research in the hydrological, atmospheric, ecological, agrologic, and remote sensing communities in the recent decades. With the implementation of a number of international observation experiments since the 1980s, such as the Global Energy and Water Cycle Experiment (GEWEX) Asian Monsoon Experiment (GAME), Hydrological and Atmospheric Pilot Experiment (HAPEX), First ISLSCP (International Satellite Land Surface Climatology Project) Field Experiment (FIFE), and Boreal Ecosystems Atmosphere Study (BOREAS), abundant observed data have been accumulated, which effectively promote studies on land–atmospheric interactions and the development of LSMs. Yang (2004) provided a comprehensive summary on LSMs. In this section, we will briefly introduce the development of LSMs based on their review and on other documents.

Early LSMs, such as the "bucket" model (Manabe, 1969) assumed that the global soil had fixed soil properties and constant soil depth, in which the "bucket" is filled by precipitation and emptied by evaporation. The excess, above its capacity or a critical value, is termed runoff. The evaporation rate is a product of a coefficient that is assumed to be a linear function of soil moisture content, and potential evaporation. The process of heat transfer in soils is ignored in the "bucket" model. The Project for Inter-comparison of Land-surface Parameterization Schemes (PILPS) showed that the "bucket" model was inadequate for diurnal to multi-annual scale surface hydrological representation (Henderson-Sellers *et al.*, 1993).

Deardorff (1978) first introduced vegetation impacts within land surface modeling. This model considered the evaporation from the soil layer and the wet canopy as well as the interception and transpiration from the dry parts of the canopy. This formulation facilitated subsequent developments toward constructing advanced LSMs. Since then, active research on land–atmospheric interactions has led to the construction of more than two dozen LSMs for use in GCMs (e.g., Henderson-Sellers *et al.*, 1993; Slater *et al.*, 2001). The general characteristics of these LSMs are as follows: consideration of the effects of vegetation on momentum, energy, and water transfer; verifying the existence of at least two soil layers; application of the Richards equation-based water transfer scheme; and adoption of the saturation/infiltration excess surface runoff generation regime. Two well-known examples of this type of LSM are the Biosphere Atmosphere Transfer Scheme (BATS) (Dickinson *et al.*, 1986) and the Simple Biosphere (SiB) Model (Sellers *et al.*, 1986).

The application of these LSMs improved the modeling of land–atmospheric interactions on the timescale of days. For example, there was improvement in the precipitation weather forecast (Beljaars *et al.*, 1996) and in European soil temperature prediction (Viterbo and Betts, 1999). However, some issues were identified with respect to these models: (1) they ignore the horizontal interactions of land surface processes between adjacent grid squares; (2) vegetation is treated as a "big leaf", and hence, a single stoma is associated with the big leaf; (3) only three land components (soil, snow, and vegetation) are treated explicitly, whereas land ice and lakes are neglected; and (4) the number and areal coverage of vegetation types need to be prescribed.

In the past two decades, advances in related disciplines, e.g., ecology and remote sensing, have effectively promoted the development of LSMs. Previous LSMs were revised to model the carbon balance by a semi-empirical representation of vegetation conductance. A typical example is the revised Simple Biosphere Model (SiB2, Sellers *et al.*, 1996). Other newly developed LSMs, e.g., the Common Land Model (CoLM, Dai *et al.*, 2003) and Community Land Model version 3.5 (CLM 3.5, Oleson *et al.*, 2004), led to further advancement in model structures. For example, the CLM3.5 was expanded to include the carbon cycle, vegetation dynamics, and river routing module; the CoLM implemented a two-big-leaf model for leaf temperatures and vegetation conductance, and a two-stream approximation for the calculation of canopy albedos and radiation for the sunlit and shaded canopy parts. The recently released CLM 4.0 (Oleson *et al.*, 2010) has been extended to include a carbon–nitrogen biogeochemical model, an urban canyon model, and transient land cover and land-use change capability, in addition to other improvements (Lawrence *et al.*, 2010).

Nevertheless, unresolved uncertainties remain in vegetation and soil datasets, in model parameterization schemes, and in forcing datasets. In this chapter, we focus on the uncertainties of land surface modeling in drylands and high-elevation regions.

4.2 Issues in Land Surface Modeling of Drylands and High-Elevation Regions

Drylands, including arid and semi-arid regions, occupy approximately 41% of the Earth's land area and are home to more than 38% of the total global population of 6.5 billion (Millennium Ecosystem Assessment, 2005; Global Land Project, 2005). However, increasing desertification due to climatic change and human activities is severely threatening these regions (e.g., Puigdefabregas, 1995). In China, the arid and semi-arid regions in the north have experienced significant environmental changes, such as an enhanced drying trend (e.g., Wang and Zhai, 2003), an enhanced warming signal (e.g., Ren *et al.*, 2005), grassland degradation and desert encroachment (e.g., Zhu and Chen, 1994; Zhao and Huang, 1993; Fu and Wen, 2002), reduced discharge in the Yellow River (e.g., Yang *et al.*, 2004; Tang *et al.*, 2008), and an extended drought area in principal farming areas (e.g., Wang and Zhai, 2003). Land–atmosphere interactions are very strong in these areas, and understanding their responses to climate change will help to confront severe environment problems.

High-elevation regions are also characterized by strong land–atmosphere interactions due to strong solar heating, which play a very important role in regional and/or global climate systems. In China, the Tibetan Plateau (TP) is a key region affecting the Asian summer monsoon. TP land processes are generally characterized by the following features. The first is apparent diurnal variations of weather conditions due to strong solar radiation and low air density. The solar irradiance over the Plateau is often observed to exceed 1200 W m^{-2} near noon (Ma *et al.*, 2005), which results in a very strong diurnal change of the surface energy budget and near-surface meteorological variables. For example, the diurnal range of the surface skin temperature can exceed 60 K. The second is the distinct seasonal march of the surface water and energy budget in the central and eastern TP (CE-TP). Before the onset of the monsoon, the surface is relatively dry and the sensible heat flux dominates the surface energy budget; during the monsoon season, the land surface becomes wet and the latent heat flux dominates the energy budget. The third is the contrast between the dry western region and the wet eastern region. Annual precipitation is approximately 400 mm or more in most of the CE-TP, compared with around 100 mm or less in the western TP (W-TP). Under the unique plateau climate, land surfaces are typically characterized by alpine meadows and grasslands in CE-TP and by alpine deserts in W-TP. It has been widely accepted that the TP provides a huge elevated heat source to the middle-troposphere and that land–atmosphere interactions play an important role in the formation of the Asian monsoon (e.g., Ye and Gao, 1979; Yanai and Wu, 2006). Therefore, it is essential to accurately represent these land processes in current LSMs.

4.2.1 *Thermal coupling between land and atmosphere in drylands*

The physical representation of land–atmosphere interactions in models plays a key role in improving the predictability of weather and climate (e.g., Pielke *et al.*, 1999; Koster *et al.*, 2004). However, issues concerning the

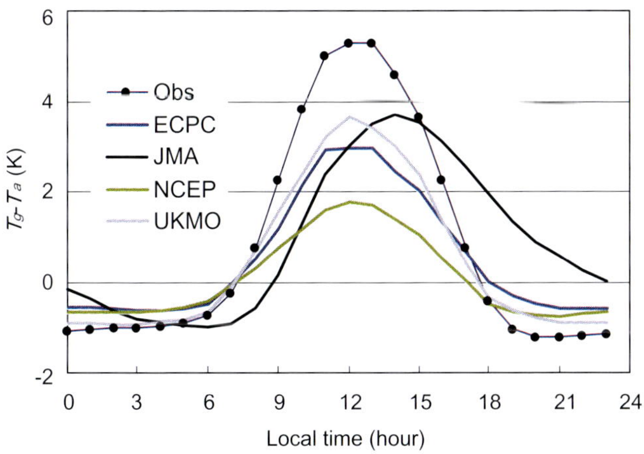

Fig. 4.1. Composite diurnal variation of surface-air temperature difference from *in situ* data and GCMs (Experimental Climate Prediction Center [ECPC] of the Scripps Institution of Oceanography, Japan Meteorological Agency [JMA], NOAA National Centers for Environmental Prediction [NCEP], UK Meteorological Office [UKMO]) at 14 CEOP sites for the period of October 2002–September 2003 (after Yang *et al.* 2007, Fig. 8). Copyright 2007, Meteorological Society of Japan.

exchange efficiencies of energy and water vapor between the land surface and the atmosphere remain poorly understood (Chen and Zhang, 2009), leading to uncorrected LSM predictions of land surface temperature as well as surface energy partition in drylands.

Yang *et al.* (2007) evaluated five operational global circulation models (GCMs) against the Coordinated Enhanced Observing Period (CEOP) observations (Koike, 2004), and pointed out that these models systematically underestimate the diurnal range of surface-air temperature differences (see Fig. 4.1), particularly in arid and semi-arid regions, due to the underprediction of heat transfer resistances. Hogue *et al.* (2005) found that the Noah LSM tended to overestimate the sensible heat flux and underestimate surface temperature in the dry season. LeMone *et al.* (2008) demonstrated that the default Noah model tended to overestimate sensible heat flux and underestimate surface temperature in relatively dry conditions. Yang *et al.* (2009) further evaluated three offline LSMs and reported that all three models significantly underestimated the daytime surface temperature. Chen and Zhang (2009) reviewed the observed surface heat exchange coefficient (C_h) in various surface vegetation types and suggested that the coefficient in the thermal roughness length (z_{0h}) scheme should be dependent on the vegetation type.

Ma *et al.*, 2002; Tanaka *et al.*, 2003; Yang *et al.*, 2005; Hu *et al.*, 2006; Li and Sun, 2008). Based on a strict analysis of the quality-controlled data obtained from these field experiments, a novel z_{0h} scheme (i.e., Y08) was developed by Yang *et al.* (2002) and further validated in Yang *et al.* (2008).

Previous works (see Table 4.1) have related z_{0h} directly to z_{0m} through the parameter kB^{-1}. It is well known that momentum transport is generally more efficient than heat transport on account of the influence of pressure fluctuation, because individual roughness elements may enhance the momentum flux through form drag with little contribution to the area-averaged heat flux (Mahrt, 1996). Therefore, z_{0h} is typically less than z_{0m}, especially over a surface with bluff roughness elements, and a higher z_{0m} usually corresponds to a lower z_{0h}.

Following this reasoning, Yang *et al.* (2002) correlated z_{0h} to a physical height (h_T), which is related to a height to separate the fully turbulent layer and the transitional layer. The value of h_T is determined by the critical Reynolds number ($\mathrm{Re}_{\mathrm{crit}}$):

$$h_T = \frac{\nu \mathrm{Re}_{\mathrm{crit}}}{u_*} \tag{4}$$

where $\mathrm{Re}_{\mathrm{crit}}$ equals 70, $\nu[= 1.328 \times 10^{-5}\,(p_0/p)(T/T_0)^{1.745}$, for $p_0 = 1.013 \times 10^{-5}$ Pa and $T_0 = 273.15\,\mathrm{K}]$ is the fluid kinematic viscosity, and u_* ($\mathrm{m\,s^{-1}}$) is the friction velocity.

For a surface with bluff roughness elements, u_* is quite large due to form drag, which gives a small h_T. Therefore, the variation in h_T is similar to that in z_{0h}, making it reasonable to use h_T as a length to scale z_{0h}. Accordingly, Yang *et al.* (2002) defined the parameter

$$kA^{-1} = \ln \frac{h_T}{z_{0h}} \tag{5}$$

Based on a data analysis for three GAME-Tibet sites, they found the typical diurnal variation in z_{0h}, and then parameterized kA^{-1} with both the frictional velocity and a temperature scale $[\theta_*, \equiv -H/(\rho c_p u_*)]$,

$$kA^{-1} = \beta u_*^m |\theta_*|^n \tag{6}$$

where β, m, and n are empirical coefficients. The data analysis indicates that $m = 1/2, n = 1/4$ are reasonable values. The value $\beta = 7.2\,\mathrm{m^{-1/2}\,s^{1/2}}$ $\mathrm{K^{-1/4}}$ was suggested in Yang *et al.* (2008). By combining Eqs. (4)–(6), a general expression of z_{0h} for bare-soil or short-vegetation surfaces was

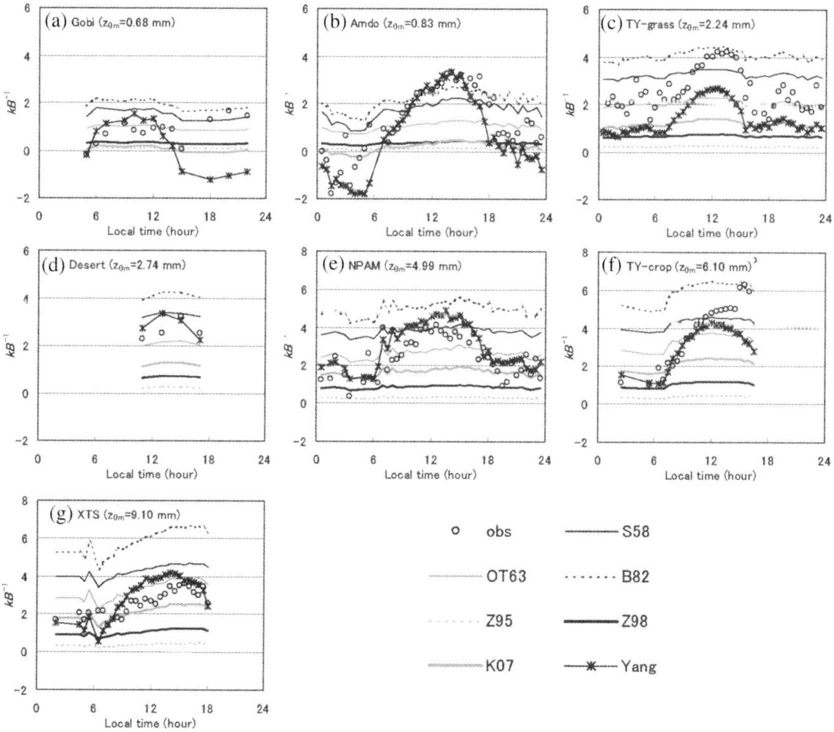

Fig. 4.5. Comparisons between composite diurnal variations of observed kB^{-1} and the scheme-predicted one for the seven bare-soil sites: Gobi, Amdo, TY-grass, Desert, NPAM, TY-crop, XTS (after Yang *et al.*, 2008, Fig. 6).

obtained, as follows

$$z_{0h} = \frac{70\nu}{u_*} \exp(-\beta u_*^{\frac{1}{2}} |\theta_*|^{\frac{1}{4}}) \tag{7}$$

Yang *et al.* (2008) evaluated several widely used schemes (see Table 4.1) against turbulent flux data collected at seven dryland stations. As shown in Fig. 4.5, they indicated that a common feature of z_{0h} over drylands is diurnal variation, and this has been reported in several studies over bare ground and grassland (Verhoef *et al.*, 1997; Sun, 1999; Ma *et al.*, 2002; Yang *et al.*, 2003) and has not been reproduced by most of the parameterization schemes; one exception is the scheme given by Eq. (7), which is able to reproduce the diurnal variation in z_{0h} through the diurnal change in θ_*. Figure 4.6 shows the averaged diurnal variations in relative mean biases in heat flux prediction for the seven sites (listed in the caption of Fig. 4.5).

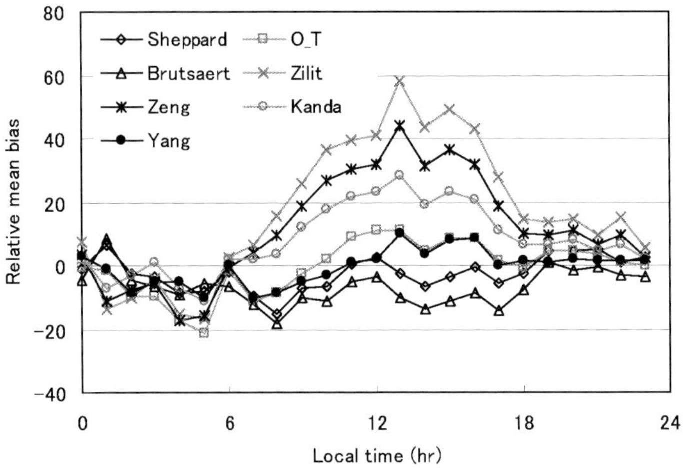

Fig. 4.6. The diurnal variation of the seven sites' composite relative mean biases in sensible heat flux. Diurnal variation of relative mean biases for each site was made first, and then averaged over the seven sites to obtain the composite one (after Yang *et al.*, 2008, Fig. 9).

Schemes OT63, Z95, Z98, and K07 produce a clear diurnal variation in errors of sensible heat flux (positive biases during the day and negative biases at night). The S58 and B82 schemes underestimate fluxes for almost all time slots. The Y08 scheme shows no more than a 10% mean bias for all time slots, indicating its capability of reflecting the diurnal variation in kB^{-1}. In summary, it is important for a scheme to provide not only the mean value but also the diurnal variation in kB^{-1}.

The applicability of scheme Y08 is further evaluated for stable boundary conditions based on data collected from a glacier experiment that was carried out from May 20 to September 9, 2009, at Palong-Zangbu No. 4 Glacier on the south-east TP, by the hydrometeorological research group at the Institute of Tibetan Plateau Research, Chinese Academy of Sciences. This valley glacier covers an area of approximately 12.8 km^2 and spans an altitude of 4650–5800 m (see Xu *et al.* [2009] for a topographic map). This site was located in the glacier's ablation zone, namely at 4800 m above sea level (29°15′N, 96°55′E; approximately 1500 m from the glacier terminus). Meteorological instruments included an eddy-covariance system for measuring three components of turbulent wind velocity, virtual temperature, and water vapor concentrations. Also available were CNR1 pyranometers and pyrgeometers for measuring the four radiation components. Auxiliary measurements of the ambient air were recorded by an HMP45C temperature and relative humidity sensor housed in a radiation shield. It was observed that the surface cover showed a marked evolution during the ablation season (i.e., snow, bare ice, and ice hummocks in chronological order). The ice/snow melt led to substantial darkening of the surface, and ice hummocks and meltwater streams emerged due to differential ice melt. Using height records of nine mass balance stakes, monthly surface melt was quantified as -1.57 ± 0.03 m from June 3 to September 9.

Based on the quality-controlled data, Guo *et al.* (2011) carried out a critical evaluation of three physical z_{0h} parameterizations (i.e., Andreas, 1987, hereinafter A87; Yang *et al.*, 2002; Smeets and van den Broeke, 2008, hereinafter SvdB) for energy flux estimations using the bulk aerodynamic method. Figure 4.7 compares the estimated sensible heat fluxes (H_{BA}) with eddy-covariance measurements (H_{EC}); linear regressions of H_{BA} versus H_{EC} are made for overall evaluations. In Figs. 4.7(a) and 4.7(c), the sensible heat exchange is notably underestimated by the A87 and overestimated by the SvdB schemes. One explicit cause is the significant underestimation and overestimation of z_{0h} produced by the A87 and SvdB schemes, respectively. The Y08 scheme tends to slightly underestimate high values of H and overestimate low values. However, its overall performance is better than that of the other two schemes.

Figure 4.8 compares the estimated latent heat fluxes (LE_{BA}) with eddy-covariance measurements (LE_{EC}). In Figs. 4.8(a) and 4.8(c), the A87 and SvdB schemes generally underestimate and overestimate the LE (absolute values), respectively. One major contributor to this outcome is the notable underestimation and overestimation of z_{0h} produced by the A87 and SvdB schemes, respectively. In comparison, the Y08 scheme seems promising, which produces the lowest errors over the entire ablation season (see Table 3 by Guo *et al.*, 2011). A comparison of the linear regressions in Fig. 4.8 lends further support to the Y08 scheme, as the results compare well with eddy-covariance measurements, with a perfect slope of unity in Fig. 4.8(b).

The applicability of this scheme to LSMs is addressed in Sec. 4.4.

4.3.2 *Inverse analysis of the role of soil vertical heterogeneity*

Surface soil moisture and temperature play a key role in land surface energy and water budgets. However, it is rather difficult for some LSMs to reproduce the surface soil sate in areas with high soil vertical heterogeneity, because these models use a single parameter set to characterize soil hydraulic and thermal processes.

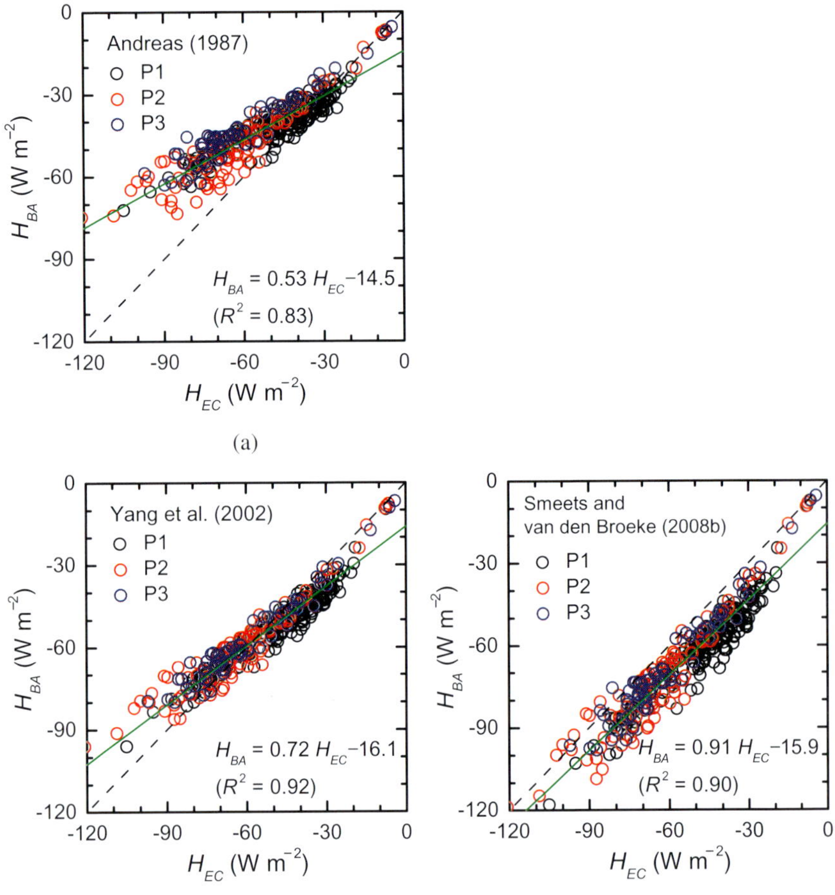

Fig. 4.7. Comparison in sensible heat fluxes between the bulk aerodynamic method (H_{BA}) and eddy-covariance method (H_{EC}) (after Guo *et al.*, 2011, Fig. 8). Data points distinguish among the three melt phases (P1, P2, and P3); least-squares linear regressions (green line) are made for an overall evaluation without distinguishing the data (regression equations given with coefficients of determination, i.e., R^2). Copyright 2011, Springer Science and Business Media.

Yang *et al.* (2005) developed an LSM to parameterize this heterogeneity, including a multi-layer soil model with the thermal diffuse equation for soil temperature, Richards' equation for soil moisture, and bulk heat transfer equations for surface turbulent fluxes. The model approximates a vertically heterogeneous soil column in a sandwich-like structure: a uniform top domain, a transitional middle domain, and a uniform bottom domain, as

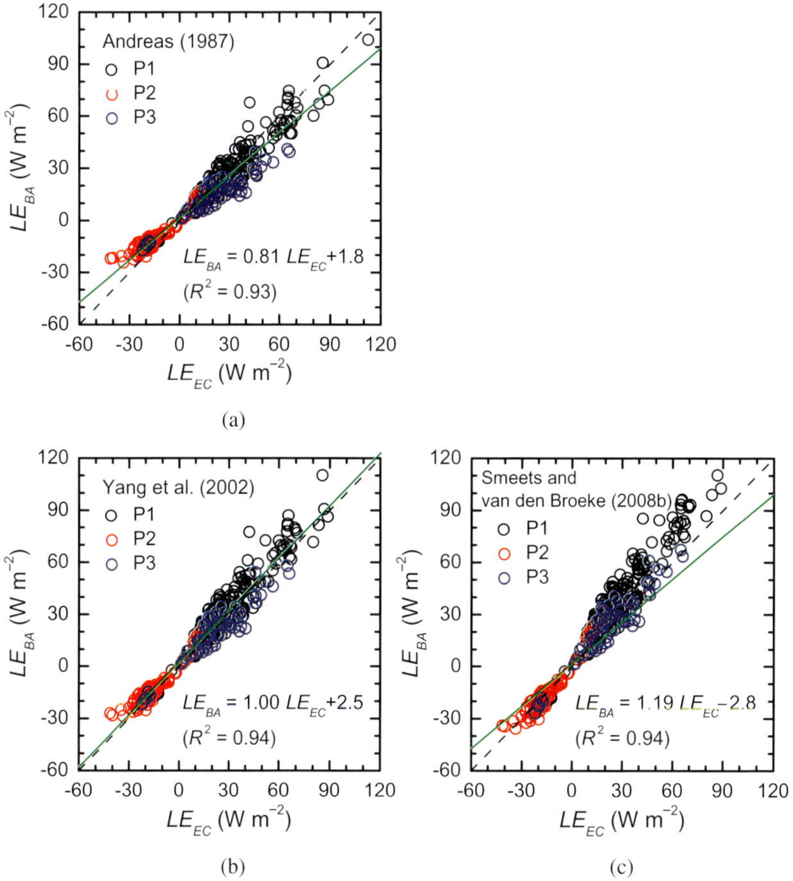

Fig. 4.8. Comparison in latent heat fluxes between the bulk aerodynamic method (LE_{BA}) and eddy-covariance method (LE_{EC}) (after Guo *et al.*, 2011, Fig. 11). Data points distinguish among the three melt phases (P1, P2, and P3); least-squares linear regressions (green line) are made for an overall evaluation without distinguishing the data (regression equations given with coefficients of determination, i.e., R^2). Copyright 2011, Springer Science and Business Media.

shown in Fig. 4.9. The transitional domain is bounded by depths d_1 and d_2. The top domain and the bottom domain are characterized by distinct soil parameters. The soil properties for the transitional domain are interpolated from those for the top and bottom domains. In order to account for the non-linearity of soil properties, they interpolate soil functions rather than soil parameters, following an exponential variation in the transitional domain.

$$A(w) = A_1(w)^x A_2(w)^{1-x} \tag{8}$$

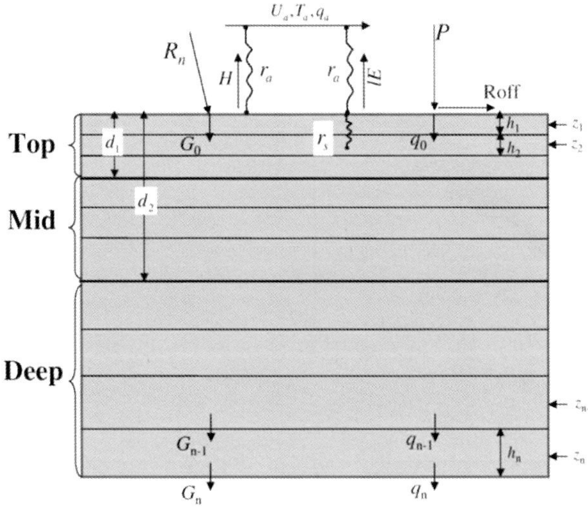

Fig. 4.9. Schematic of the single-source land surface model used to parameterize the soil vertical heterogeneity. There are three soil domains (top, mid, deep), each of which contains multiple computational layers. All the symbols have their common meanings (after Yang *et al.*, 2005, Fig. 1). Copyright 2005, John Wiley & Sons, Inc.

where $A(w)$ denotes any hydraulic function or thermal function; subscripts 1 and 2 represent the top domain and the deep domain, respectively; and $x = (d_2 - z)/(d_2 - d_1)$.

The soil hydraulic and thermal parameters are inversely estimated simultaneously by minimizing a cost function that is objectively determined by the discrepancy between observed and model-predicted values of soil moisture and temperature. The Shuffled Complex Evolution method developed at The University of Arizona (SCE-UA) (Duan *et al.*, 1992, 1993) is used as an optimization algorithm.

4.3.3 *A soil surface resistance scheme for evaporation*

A new parameterization for surface resistance of evaporation (r_{soil}) was developed by Yang *et al.* (2009). According to the condition of mass continuity, the water supply from the soil and the water demand by the air must be satisfied. Therefore, actual evaporation must be the minimum of the two fluxes, which may be expressed as:

$$E = \min(\rho_w q_{\mathrm{supply}}, E_{\mathrm{demand}}), \tag{9}$$

$$q_{\mathrm{supply}} = -\left(K - \frac{\partial \phi}{\partial z}\right) \tag{10}$$

$$E_{\text{demand}} = \rho \frac{rh_{\text{eq}} q_{\text{sat}}(T_{\text{sfc}}) - q_a}{(r_{ah} + r_{\text{soil}})} \tag{11}$$

$$rh_{\text{eq}} = \exp\left(\frac{\psi(\theta_{\text{sfc}})}{R_v T_{\text{sfc}}} g\right) \tag{12}$$

where E ($\text{mm}\,\text{s}^{-1}$) is the actual evaporative flux; q_{supply} ($\text{m}\,\text{s}^{-1}$; positive if upward) is the maximum soil water flux from the first node to a very dry surface; ϕ is the soil flux potential; E_{demand} ($\text{mm}\,\text{s}^{-1}$) is the demand water flux by the air; rh_{eq} is the equilibrium relative humidity in the air space of the soil; θ_{sfc} is the near-surface soil water content; $q_{\text{sat}}(T_{\text{sfc}})$ is the saturated specific humidity; q_a is the reference level air specific humidity; r_{ah} is the heat transfer resistance; $\rho(\text{kg}\,\text{m}^{-3})$ is the density of air; ρ_w ($\text{kg}\,\text{m}^{-3}$) is the density of water; $R_v = 461.5\,\text{J}\,\text{K}^{-1}\text{kg}^{-1}$; and $g = 9.81\,\text{m}\,\text{s}^{-1}$.

Equations (9)–(11) can be rearranged to give

$$r_{\text{soil}} = \max\left(0, \rho \frac{rh_{\text{eq}} q_{\text{sat}}(T_{\text{sfc}}) - q_a}{\rho_w q_{\text{sup}}} - r_{ah}\right) \tag{13}$$

In addition, an advanced scheme for soil water flow simulations, developed by Ross (2003), was implemented in SiB2 by Yang *et al.* (2009). This scheme adopts the Kirchhoff transformation to calculate soil water flux to avoid calculating the soil water content at the interface. This is beneficial for precisely obtaining the soil hydraulic conductivity.

4.4 Land Surface Modeling Improvements

This section introduces the land surface modeling improvements achieved by the aforementioned efforts, including improvements in the simulation of surface temperature as well as surface energy budget in drylands; in modeling soil moisture of topsoils in alpine grasslands; and in simulation of the soil moisture in dry conditions.

4.4.1 *Modeling improvements in drylands*

In a preliminary evaluation, Yang *et al.* (2009) implemented the Y08 scheme into SiB2 model, which was then applied to two alpine desert sites. Figure 4.10 shows the results of SiB2 with and without this revision to account for the excess resistance for heat transfer. Figure 4.10 also shows that the ground–air temperature difference is well simulated by SiB2 when this scheme is included in the model, whereas the simulation without it yields higher H although the T_{sfc} is underestimated. The lower H in

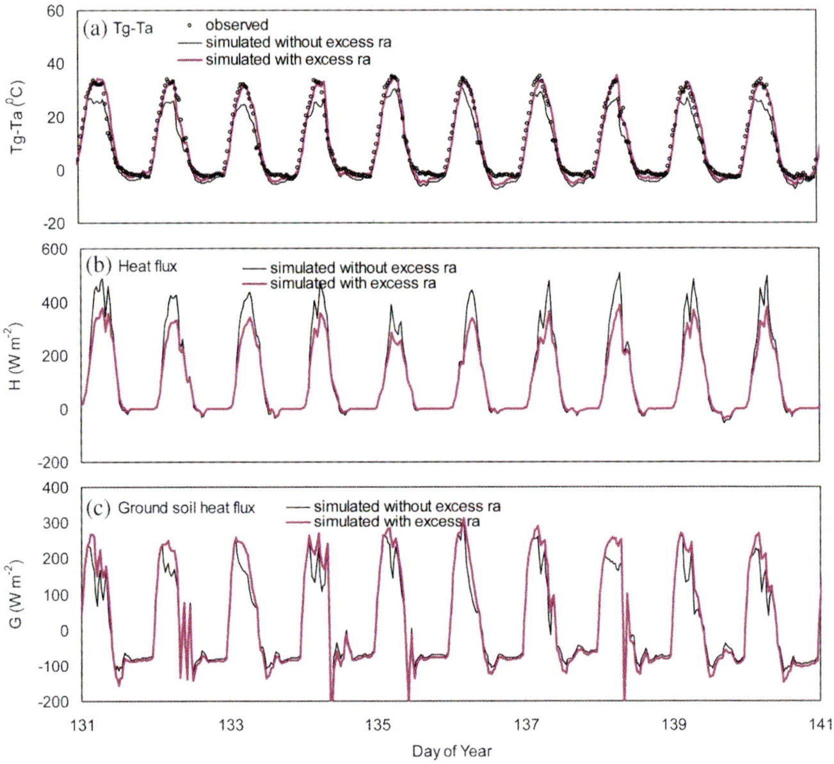

Fig. 4.10. Comparisons between two SiB2 simulations with and without the Y08 thermal roughness length scheme for an alpine desert site (SQH) in 1998. Panel (a) ground-air temperature gradient, (b) sensible heat flux, (c) soil heat flux (after Yang *et al.*, 2009, Fig. 12).

the revised scheme is consistent with enhanced upward longwave cooling and lower R_{net} amounts during the daytime. Meanwhile, the higher T_{sfc} would directly result in higher ground soil heat fluxes (G_0), which are also consistent with the lower H. The results in Fig. 4.10 verify this reasoning.

Chen *et al.* (2010) implemented this scheme in the Noah LSM, which was then evaluated at several dryland sites, including Audubon in USA and Dunhuang, Shiquanhe (SQH), and Gaize in China. Figures 4.11(a)–4.11(d) compare the simulated diurnal variation in T_{sfc}, R_{net}, H, and G_0, respectively, between the revised Noah LSM and the original model using data collected at the Audubon Research Ranch. The results confirmed that T_{sfc} and H as well as R_{net} were properly simulated by the revised model, while the original model produced higher H and lower T_{sfc}, and thus, higher R_{net}.

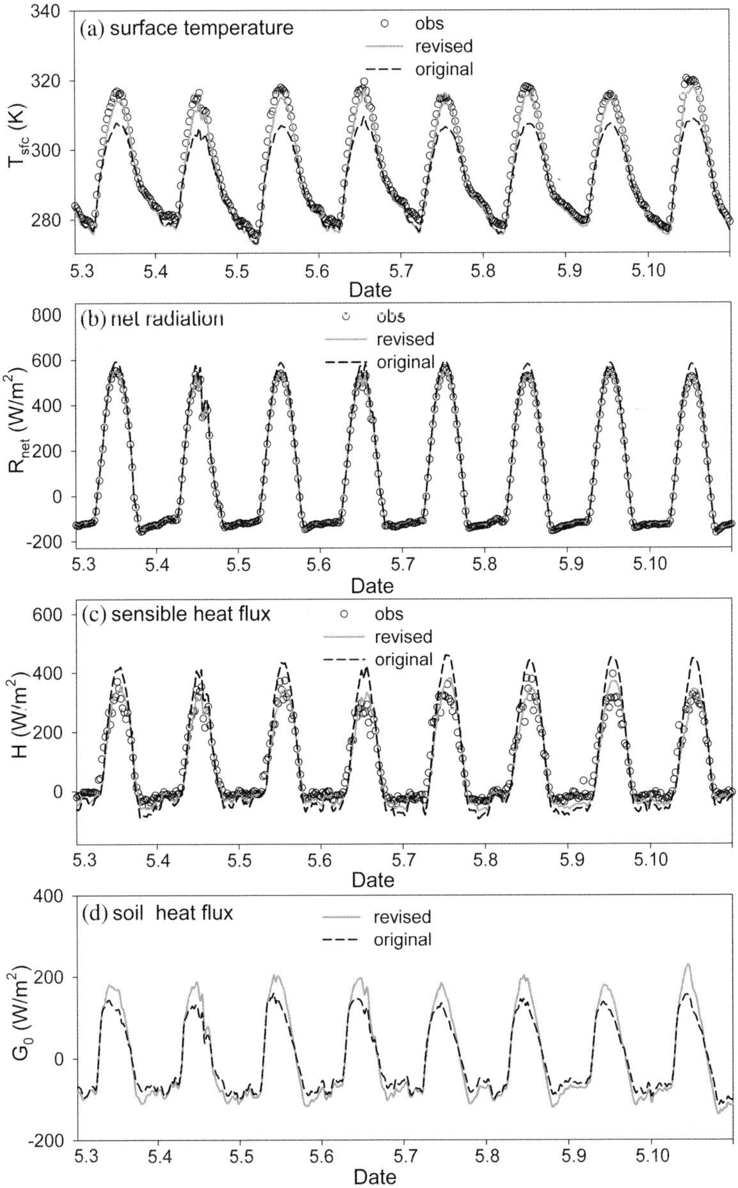

Fig. 4.11. Comparison of the diurnal variation of (a) surface temperature ($T_{\rm sfc}$), (b) net radiation ($R_{\rm net}$), (c) sensible heat flux (H), and (d) ground soil heat flux (G_0) between two simulations by the revised Noah and the original one against observations for Audubon site (after Chen *et al.*, 2010, Fig. 2). Circles represent the observations; dark grey line represents the simulations by revised Noah; dashed black line represents the simulations by original model.

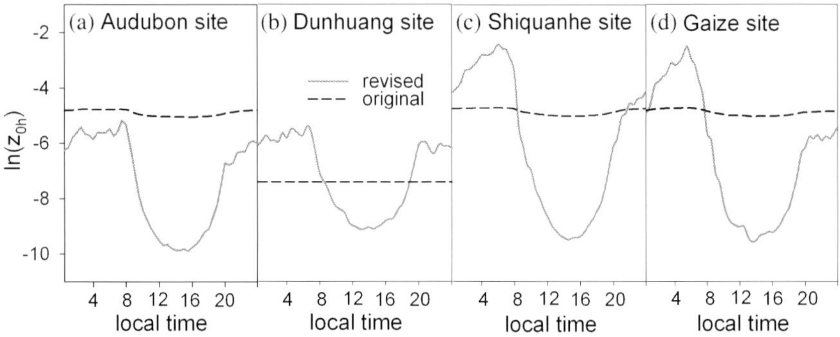

Fig. 4.12. Simulated mean diurnal variation of $\ln(z_{0h})$ throughout the simulation period at (a) Audubon Research Ranch, (b) Dunhuang site, (c) Shiquanhe site, and (d) Gaize site (after Chen *et al.*, 2010, Fig. 6).

Table 4.2 shows the error metrics using data for the period 0900–1600 hours (local time), which indicates that the revised model significantly reduced simulation errors. The lower T_{sfc} in the original model would directly result in a lower G_0, corresponding to a higher H. The outcomes are similar for other sites (see error metrics in Table 4.2). However, the figures for other sites are not shown here.

Figure 4.12 shows the simulated mean diurnal variation of $\ln(z_{0h})$ throughout the simulation period at four sites. This helps to understand the difference in performance between the revised Noah LSM and the original model. It is clear that the values of z_{0h} simulated by the revised model exhibit evident diurnal variations. However, this diurnal variation was unsatisfactorily simulated by the original Noah LSM. Figure 4.12 shows that the original model produces a rather high z_{0h}, and thus, a fairly high C_h, compared with those produced by the revised model in the daytime. Moreover, the original simulation of T_{sfc} is lower because the overestimated C_h results in an excessive transfer of H away from the surface. Two direct effects arise from the underestimation of T_{sfc}: one is the overestimation of R_{net} due to the reduced upward longwave radiation flux; the other is the underestimation of G_0 due to an underestimated soil temperature gradient. Figure 4.12 also shows that the values of z_{0h} produced by the revised model have larger diurnal ranges at two TP sites than elsewhere. This is consistent with the results of Yang *et al.* (2008), who found that the diurnal variations in z_{0h} are more evident at TP sites than at other sites. This phenomenon may be attributed to strong diurnal changes of near-surface meteorological variables and a higher land–atmosphere temperature difference caused by

Table 4.2. The error metrics of the difference between the half-hourly observations and the simulated ones for four sites by the revised Noah and the original one during daytime (09:00–16:00, local time), respectively (after Chen *et al.*, 2010, Table 7).

		T_{sfc}			H			R_{net}		
		BIAS (K)	*ME* (K)	*RMSE* (K)	*BIAS* (Wm^{-2})	*ME* (Wm^{-2})	*RMSE* (Wm^{-2})	*BIAS* (Wm^{-2})	*ME* (Wm^{-2})	*RMSE* (Wm^{-2})
Audubon	Original	−8.08	8.09	8.39	63.42	69.26	80.22	38.39	38.62	41.47
	Revised	−2.19	2.28	2.68	−1.91	34.23	41.62	−0.23	5.96	7.69
Dunhuang	Original	−4.99	5.15	5.93	38.36	47.84	57.29	19.48	22.11	27.89
	Revised	−2.44	2.96	3.86	14.48	30.52	38.4	1.22	16.08	19.97
Shiquanhe	Original	−11.15	11.29	12.21	—	—	—	99.54	100.92	107.63
	Revised	−1.66	3.82	5.18	—	—	—	35.59	41.07	46.24
Gaize	Original	−10.18	10.19	10.91	—	—	—	64.01	64.01	68.44
	Revised	−2.62	3.73	4.76	—	—	—	15.48	20.78	25.53

Note: The observed sensible heat flux was not available at Shiquanhe and Gaize sites.
BIAS, *ME*, and *RMSE* are defined as:

$$BIAS = \frac{\sum_{i=1}^{N}(X_i - Obs_i)}{N}, \quad ME = \frac{\sum_{i=1}^{N}|X_i - Obs_i|}{N}, \quad RMSE = \sqrt{\frac{\sum_{i=1}^{N}(X_i - Obs_i)^2}{N}},$$

where N is the sample number.

the high elevation and the consequently strong solar radiation. As the differences between the two simulated z_{0h} values at the TP sites are also larger than those at Dunhuang and Audubon during the daytime, the improvements in the revised Noah LSM are believed to be particularly meaningful for high-elevation regions, as summarized in Table 4.2.

The revised Noah LSM was applied by Chen *et al.* (2011) to simulate T_{sfc} as well as the surface energy budget of dry regions in China in order to verify its effectiveness for a regional scale. They presented two improvements of Noah land surface modeling for China's dry areas, which are characterized by bare ground surfaces and grasslands. The first improvement is the replacement of the model's z_{0h} scheme with the Y08 scheme (i.e., applying the revised Noah LSM), as introduced above. Both the original and revised models are driven by the Global Land Data Assimilation System (GLDAS) forcing data (Rodell *et al.*, 2004). A comparison between the simulations of the two models and the daytime MODIS/Aqua T_{sfc} products indicates that the original Noah LSM produces a mean bias of approximately -6 K in the early afternoon (around 13:30 local solar time) for drylands, whereas the revised Noah LSM reduces the mean bias to approximately -3 K, as shown in Figs. 4.13(a) and 4.13(b) and 4.14(a) and 4.14(b). The second improvement is the use of newly developed forcing data for China to drive the revised model (ITPCAS forcing data, named after the authors' institute; He, 2010). The mean bias in early afternoon is further reduced by more than 2 K, as shown in Figs. 4.13(c) and 4.14(c). Table 4.3 gives the error metrics for simulated T_{sfc} against daytime MODIS/Aqua MYD11C1 T_{sfc} products, which further supports

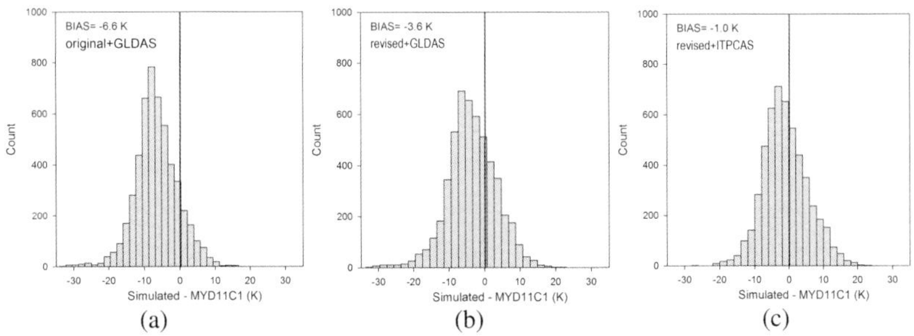

Fig. 4.13. The histograms for the difference between the simulated surface temperature (T_{sfc}) and the aggregated daytime MODIS/Aqua MYD11C1 T_{sfc} on bare ground surfaces (corresponding to arid regions): (a) for the result of the original Noah forced by GLDAS forcing data, (b) for the result of the revised Noah forced by GLDAS data, and (c) for the result of the revised Noah forced by ITPCAS forcing data (after Chen *et al.*, 2011). Copyright 2011, John Wiley & Sons, Inc.

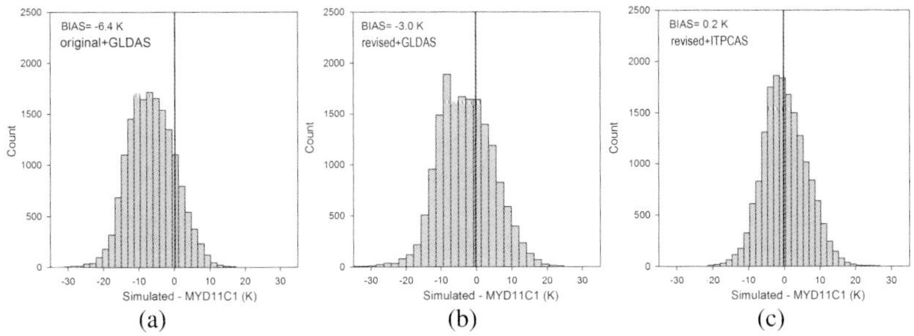

Fig. 4.14. Similar to Fig. 4.13, but for the grassland (after Chen *et al.*, 2011). Copyright 2011, John Wiley & Sons, Inc.

Table 4.3. Determination coefficient (R^2), mean bias ($BIAS$), and root mean square error ($RMSE$) for simulated surface temperature ($T_{\rm sfc}$) against daytime MODIS/Aqua MYD11C1 $T_{\rm sfc}$ products (after Chen *et al.*, 2011, Table 1). Copyright 2011, John Wiley & Sons, Inc.

Land type	Model	Forcing data	R^2	$BIAS$ (K)	$RMSE$ (K)
Bare ground	Original	GLDAS	0.871	−6.6	9.0
	Revised	GLDAS	0.861	−3.6	7.8
	Revised	ITPCAS	0.889	−1.0	6.6
Grassland	Original	GLDAS	0.809	−6.4	9.2
	Revised	GLDAS	0.792	−3.0	8.2
	Revised	ITPCAS	0.855	0.2	6.2

Note: The sample number for bare ground is 5171, the sample number for grassland is 16,852.

these results. A similar reduction is also found when the original Noah LSM is driven by the new dataset. Furthermore, the authors found that the original Noah LSM, when driven by the new forcing data, satisfactorily reproduces the $T_{\rm sfc}$ for forest, shrubland, and cropland (not shown here). The results suggest that it may be sensible to select the z_{0h} scheme according to the vegetation types in the Noah LSM.

4.4.2 *Improvements considering soil vertical stratification in alpine grasslands*

The single-source LSM and the inverse system for estimation of soil parameters discussed in Sec. 4.3.2 are then used at alpine grassland sites. According to the inverse analysis, the topsoil has high porosity, and consequently, high water-holding capacity; this may enhance evaporation during the wet season (not shown). However, this layer shows low heat

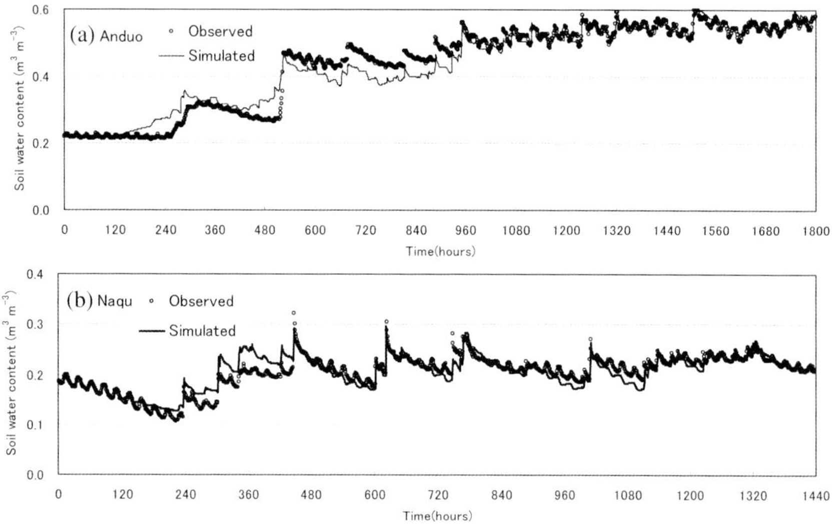

Fig. 4.15. Comparisons of near-surface soil water content between observation and simulation at Anduo site in 1998 and Naqu site in 2001. The simulations were conducted using the LSM in Yang *et al.* (2005) with stratified soil parameters (after Yang *et al.*, 2009, Fig. 8).

capacity and low thermal conductivity in the dry season; this may lead to high T_{sfc} and high H (not shown). With the incorporation of stratified soil parameters, the soil moisture profile can be simulated suitably. Figure 4.15 shows examples at Anduo and Naqu sites, where near-surface soil moistures were simulated suitably by using a sandwich structure to delineate the soil stratification.

In some models (e.g., CoLM), it is possible to specify soil parameters for each computational layer. Further, some previous studies attempted to formulate the effect of SOMs on soil parameters (Beringer *et al.*, 2001; Lawrence and Slater, 2008). However, our knowledge of the Tibetan Plateau soils is still very limited. In order to develop the parameterization for soil properties in this region, laboratory soil experiments are required to measure basic parameters such as the content of grassroots and SOC.

More than 40 soil profiles (0–50 cm) were collected from alpine grassland sites in the central and eastern TP. The soil porosity, bulk density, gravel proportion, soil thermal conductivity at several soil wetness levels, soil texture, organic carbon content, and soil water retention curve were measured in laboratory. Preliminary analysis confirmed that higher SOC content in the topsoil of alpine grassland leads to higher porosity, with lower

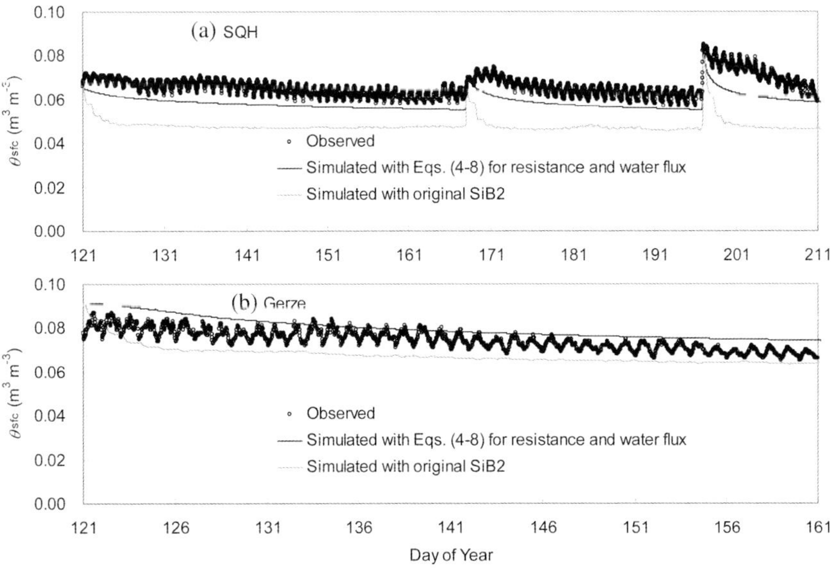

Fig. 4.16. Comparison of soil water content in the top 15 cm between observation and simulation at an alpine desert site (SQH). The simulation is conducted using SiB2 with or without Ross's scheme as well as the surface resistance for calculating soil water flow and evaporation (after Yang *et al.*, 2009, Fig. 11).

bulk density and thermal conductivity, thereby producing the vertical soil stratification.

4.4.3 *Improvements by introduction of soil surface resistance for evaporation in LSM*

The proposed parameterization for the resistance of evaporation was implemented in SiB2 by Yang *et al.* (2009) and the revised SiB2 was then applied to the desert sites with default soil hydraulic parameters. As indicated in Fig. 4.16, after introducing Ross's scheme and the new parameterization for soil surface resistance into SiB2, the simulated soil moisture in the top 15 cm at the alpine desert sites was closer to the observed value. Although it remains difficult to interpret the result due to the uncertainties in the specified model parameters in the simulations, the differences in the simulated soil moisture suggest that implementing physically or mathematically advanced schemes in an LSM is an important aspect that needs to be considered for improving soil moisture simulations.

4.5 Summary and Remarks

This chapter reviewed the progress toward improving LSMs for drylands and high-elevation regions.

First, the daytime land surface temperature for drylands is evidently underestimated by current LSMs. This underestimation actually corresponds to the overestimation of sensible heat flux and the underestimation of soil heat flux. These biases result from the underestimation or neglect of an excess resistance for heat transfer. After incorporating an excess-resistance scheme into LSMs, as recommended by Yang *et al.* (2008), the surface temperature and surface energy partitioning for drylands can be simulated well. This was also verified by a recent study (Zhang, 2011), confirming that this scheme is applicable to bare soils, short vegetation, and even glacier surfaces.

Second, water content within the topsoil of alpine grasslands is usually under-predicted by LSMs due to SOC-induced soil stratification, where the topsoil contains dense grassroots and soil organic matter. Limited experiments have shown that this layer exhibits significantly different properties than the deep soil, and, in general, high soil water content is observed in the topsoil. A single-source model and an inverse estimation system for soil properties were developed to account for the effects of soil vertical heterogeneity. Although the topsoil interacts with the atmosphere directly (due to short vegetation) and strongly (due to high radiation), we have limited knowledge of its hydraulic and thermal properties. Future studies should address this issue in order to develop proper parameterization schemes of the topsoil characteristics.

Third, to improve soil water simulations, we implemented the scheme proposed by Ross (2003) for soil water flow into an LSM and a new parameterization for soil surface resistance for evaporation. The Ross scheme can handle the high nonlinearity of the soil water flow equation. It is shown that the modeling of soil water content can be improved through the implementation of physically or mathematically advanced schemes.

References

Andreas, E. L. (1987). A theory for the scalar roughness and the scalar transfer coefficients over snow and sea ice. *Boundary-Layer Meteorology*, **38**, 159–184.

Beljaars, A. C. M. and Holtslag, A. A. M. (1991). Flux parameterization over land surfaces for atmospheric models. *Journal of Applied Meteorology*, **30**, 327–341.

Beljaars, A. C. M., Viterbo, P., Miller, M. and Betts, A. K. (1996). The anomalous rainfall over the USA during July 1993: Sensitivity to land surface parametrization and soil moisture anomalies. *Monthly Weather Review*, **124**, 362–383.

Beringer, J., Lynch, A. H., Chapin III, F. S., Mack, M. and Bonan, G. B. (2001). The representation of Arctic soils in the land surface model: The importance of mosses. *Journal of Climate*, **14**, 3321–3335.

Brutsaert, W. H. (1982). *Evaporation Into the Atmosphere: Theory, History, and Applications.* 299pp. Dordrecht: Springer.

Camillo, P. J. and Gurney, R. J. (1986). A resistance parameter for bare soil evaporation models. *Soil Scicncc*, **141**, 95 105.

Chen, F. and Zhang, Y. (2009). On the coupling strength between the land surface and the atmosphere: From viewpoint of surface exchange coefficients. *Geophysical Research Letters*, **36**, L10404.

Chen, F., Pielke, R. A. and Mitchell, K. (2001). Development and application of land-surface models for mesoscale atmospheric models: Problems and promises. In *Land-Surface Hydrology, Meteorology, and Climate: Observations and Modeling Water Science and Application*, Lakshmi, V., Alberston, J. and Schaake, J. (eds.), Vol. 3, pp. 107–135. Wahington, DC: American Geophysical Union.

Chen, Y., Yang, K., He, J., Qin, J., Shi, J., Du, J. and He, Q. (2011). Improving land surface temperature modeling for dry land of China. *Journal of Geophysical Research*, p. 116, D20104.

Chen, Y.-Y., Yang, K., Zhou, D.-G., Qin, J. and Guo, X.-F. (2010). Improving the Noah land surface model in arid regions with an appropriate parameterization of the thermal roughness length. *Journal of Hydrometeorology*, **11**, 995–1006.

Dai, Y. *et al.* (2003). The common land model (CLM). *Bulletin of the American Meteorological Society*, **84**, 1013–1023.

Deardorff, J. W. (1978). Efficient prediction of ground surface temperature and moisture, with inclusion of a layer of vegetation. *Journal of Geophysical Research*, **83**, 1889–1903.

Dickinson, R. E. (1992). Land surface. In *Climate System Modeling*, Trenberth, K. E. (ed.), pp. 149–171. Cambridge: Cambridge University Press.

Dickinson, R. E., Henderson-Sellers, A., Kennedy, P. J. and Wilson, M. F. (1986). Biosphere Atmosphere Transfer Scheme (BATS) for the NCAR Community Climate Model. NCAR Technical Note, NCAR/TN-275 + STR.

Duan, Q., Sorooshian, S. and Gupta, V. K. (1992). Effective and efficient global optimization for conceptual rainfall-runoff models, *Water Resources Research*, **28**, 1015–1031.

Duan, Q., Gupta, V. K. and Sorooshian, S. (1993). A shuffled complex evolution approach for effective and efficient global minimization. *Journal of Optimization Theory and Applications*, **76**, 501–521.

Foley, J. A., Prentice, I. C. *et al.* (1996). An integrated biosphere model of land surface processes, terrestrial carbon balance, and vegetation dynamics, *Global Biogeochemical Cycles*, **10**, 603–628.

Fu, C. B. and Wen, G. (2002). Several issues on aridification in the northern China. *Climatic and Environmental Research*, **7**(1), 22–29 (in Chinese).

Garratt, J. R. and Francey, R. J. (1978). Bulk characteristics of heat transfer in the unstable, baroclinic atmospheric boundary layer. *Boundary-Layer Meteorology*, **15**, 399–421.

Global Land Project (GLP) (2005). Science Plan and Implementation Strategy. IGBP Report No. 53/IHDP Report No. 19. IGBP Secretariat, Stockholm. 64pp.

Guo, X. F., Yang, K., Zhao, L., Yang, W., Li, S. H., Zhu, M. L., Yao, T. D. and Chen, Y. Y. (2011). Critical evaluation of scalar roughness length parametrizations over a melting valley glacier. *Boundary-Layer Meteorology*, **139**(2), 307–322.

He, J. (2010). Development of surface meteorological dataset of China with high temporal and spatial resolution, M.S. Thesis, Institute of Tibetan Plateau Research, Chinese Academy of Sciences, Beijing, China.

Henderson-Sellers, A., Yang, Z.-L. and Dickinson, R. E. (1993). The project for intercomparison of land-surface parameterization schemes. *Bulletin of the American Meteorological Society*, **74**, 1335–1349.

Hogue, T. S., Bastidas, L. Gupta, H., Sorooshian, S., Mitchell, K. and Emmerich, W. (2005). Evaluation and transferability of the Noah LSM in semiarid environments. *Journal Hydrometeorology*, **6**, 68–83.

Hu, H.-P., Ye, B.-S., Zhou, Y.-H. and Tian, F.-Q. (2006). A land surface model incorporated with soil freeze/thaw and its application in GAME/Tibet. *Science in China, Series D*, **49**(12), 1311–1322.

Jarvis, P. G. (1976). The interpretations of the variation in leaf water potential and stomatal conductance found in canopies in the field. *Philosophical Transactions of the Royal Society of London, Series B, Biological Sciences*, **273**, 593–610.

Kanda, M., Kanega, M., Kawai, T., Moriwaki, R. and Sugawara, H. (2007). Roughness lengths for momentum and heat derived from outdoor urban scale models. *Journal of Applied Meteorology and Climatology*, **46**, 1067–1079.

Koike, T. (2004). The coordinated enhanced observing period — An initial step for integrated global water cycle observation. *WMO Bulletin*, **53**(2), 1–8.

Koike, T., Yasunari, T., Wang, J. and Yao, T. (1999). GAME–Tibet IOP summary report. *Procceedings of First International Workshop on GAME–Tibet*, Xi'an, China, Chinese Academy of Sciences and Japanese National Committee for GAME, pp. 1–2.

Koster, R. D. *et al.* (2004). Regions of strong coupling between soil moisture and precipitation. *Science*, **305**, 1138–1141.

Kustas, W. P., Choudhury, B. J., Moran, M. S., Reginato, R. J., Jackson, R. D., Gay, L. W. and Weaver, H. L. (1989). Determination of sensible heat flux over sparse canopy using thermal infrared data. *Agricultural and Forest Meteorology*, **44**, 197–216.

Lawrence, D. M. and Slater, A. G. (2008). Incorporating organic soil into a global climate model. *Climate Dynamics*, **30**, 145–160.

Lawrence, D. M. *et al.* (2011). Parameterization improvements and functional and structural advances in version 4 of the Community Land Model. *Journal of Advances in Modeling Earth Systems*, **3**, M03001, 27pp.

LeMone, M. A., Tewari, M., Chen, F., Alfieri, J. and Niyogi, D. (2008). Evaluation of the Noah land-surface model using data from a fair-weather IHOP_200 day with heterogeneous surface fluxes. *Monthly Weather Review*, **136**, 4915–4941.

Li, Q. and Sun, S.-F. (2008). Development of the universal and simplified soil model coupling heat and water transport. *Science in China, Series D, Earth Sciences*, **51**, 88–102.

Los, S. O., Weedon, G. P., North, P. R. J., Kaduk, J. D., Taylor, C. M. and Cox, P. M. (2006). An observation-based estimate of the strength of rainfall — Vegetation interactions in the Sahel. *Geophysical Research Letters*, **33**, L16402.

Luo, S. Q., Lü, S. H., Zhang, Y., Hu, Z. Y., Ma, Y. M., Li, S. S. and Shang, L. Y. (2008). Simulation analysis on land surface process of BJ site of central Tibetan Plateau using CoLM. *Plateau Meteorology*, **27**(2), 259–271 (in Chinese).

Ma, Y.-M., Fan, S.-L., Ishikawa, H., Tsukamoto, O., Yao, T.-D., Koike, T., Zuo, H., Hu, Z.-Y. and Su, Z. (2005). Diurnal and intermonthly variation of land surface heat fluxes over the central Tibetan Plateau area. *Theoretical and Applied Climatology*, **80**, 259–273.

Ma, Y.-M., Kang, S.-C., Zhu, L.-P., Xu, B.-Q., Tian, L.-D. and Yao, T.-D. (2008). Tibetan observation and research platform (Torp): Atmosphere-land interaction over a heterogeneous landscape. *Bulletin of the American Meteorological Society*, **89**, 1487–1492.

Ma, Y.-M., Tsukamoto, O., Wang J., Ishikawa, H. and Tamagawa, I. (2002). Analysis of aerodynamic and thermodynamic parameters over the grassy marshland surface of Tibetan Plateau. *Progress in Natural Science*, **12**, 36–40.

Mahrt, L. (1996). The bulk aerodynamic formulation over heterogeneous surfaces. *Boundary-Layer Meteorology*, **78**, 87–119.

Mahrt, L. and Vickers, D. (2004). Bulk formulation of the surface heat flux. *Boundary-Layer Meteorology*, **110**, 357–379.

Manabe S. (1969). Climate and ocean circulation, the atmospheric circulation and the hydrology of the earth's surface. *Monthly Weather Review*, **91**, 739–774.

Millennium Ecosystem Assessment (MEA) (2005). *Ecosystems and Human Well-Being: Desertification Synthesis*. Washington, DC: World Resources Institute.

Oleson, K. W. *et al.* (2004). Technical description of the Community Land Model (CLM). NCAR Technical Note NCAR/TN-461+STR. National Center for Atmospheric Research, Boulder, CO, 173pp.

Oleson, K. W. *et al.* (2010). Technical description of version 4.0 of the Community Land Model. NCAR Technical Note, NCAR/TN-478+STR, 257pp.

Pielke, R. A. and Avissar, R. *et al.* (1998). Interactions between the atmosphere and terrestial ecosystems: Influence on weather and climate. *Global Change Biology*, **4**, 461–475.

Pielke, R. A., Liston, G. E., Eastman, J. L., Lu, L. and Coughenour, M. (1999). Seasonal weather prediction as an initial value problem. *Journal of Geophysical Research*, **104**, 19463–19479.

Pitman, A. J. (2003). The evolution of, and revolution in land surface schemes designed for climate models. *International Journal of Climatology*, **23**(5), 479–510.

Puigdefabregas, J. (1995). Desertification: Stress beyond resilience, exploring a unifying process structure. *Ambio*, **24**, 311–313.

Ren, G. Y., Xu, M. Z., Chu, Z. Y., Guo, J., Li, Q. X., Liu, X. N. and Wang, Y. (2005). Changes of surface air temperature in China during 1951–2004. *Climatic and Environmental Research*, **10**(4), 711–727 (in Chinese).

Rodell, M. *et al.* (2004). The global land data assimilation system. *Bulletin of the American Meteorological Society*, **85**(3), 381–394.

Ross, P. J. (2003), Modeling soil water and solute transport — Fast, simplified numerical solutions. *Agronomy Journal*, **95**, 1352–1361.

Schelde, K. (1996). Modelling the forest energy and water balance. Series Paper No. 62, Department of Hydrodynamics and Water Resources, Technical University of Denmark, 2800 Lyngby, 263pp.

Sellers, P. J., Randall, D. A., Collatz, G. J., Berry, J. A., Field, C. B., Dazlich, D. A., Zhang, C., Collelo, G. D. and Bounoua, L. (1996). A revised land surface parameterization (SiB2) for atmospheric GCMs, Part I: Model formulation. *Journal of Climate*, **9**, 676–705.

Sellers, P. J. (1992). Biophysical models of land surface processes. In *Climate System Modeling*, Trenberth, K. E. (ed.), pp. 451–490. Cambridge: Cambridge University Press.

Sellers, P. J., Mintz, Y., Sud, Y. C. and Dalcher, A. (1986). A simple biosphere model (SiB) for use within general circulation models. *Journal of the Atmospheric Sciences*, **43**, 305–331.

Sheppard, P. A. (1958). Transfer across the earth's surface and through the air above. *Quarterly Journal of the Royal Meteorological Society*, **84**, 205–224.

Slater, A., Schlosser, C. A. *et al.* (2001). The representation of snow in land-surface schemes: Results from PILPS 2(d). *Journal of Hydrometeorology*, **2**, 7–25.

Smeets, C. J. P. P. and van den Broeke, M. R. (2008). The parameterisation of scalar transfer over rough ice. *Boundary-Layer Meteorology*, **128**, 339–355.

Sun, J. (1999). Diurnal variations of thermal roughness height over a grassland. *Boundary-Layer Meteorology*, **92**, 407–427.

Sun, S.-F. (1982). Moisture and heat transport in a soil layer forced by atmospheric conditions. M.S. Thesis, Department of Civil Engineering, University of Connecticut, 72pp.

Tanaka, K., Tamagawa, I., Ishikawa, H., Ma, Y.-M. and Hu, Z. (2003). Surface energy and closure of the eastern Tibetan Plateau during the GAME-Tibet IOP 1998. *Journal of Hydrology*, **283**, 169–183.

Tang, Q., Oki, T., Kanae, S. and Hu, H. (2008). Hydrological cycles change in the Yellow River basin during the last half of the 20th century. *Journal of Climate*, **21**(8), 1790–1806.

van de Griend, A. A. and Owe, M. (1994). Bare soil surface resistance to evaporation by vapor diffusion under semiarid. *Water Resoures Research*, **30**, 181–188.

Van der Velde, R., Su, Z., Ek, M., Rodell, M. and Ma, Y. (2009). Influence of thermodynamic soil and vegetation parameterizations on the simulation of soil temperature states and surface fluxes by the Noah LSM over a Tibetan Plateau site. *Hydrology and Earth System Sciences*, **13**, 759–777.

Verhoef, A., de Bruin, H. A. R. and van den Hurk, B. J. J. M. (1997). Some practical notes on the parameter kB^{-1} for sparse vegetation. *Journal of Applied Meteorology*, **36**, 560–572.

Viterbo, P. and Betts, A. K. (1999). The forecast impact of the albedo of the boreal forests in the presence of snow. *Journal of Geophysical Research*, **104D**, 27, 803–827, 810.

Wang, Z. W. and Zhai, P. M. (2003). Climate change in drought over Northern China during 1950-2000. *Acta Geographica Sinica*, **58**(suppl), 61–68 (in Chinese).

Xu, B.-Q., Wang, M., Joswiak, D. R., Cao, J.-J., Yao, T.-D., Wu, G.-J., Yang, W. and Zhao, H.-B. (2009). Deposition of anthropogenic aerosols in a southeastern Tibetan glacier. *Journal of Geophysical Research*, **114**, D17209.

Xu, X.-D., Zhang, R.-H., Koike, T., Lu, C.-G., Shi, X.-H., Zhang, S.-J., Bian, L.-G., Cheng, X.-H., Li, P.-Y. and Ding, G.-A. (2008). A new integrated observational system over the Tibetan Plateau (NIOST). *Bulletin of the American Meteorological Society*, **89**(10), 1492–1496.

Yanai, M. and Wu, G.-X. (2006). Effect of the Tibetan Plateau. In *The Asian Monsoon*, Wang, B. (ed.), pp. 513–549. Dordrecht: Springer.

Yang, D., Li, C., Hu, H., Lei, Z., Yang, S., Kusuda, T., Koike, T. and Musiake, K. (2004). Analysis of water resources variability in the Yellow River of China during the last half century using historical data. *Water Resources Research*, **40**, W06502.

Yang, K. *et al.* (2007). Initial CEOP-based review of the prediction skill of operational general circulation models and land surface models. *Journal of the Meteorological Society of Japan*, **85**, 99–116.

Yang, K., Koike, T. and Yang, D. (2003). Surface flux parameterization in the Tibetan Plateau. *Boundary-Layer Meteorology*, **106**, 245–262.

Yang, K., Koike, T., Ye, B.-S. and Bastidas, L. (2005). Inverse analysis of the role of soil vertical heterogeneity in controlling surface soil state and energy partition. *Journal of Geophysical Research*, **110**, D08101.

Yang, K., Koike, T., Fujii, H., Tamagawa, K. and Hirose, N. (2002). Improvement of surface flux parameterizations with a turbulence-related length. *Quarterly Journal of the Royal Meteorological Society*, **128**, 2073–2087.

Yang, K., Koike, T., Ishikawa, H., Kim, J., Li, X., Liu, H.-Z., Liu, S.-M., Ma, Y.-M. and Wang, J.-M. (2008). Turbulent flux transfer over bare soil surfaces: Characteristics and parameterization. *Journal of Applied Meteorology and Climatology*, **40**, 276–290.

Yang, K., Chen, Y. Y. and Qin, J. (2009). Some practical notes on the land surface modeling in the Tibetan Plateau. *Hydrology and Earth System Sciences*, **13**, 687–701.

Yang, Z. L. (2004). Modeling land surface processes in short-term weather and climate studies. In *Observation, Theory and Modeling of Atmospheric Variability*, Zhu, X., Li, X., Cai, M., Zhou, S., Zhu, Y., Jin, F.-F., Zou, X. and Zhang, M. World Scientific Series on Meteorology of East Asia, pp. 288–313. New Jersey: World Scientific.

Ye, D.-Z. and Gao, Y. (1979). *The Meteorology of the Qinghai-Xizang (Tibet) Plateau.* Beijing: Science Press (in Chinese).

Zeng, X. and Dickinson, R. E. (1998). Effect of surface sublayer on surface skin temperature and fluxes. *Journal of Climate*, **11**, 537–550.

Zhang, X. (2011). Simulation of the bare soil surface energy balance at the Tongyu reference site in semiarid area of North China. *Atmospheric and Oceanic Science Letter*, **3**, 330–335.

Zhao H.-L. and Huang, X.-W. (1993). *Mechanism Study on Cropland Desertification in Horqin Sandland: Change of Soil Physical and Chemical Properties in Desertifcation Ecosystem*, Liu, X. (ed.), pp. 164–173. Lanzhou: Gansu Science and Technology Press (in Chinese).

Zhu, Z.-D. and Chen, G.-T. (1994). *Sandy Desertification in China.* Beijing: Science Press (in Chinese).

Zilitinkevich, S. S. (1995). Non-local turbulent transport: Pollution dispersion aspects of coherent structure of convective flows. In *Air Pollution Theory and Simulation*, Power, H., Moussiopoulos, N. and Brebbia, C. A. (eds.), Air Pollution III, Vol. I, pp. 53–60. Southampton: Computational Mechanics Publications.

Chapter 5

Review of Parameterization and Parameter Estimation for Hydrologic Models

Soroosh Sorooshian and Wei Chu

Center for Hydrometeorology and Remote Sensing,
University of California, Irvine

Hydrologic modeling of the rainfall runoff processes for predictions of future flow events requires a modeling system composed of three elements: (1) selection of an appropriate mathematical rainfall-runoff model; (2) a suitable calibration system, and (3) the required observations by the model. Over the past 40 years and with the advent of digital computers, hydrologic models of various levels of sophistication have been developed. Progress toward development of more advanced parameter estimation methods for model calibration has also been made. This chapter provides an overview of the recent developments in modeling and parameter estimation methods available for model calibration purposes.

5.1 Overview

Hydrologic modeling is of great importance in many practical problems such as water resource management, flood forecast and control, civil infrastructure design, and water quality assessing. Unfortunately, there is a fundamental barrier in understanding and studying the hydrologic process. Most of the process takes place below the surface and may extend deeply in the ground, which poses tremendous difficulties.

However, this barrier has never intimidated researchers from developing hydrologic models. On the contrary, there are so many models that have been developed that the word "plethora" sprung to Beven's mind (2001). Despite the great varieties of realizations of different models, the underlying rationale is the same: using the tractable simplification of the hydrologic process to simulate the intractable complexity and therefore

S. Liang, X. Li & X. Xie (eds.), *Land Surface Observation, Modeling and Data Assimilation*, 125–138.

make the infeasible problem feasible. In this attempt, efforts have been focused on two major types of models: conceptual models and process-based models. The conceptual model is characterized by using conceptual components (especially the water storage) to represent the process, whereas the process-based model features simplifying the physical processes, such as water movement, to simulate the system. Nevertheless, both types of models are just approximations of the true system, and there remains substantial discrepancy from the truth and uncertainty within these models. Through calibration of model parameters, hopefully the discrepancy can be reduced and the uncertainty characterized with the light shed by real observations.

Parameter calibration of hydrologic models is among the pioneering applications that have been utilizing direct search optimization techniques (Nash and Sutcliffe, 1970; Johnston and Pilgrim, 1976; Pickup, 1977; Hendrickson *et al.*, 1988). However, calibration of hydrologic models proposes substantial difficulties. As presented by Duan *et al.* (1992), the parameter space response surfaces of models usually possess very complex nature: (a) nonconvexity with multiple optima that have broad attractive regions; (b) discontinuous derivatives and roughness on the response surface, where the roughness refer to the numerous minor local optima; and (c) interaction between parameters and the resulted correlation. These difficulties deter traditional analytical optimization algorithms, such as quasi-Newton method, and on the other hand, suit the heuristic and more flexible direct search strategies. As a successful attempt to automatically calibrate CRR models, the shuffled complex evolution (SCE-UA) method (Duan *et al.*, 1992, 1993) was developed and has exhibited its effectiveness and efficiency compared with other global search procedures. Therefore, it has been widely used in the community of hydrological modeling.

5.2 Review of Hydrologic Models

5.2.1 *Basic concepts of a hydrologic model*

The history of modern hydrologic models stems back to the early 1970s. A milestone, the Sacramento Soil Moisture Accounting Model (SAC-SMA), was introduced by Burnash *et al.* (1973) and has been frequently used ever since. This model is a conceptual lumped-parameter model. It simulates the rainfall-runoff process by processing precipitation through a computational program, in which the uppermost of the soil mantle and the deeper portion of soil are represented as upper zone and lower zone, respectively.

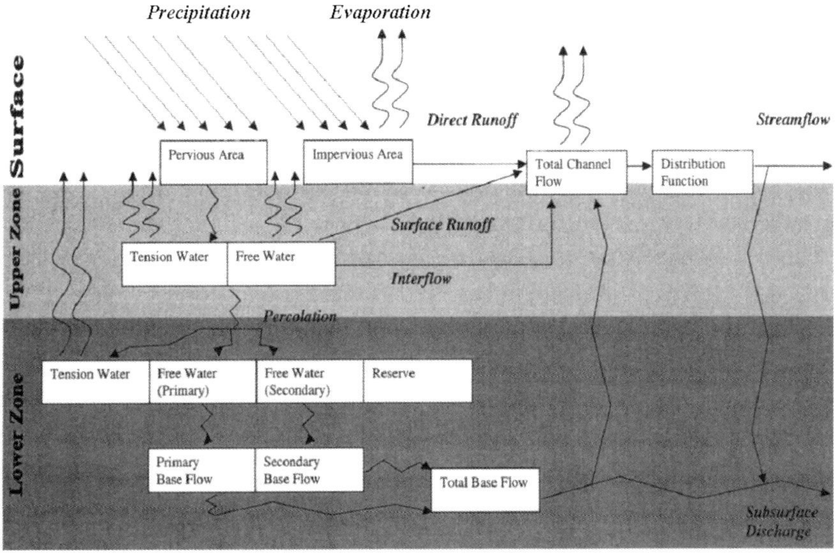

Fig. 5.1. Conceptual flow diagram of the SAC-SMA model (Thiemann *et al.*, 2001). Copyright 2001, John Wiley & Sons, Inc.

The SAC-SMA model utilizes a series of conceptual storages to approximate the soil moisture conditions and to control the production of streamflow. In the upper zone, there are two storages: the upper zone tension water storage and the upper zone free water storage; in the lower zone, storage is divided into four parts: tension water, primary free water, secondary free water, and reserved water. A very brief description of the SAC-SMA model is derived from more detailed introduction of Singh (1995) and is presented here to illustrate the basic concepts of hydrologic models.

Figure 5.1 illustrates how the water is routed through the whole system starting from the precipitation. When rainfall occurs over the surface of the catchment, the portion that falls on the impervious area forms the direct runoff. Rainfall infiltrates into the soil in the pervious area. However, the pervious area starts to produce surface runoff when the rainfall rate is sufficiently heavy and part of the pervious area becomes actively impervious.

The first portion of the infiltration goes to the upper zone tension water storage, which represents the moisture storage that is held in place by the molecular attraction between soil particles and water. Once the tension water is totally satisfied, the excess water is temporarily accumulated as upper zone free water storage which is not bounded to soil particles. The free water feeds to both percolation and interflow. However, the interflow only forms after immediate percolation requirements have been met and is proportional to the residual upper zone free water volume. Water percolated from the upper soil to the lower soil goes to three storages simultaneously according to allocation parameters: lower zone tension water, which is the water held by lower zone soil particles after saturation and drainage and is only removed by evapotranspiration; and two lower zone free water storages, primary and secondary, which represent volumes that are available for drainage as base flows. These two free water storages drain independently at different rates, producing a variable ground water recession.

The heart of the SAC-SMA model is the percolation formula, which is designed based on observed characteristics of the motion of moisture through the soil, including the formation and transmission characteristics. In the model, the percolation, PERC is defined as:

$$\text{PERC} = \text{PBASE}\left(1 + \text{ZPERC}\left(\frac{\text{Lower Zone Deficiency}}{\text{Lower Zone Capacity}}\right)^{\text{REXP}}\right)\frac{\text{UZFWC}}{\text{UZFWM}} \tag{1}$$

where PBASE is the saturated drainage from the lower zone, which is equal to the percolation when the lower zone is totally saturated. PBASE is defined as the sum of the products of each of the two lower zone free water storages and its respective drainage rate:

$$\text{PBASE} = (\text{LZFPM} \times \text{LZPK} + \text{LZFSM} \times \text{LZSK}) \tag{2}$$

In Eq. (1), UZFWC is the upper zone free water contents, whereas ZPERC, REXP, UZFWM, LZFPM, LZPK, LZFSM, and LZSK are all model parameters whose definitions can be found in Table 5.1.

5.2.2 *Trends of modern hydrologic modeling*

With the increasing demands from both research and operation, the hydrologic modeling community is making efforts toward two objects: distributed model and physically-based model.

5.2.2.1 *Distributed hydrologic modeling*

With a lumped hydrologic model, the entire watershed is treated as a spatially homogeneous unit. This assumption is sufficient, if only the variables at the outlet or at the entire watershed level are of interest.

Table 5.1. Parameters of the SAC-SMA model subjected to optimization.

Parameter	Description
UZTWM	Upper zone tension water capacity
UZFWM	Upper zone free water capacity
UZK	Fractional upper zone free water withdraw rate
PCTIM	Minimum imperious area
ADIMP	Additional impervious area
ZPERC	Maximum percolation rate coefficient
REXP	Percolation equation exponent
LZTWM	Lower zone tension water capacity
LZFSM	Lower zone supplemental free water capacity
LZFPM	Lower zone primary free water capacity
LZSK	Fractional supplemental withdrawal rate
LZPK	Fractional primary withdrawal rate
PFREE	Fraction of percolated water into lower zone free water

However, in many situations, we are interested in the hydrologic variables at the interior points of the watershed. Then, the heterogeneity of the watershed and the spatial variance of precipitation play critical roles and need to be correctly simulated. Motivated by this demand, many attempts have been made to develop semi-distributed or fully distributed models. So many models have emerged in recent literature that we cannot afford to enumerate all of them here. However, a sample of models can be found in the Distributed Model Intercomparison Project (DMIP) (http://www.weather.gov/oh/hrl/dmip), a campaign organized by US National Weather Service. But, it is important to notice that distributed hydrologic modeling still remains a great challenge, which means that there still exist many open questions. As summarized in the introduction of DMIP (Smith *et al.*, 2004), some of the most important questions are:

(1) What are the characteristics of a basin that is more likely to benefit from distributed modeling (i.e., accounting for the spatial variability of precipitation and model parameters)? Can these characteristics be identified?
(2) What is the optimal choice of computational element size to capture the essential spatial variability of precipitation in runoff generation and of flow in routing runoff to stream channels?
(3) What level of complexity is required in distributed models to improve basin outlet simulations?
(4) What is the potential for distributed models set up for basin outlet simulations to generate hydrographs at interior locations for flash flood forecasting?

(5) Which approaches work well for handling sub-grid heterogeneity of hydrologic variables?
(6) Does ease of parameterization/calibration of a physically-based distributed parameter model warrant its use, even when it might not provide improvements over simpler (but harder to calibrate) lumped conceptual models?

Answers to these questions will pave a clear pathway to efficient and effective distributed modeling for the future.

5.2.2.2 *Physically-based hydrological modeling*

Certain physical features, such as soil type, vegetation type, and topography, largely affect or shape the hydrologic process of a watershed. If the information of these features is available, it is necessary and beneficial to simulate their effects.

There are many ways of integrating the geophysical features in modeling the hydrologic process, depending on the availability of the data and the object of the simulation. One of the most common practices is improving the water routing through the utilization of high-resolution digital elevation maps (DEMs). Meanwhile, more and more sophisticated schemes have been proposed, developed, and tested. Two most popular types of these schemes are discussed as follows.

The first type is deriving the parameter values from geophysical properties of the studied watershed. For instance, Koren *et al.* (2000, 2003) demonstrated the feasibility of estimating the parameters of a conceptual rainfall-runoff model using soil property maps. Sometimes, though the parameters derived are not so accurate to be used directly, they still provide very important prior information, which will facilitate the derivation of the parameters.

The other type is taking account of the effect of vegetation dynamics. For example, Ivanov *et al.* (2008a, b) introduced a dynamic ecohydrological model [tRIBS + VEGGIE]. It simulates essential water and energy processes over the complex topography of a river basin and links them to the basic plant life regulatory processes. The framework emphasizes ecohydrology of semi-arid environments, exhibiting abundant input of solar energy but limiting soil water that correspondingly affects vegetation structure and organization.

Physically-based hydrologic models not only simulate the rainfall-runoff process but also try to depict the linkage between the surface hydrology and the dynamic environment surrounding it.

5.3 Review of Parameter Estimation Methods

The usefulness of a hydrologic model largely depends on how well the model is calibrated. Therefore, the calibration procedure must be conducted carefully and properly to maximize the reliability of the model. In general, manual procedures for calibration can be extremely time-consuming and this has been a major factor inhibiting the widespread use of the more sophisticated and complex hydrologic models. Fortunately, with the emerging and rapid developing of computers, it is feasible to conduct automatic calibration with powerful global optimization algorithms. Gupta *et al.* (1999) compared the effectiveness of automatic calibration and expert calibration and concluded that the state-of-the-art in automatic calibration now can be expected to perform with a level of skill approaching that of a well-trained hydrologist.

5.3.1 *Automatic calibration requirements*

The concept of automatic calibration (Fig. 5.2) requires three essential components: observational data, models, and optimization algorithms.

Calibration, the identification of hydrologic models, requires proper observational data. What observational data are proper? First of all, it is a common sense that the uncertainty or error of the data should be reasonable, which means the magnitude level and, ideally, the distribution of errors are tractable. Another property of proper data is that the available data can really activate all the model functionalities. For instance,

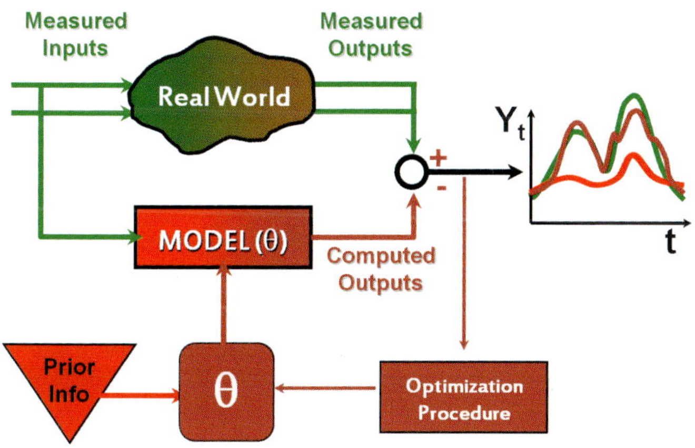

Fig. 5.2. The general scheme of model parameter estimation.

Chu *et al.* (2010) demonstrated that if the forcing data cannot activate certain part of a hydrologic model, then the identification of parameters of that part is consequently poor.

In order to implement automatic calibration, the hydrologic model must be executable by computers. Models should be coded in ways such that the following requirements are met:

(1) The computation time of a model should be feasible. More complexity does not necessarily lead to more effectiveness. Models should be designed as simple as possible. Overfitting not only increases the computation burden but also inflates the uncertainty of predictions. Another way of reducing the computation cost is utilizing parallel processing, especially for distributed models.
(2) Models should have well-defined interfaces for implementing optimization algorithms. It means that the parameters are easily accessible and modifiable. This allows the model developers or practitioners to easily implement optimization algorithms, which are usually generic and not oriented to any specific models.

The response surfaces of hydrologic models in the parameter space usually are very complex, which prohibits traditional analytical optimization algorithms. On the other hand, evolutionary global optimization method, the modern class of direct search methods, can cope with the difficulties resulting from the complex nature of hydrologic models. Therefore, many evolutionary algorithms have developed or alternated for optimization of hydrological models in the past two decades. Some of the state-of-the-art algorithms will be introduced in Sec. 5.3.3.

5.3.2 *Choice of calibration criteria*

Choices of criteria, namely the cost functions, are critical in the success of model calibration. It is determined by both the objectives of the calibration and the properties of the model to be calibrated.

The cost functions must reflect the expectations of the calibration. For instance, root-mean-squared error (RMSE) is one of the most popular criteria in calibrating rainfall-runoff models. It is well known that RMSE emphasizes high flows. If one wants to calibrate a model in hope that the model can simulate peak flow well in order to improve flash flood forecast, RMSE is a good candidate. However, if the expectation of calibrating a model is to simulate the water balance changes in a watershed, the low flow part is not trivial, and RMSE is not a proper statistics in this case. A more sophisticated criterion, which can characterize the total water volume change, is needed.

The properties of models also guide the choice or design of cost functions. If a model is constructed in such a way that its parameters are well defined at certain deterministic values, the cost function usually should be deterministic as well. On the other hand, many models have parameters are not well-defined or have stochastic nature. As a consequence, the calibration criterion should be able to reflect the distributions of the uncertainties associated with the parameters.

5.3.3 *State-of-the-art algorithms of optimization for hydrologic models*

In this section, we introduce several popular global optimization algorithms for calibrating hydrologic models. This is far from a comprehensive review of all types of algorithms used by hydrologic model researchers and practitioners, which can be a subject of a standalone book. Here we only select several algorithms that are well recognized in the community of hydrologic modeling, and these algorithms delineate the pathway of the development of automatic calibration techniques.

The SCE-UA algorithm (Duan *et al.*, 1992, 1993) employs the Nelder–Mead simplex scheme (Lagarias *et al.*, 1998) to make particle replacements for population evolution. However, SCE-UA only adopts the reflection and inside contraction movements and excludes expansion, outside contraction, and shrinkage. If the simplex finds a new point with improved fitness, the worst vertex is replaced by the new point; otherwise, the worst vertex is replaced by a random point in the search domain, which is the mutation procedure. In this scheme, information from a simplex's vertices on the response surface of the objective function is used to approximate the direction of the steepest descent. Driven by the direction of steepest descent, the simplex can effectively find a better offspring. To prevent simplexes from converging to local minima and to enhance the chances of finding the global optimum, SCE-UA utilizes the shuffled-complex process. At the start, the particle population is partitioned evenly into several complexes. Each complex includes $2n + 1$ (n is the dimensionality of the problem) particles and evolves independently using the simplex method. At the beginning of every iteration, a simplex is formed within each complex by randomly selected $n+1$ out of the $2n+1$ particles in order to perform simplex search. Once a worst vertex is replaced, the simplex will be broken down and its particles return to the complex. We refer to this procedure as simplex sub-routine. After each complex completes a certain number of simplex sub-routines, particles in all complexes are mixed and redivided into new complexes through the shuffling procedure, and one iteration ends. By shuffling, the new complex is likely to contain particles from all of the previous complexes, hence having information about the function over the region covered by the entire population instead of a single complex. This gathering of information from all of the previous complexes results in the new complexes to have a better chance of moving toward the global minimum.

Practical experience with model calibration suggests that any single-objective function, no matter how carefully chosen, may not adequately measure the ways in which the model fails to match the important characteristics of the observed data. This is reflected in the fact that the US National Weather Service typically uses as many as 10 different objective functions to measure the goodness-of-fit of the SAC-SMA model during a multi-stage semi-automated calibration procedure (Brazil, 1988). Furthermore, many of the latest hydrologic models simulate several of the watershed output fluxes (e.g., water, energy, chemical constituents, etc.) for which measurement data are available, and all these data must be properly utilized to ensure a proper modeling. In particular, land-surface hydrology models are designed for coupling with General Circulation Models and typically simulate several energy and water fluxes and state variables, including latent heat, sensible heat, temperature, runoff, and soil moisture at various depths. Therefore, procedures for the proper calibration of complex hydrologic models must effectively and efficiently utilize the various measurement data time series that provide useful information about the physical system. The essential characteristic of a multi-objective problem is that the solution will not, in general, be unique, if the individual objectives are non-commensurate. Thus, the goal of optimizing a multi-objective problem is to approximate the Pareto set (or Pareto frontier) in the objective space. Inspired by the success of population-based evolution strategies, such as the Genetic Algorithm (Holland, 1975) and SCE-UA, Yapo *et al.* (1998) developed the multi-objective complex evolution (MOCOM-UA) global optimization method. The MOCOM-UA method is a general-purpose global multi-objective optimization algorithm designed to be effective and efficient for a broad class of problems. The MOCOM-UA strategy combines the strengths "controlled random search" (Price, 1987) with the "competitive evolution" (Holland, 1975), Pareto ranking (Goldberg, 1989), and multi-objective downhill simplex search.

Another practical issue is that estimates of hydrologic model parameters are generally error-prone, because the data used for calibration contain measurement errors and because the model never perfectly represents the system or exactly fits the data. Consequently, it is often impossible to find a single point in the parameter space associated with good simulations, and the retrieving of the uncertainty distributions of parameters is more feasible and meaningful. Inspired by this consideration, Vrugt *et al.* (2003) presented an MCMC sampler entitled the Shuffled Complex Evolution Metropolis Algorithm (SCEM-UA), which is designed to infer the posterior distribution of hydrologic model parameters. The SCEM-UA algorithm is a modified version of the SCE-UA algorithm. The SCEM-UA algorithm operates by merging the strengths of the Metropolis algorithm, controlled random search, competitive evolution, and complex shuffling in order to continuously update the proposal distribution and evolve the sampler to the posterior target distribution. Case studies demonstrate that the adaptive capability of the SCEM-UA algorithm may significantly reduces the number of model simulations needed to infer the posterior distribution of the parameters when compared with the traditional Metropolis–Hastings samplers. However, MCMC methods require the *a priori* definition of a proposal or sampling distribution, which determines the explorative capabilities and efficiency of the sampler and therefore the statistical properties of the Markov Chain and its rate of convergence. How to define a proper proposal distribution for the parameters of hydrologic models remains the biggest challenge in implementing MCMC in hydrologic modeling.

The current efforts of hydrologic modeling have focus on developing distributed and physically-based models. The complexity of new models has been increasing accordingly, which means models possess more and more degrees of freedom (parameters). Then, the optimization for such models is essentially search solutions in high-dimensional parameter spaces. The features unique to high-dimensional space pose new challenges for auto-calibration procedures. As revealed by Chu *et al.* (2010), the SCE-UA method suffers from a critical problem, "population degeneration", when applied to high-dimensional problems. Population degeneration refers to the phenomenon that all searching particles are driven into a subspace (or hyperplane) of the original parameter space. Since SCE-UA uses linear operations (the Nelder–Mead simplex method) on current particles to generate new particles, the population will be confined on this plane in the remaining search. If the global optimum is not located on this plane, the

6.1 Theory of Data Assimilation

Modeling and observation are two fundamental approaches to study water and energy cycles at various scales and investigate the spatiotemporal evolution of the Earth's surface system. Modeling enables scientists to interact iteratively with reality by continuously testing the assumptions used to build models against the extent to which model predictions match reality. Observation is the way to obtain the "true" value of the model state at a certain time. It is known that each of these approaches has its own advantages and disadvantages. The main advantage of modeling is that it uses knowledge of the underlying physics and dynamics to provide a complete description of state evolution in time. The main advantage of observation is that it can provide direct information of the true state, whether it is obtained *in situ* or by remote sensing.

Both modeling and observation have multiple uncertainties. Regarding modeling, many complex land surface parameterization schemes have been developed for the existing global and regional climate models and land process models, but the accuracy of these model simulations remains low. For example, in the global soil wetness project (GSWP), 10 representative land surface models (LSMs), including the Simple Biosphere Model 2 (SiB2) (Sellers *et al.*, 1996), the Simplified Simple Biosphere Model (SSiB) (Xue *et al.*, 1991), and the Biosphere-Atmosphere Transfer Scheme (BATS) model (Dickinson *et al.*, 1993), were applied to simulate the global distribution of soil moisture in 1987 and 1988. However, it was concluded that none of these models could simulate soil moisture accurately at any location because of defects in physical process formulations and parameterization schemes in the existing models. However, it is difficult to determine the initial conditions of model state and the parameters related to soil and vegetation types.

6.1.1 *Uncertainties of modeling*

In this section, we begin our discussion of the uncertainties of modeling using the Richards equation, which represents the movement of water in unsaturated soils. The Richards equation is used by numerous LSMs and hydrological models. It can be expressed as follows:

$$\frac{\partial \theta}{\partial t} = \frac{\partial}{\partial z}\left[K(\theta)\left(\frac{\partial \psi}{\partial z} + 1\right)\right] \tag{1}$$

where θ is the water content ($\mathrm{m^3\,m^{-3}}$), ψ is the pressure head (m), K is the hydraulic conductivity ($\mathrm{m\,s^{-1}}$), z is the soil depth (m), and t is time (s).

From a mathematical viewpoint, there are three components associated with the uncertainties in the equation. The first component is the parameter, i.e., K. This parameter has strong spatial heterogeneity (more than three orders of magnitude in a short distance) and exhibits temporal variation with soil moisture. The parameter K can be observed at the microscale, but in unsaturated soil, it is nearly immeasurable in grid scale and other macroscales. The second component is the model state (θ) of the equation. It has a moderate degree of spatial heterogeneity but strong temporal variation. Although it can be observed in the microscale, obtaining reliable soil moisture in grid scale and other macroscales remains a challenge. Finally, using the SiB2 (Sellers *et al.*, 1996) as an example, we discretize the Richards equation as follows:

$$\frac{\partial \theta_1}{\partial t} = \frac{1}{D_1}\left[Pt - Q_{1,2} - \frac{1}{\rho_w}E_g\right] \tag{2}$$

$$\frac{\partial \theta_2}{\partial t} = \frac{1}{D_2}\left[Q_{1,2} - Q_{2,3} - \frac{1}{\rho_w}E_{tr}\right] \tag{3}$$

$$\frac{\partial \theta_3}{\partial t} = \frac{1}{D_3}[Q_{2,3} - Q_3] \tag{4}$$

where θ_i is volumetric soil moisture content of each layer (m^3 m^{-3}); D_i is the soil thickness of each layer (m); Pt is the precipitation rate (m s^{-1}); $Q_{1,2}$, $Q_{2,3}$, and Q_3 are soil water fluxes between layers and out of the bottom layer (m s^{-1}); E_g and E_{tr} are evaporation from the soil surface and transpiration from the vegetation canopy (m s^{-1}), respectively; and ρ_w is the water density (kg m^{-3}).

We then address the third component associated with the modeling uncertainties, i.e., the so-called atmospheric forcing of the LSM, which determines the boundary conditions of the Richards equation. In the above equations, for example, precipitation is a forcing variable. Forcing usually has a weak to moderate degree of spatial heterogeneity but strong temporal variation. Fortunately, its representative area is larger than that of the parameter or state variable so that it is measurable in the macroscale. However, an uncertainty is also associated with the spatial representativeness of the forcing variable.

Therefore, the inherent characteristics of model simulation can be summarized as follows. At the microscale, the basic processes of water cycle (and probably other land processes?) are known. However, when there is a need to scale up the hydrologic process from microscale to macroscale, for example, to the grid or watershed scale, uncertainties may arise because the parameters, boundary conditions, and initial conditions of the state variables that will influence these processes are highly heterogeneous in space and time. Therefore, many uncertainties are inherent to the modeling. Even if we assume that the physical structure of the model is perfect, the uncertainties in parameters, forcing variables, and the initial conditions will be reflected in the simulation results through complex error propagations.

6.1.2 *Uncertainties of observation*

Both *in situ* and remote sensing observations have many uncertainties. Regarding the conventional *in situ* observation, the "true" value of the observed object can be obtained in its representative space and observing time. The *in situ* observation can usually be considered as an accurate measurement of land surface states, but it has weaknesses regarding its temporal inconsistency and spatial incompleteness. Although the global observation network for land surface variables and fluxes has been enhanced (e.g., the Coordinated Enhanced Observing Period (CEOP) (Koike, 2004), which is a Global Energy and Water Cycle Experiment Project), it is still difficult to construct a complete image of the global and regional energy and water cycle accurately by objective analysis (spatial interpolation). The difficulty arises because the spatial variations of the land surface states are too strong when compared with the significant heterogeneity of the land surface; the *in situ* observations, although increased, are still sparse. Therefore, we can identify the inherent characteristics of point observation.

When the instrumental error is small, *in situ* observations can obtain only the "true value" of the observed objects in the observation time and in their representative spaces. In terms of spatial representativeness, these observations also have much greater uncertainties. If we can obtain the main state variables, parameters, and forcing variables associated with the water cycle at the watershed scale and their temporal and spatial variation characteristics (probability density function; PDF) through intensive and well-designed observation experiments, we can use these values as the model inputs for ensemble forecasting, or we can assimilate them into hydrological models. The accuracy of hydrological simulation and forecasting could be remarkably improved, and the error of the model could be estimated simultaneously.

Recently, extensive satellite remote sensing data have been used to retrieve land surface variables, greatly ameliorating the insufficiency of *in situ* observations. Remote sensing is a macro and quantitative observation approach for land surface processes. However, the relationship between remote sensing observations (such as the brightness temperature or spectral reflectance) and surface ecological or hydrological variables is implied, and observations are therefore indirect. In general, the forward model is a complex nonlinear model, which leads to a difficult or even ill-posed inversion. Additionally, remote sensing cannot directly measure many surface states, such as evapotranspiration and carbon flux; it can only estimate them indirectly via models. Generally, remote sensing observations capture instantaneous values, and they cannot provide information below the shallow surface. For example, remote sensing is not sensitive for the root zone and deep soil moisture, and ground biomass.

In summary, remote sensing itself cannot provide a continuous spatiotemporal evolution of land surface states for four reasons:

(1) The remote sensing observation is instantaneous, but the land surface process is continuous in both space and time.
(2) Remote sensing is an indirect measurement in the sense that the relationship between remote sensing observations and land surface variables is implicit. In general, we can establish a forward model of the relationship between land surface variables and satellite observations such as brightness temperatures. However, the number of observations is usually less than that of land surface variables, and the forward models are often complicated and nonlinear. These factors make the inversion difficult and an ill-posed problem. Therefore, *a priori* knowledge should be used to improve the possibility and accuracy of inversion. The LSM, as a physical constraint, can provide *a priori* knowledge for remote sensing inversions.
(3) Most remote sensors cannot detect sub-surface information. Microwave-based sensors can sense only the top few centimeters of wet soil. However, for hydrological predictions, the soil moisture in the root zone and deeper soil layer is equally important.
(4) Errors exist in remote sensing observations, including instrumental errors and representative errors caused by instrument inaccuracy and surface heterogeneity, respectively.

The inherent characteristics of remote sensing observations can be identified as follows. Remote sensing is a means for macro and quantitative observation of the terrestrial water cycle and can probably also bridge the point observation and model simulation. However, remote sensing provides an indirect observation of the terrestrial water cycle and most likely other processes of the Earth's surface. Regardless of whether inversion (such as soil moisture) or models that indirectly estimate water cycle variables (such as evapotranspiration) are used, great uncertainty remains. Therefore, remote sensing should be carefully validated and then combined with modeling to increase the observability of land surface processes.

6.1.3 *Rationales for land data assimilation*

As discussed earlier, it is important to combine model simulation with observation. Assuming that a remote sensing inverse model can be physically constrained by an LSM and, concurrently, that the trajectory of land surface modeling can be adjusted by remote sensing data to release the accumulated errors, we can make the best use of observations from multiple sources and with different spatial and temporal resolutions. We can therefore obtain high-resolution and spatiotemporally consistent datasets that can better represent the dynamic evolution of the land surface. The newly emerging land data assimilation methods provide a practical means to achieve the aforementioned goals.

Data assimilation, which originates from atmospheric and oceanographic sciences, is a method to merge all available observations with a dynamic model to produce spatiotemporally consistent system states that are as accurate as possible, by considering the observation and model errors (Talagrand, 1997). Similarly, the key objective of land data assimilation is to merge the direct and indirect observations from different sources and with different resolutions within the framework of the dynamics of land process models. By combining various types of observation operators, such as radiative transfer models, with LSMs, the trajectory of the model can be adjusted automatically by the observations to reduce the uncertainties and increase the predictability (Li *et al.*, 2007).

In the fields of land surface and hydrological sciences, data assimilation was not established as a distinct field until the mid-1990s (McLaughlin, 1995). The early work involved retrieval of soil moisture using passive microwave remote sensing data constrained by a land process model (Entekhabi *et al.*, 1994; Houser *et al.*, 1998) and evaluation of the existing data assimilation by developing and applying land data assimilation systems (LDASs) (e.g., Walker *et al.*, 2001; Reichle *et al.*, 2001; Li *et al.*, 2004; Huang *et al.*, 2008b; Jin and Li, 2009), which have become commonplace in land surface and hydrological research. Land data assimilation also displayed certain characteristics that are different from those of atmospheric and oceanographic data assimilation and permitted important progress in the theory and methodology as well as in the development of operational assimilation systems. The most representative events are the development of several operational LDASs, including the North American land data assimilation system (Mitchell *et al.*, 2004), the global land data assimilation system (Rodell *et al.*, 2004), the European land data assimilation system (van den Hurk *et al.*, 2005), and the Chinese land data assimilation system (Li *et al.*, 2007).

6.2 Methods of Data Assimilation

Based on the conclusion of Sec. 6.1, data assimilation can be defined as an important methodology in Earth system science to integrate models and observation because it can significantly improve model predictability and reduce uncertainty. As stated in "Four-Dimensional Model Assimilation of Data: A Strategy for the Earth System Sciences" (National Research Council of the United States (NRC), 1991), data assimilation is used as a strategic method for Earth system science.

The Bayesian theoretical foundation of sequential data assimilation is the focus of this section, and recursive Bayes filter forms for sequential data assimilation are discussed. Moreover, typical sequential data assimilation methods, such as the Ensemble Kalman filter (EnKF), are introduced. Bayesian filter theories provide a fundamental framework for sequential data assimilation and reveal the principles of data assimilation with basic mathematical theories.

6.2.1 *Classification of data assimilation methods*

Mathematically speaking, data assimilation is based mainly on estimation theory, control theory, optimization methods, and error estimation theory. The approach originated in the 1950s as a response to the demand for an objective analysis of weather forecasting. Early data assimilation methods included polynomial interpolation, successive corrections, nudging, and optimal interpolation (Kalnay, 2002), which was a dominant method in the 1980s and 1990s. Since 1990, modern data assimilation methods have been extensively investigated, with rapid applications (Daley, 1991; Kalnay, 2002). In this chapter, we divide the data assimilation methods into two major categories, continuous data assimilation and sequential data assimilation (Daley, 1991). Continuous data assimilation merges the model outputs with all observations in an assimilation window by iteratively and constantly adjusting the model initiative conditions using optimization algorithms. Continuous data assimilation can be further classified into variational methods and Monte Carlo global optimization methods. The variational method, represented by 3DVar and 4DVar, is used to solve objective functions, and the so-called adjoint model of the numerical model is needed to develop during the realization (Cou, 1994; Courtier, 1997; van Leeuwen, 2003). The Monte Carlo optimization method is also called as the heuristic optimization algorithm. The Monte Carlo method is generally easy to realize but has much less efficiency than the variational method (Li *et al.*, 2004; Yang *et al.*, 2007). Another advantage of the Monte Carlo global optimization method is that it can also handle nonlinearity in the dynamic model and the observation model and find the global optima of the objective function.

The sequential data assimilation methods update the model state during the running of the system using the observation when it is available, with weighted model error and observation error. Thus, the posterior optimized estimation of the model state will be obtained. After the state is updated, the system model will be reinitialized by the updated state, and then, the model will be integrated forward until the new observation is presented. Traditionally, sequential data assimilation could be represented by the Kalman filter (KF) for linear systems and the extended Kalman filter (EKF) for nonlinear systems, in which the latter requires the tangent operator of the system dynamics and the observation models to be developed (Talagrand, 1997; Kalnay, 2002). To better handle the nonlinear problem and to reasonably estimate the propagation of system error, nonlinear filter methods were gradually introduced in the field of data assimilation. Among these methods, Monte Carlo sequential data assimilation methods, such as the EnKF (Evensen, 2003), are representative. The Monte Carlo method has achieved great success and has been widely used in the fields of oceanic, land, and atmosphere data assimilation (Evensen, 2007). In recent years, the excellent performance of nonlinear/non-Gaussian filters such as the particle filter (Arulampalam *et al.*, 2002) has attracted much attention (Han and Li, 2008b; Huang *et al.*, 2008a; Nakano *et al.*, 2007; Weerts and El Serafy, 2006). In combination with EnKF, these filters are becoming vital data assimilation methods. The classification of data assimilation methods is briefly summarized in Fig. 6.1 (Li and Bai, 2010).

6.2.2 *Bayesian theoretical foundation for data assimilation*

Bayes' theorem provides a unified methodology for sequential data assimilation of both linear and nonlinear systems. Moreover, it provides a more general theoretical foundation for data assimilation. In this section, the

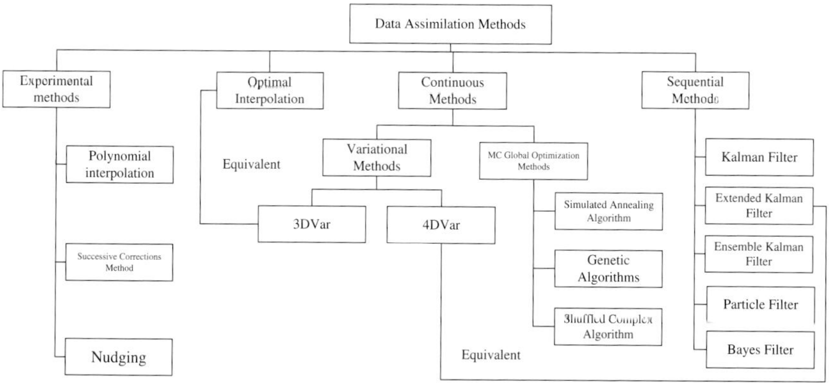

Fig. 6.1. Classification of data assimilation methods.

data assimilation of nonlinear systems is analyzed from Bayes' viewpoint using the language and standard presentation of data assimilation defined in Ide *et al.* (1997).

6.2.2.1 *Data assimilation and nonlinear dynamic systems*

To describe the estimation problems of dynamic systems, a unified framework is used in the state space method, in which it can be divided into state forecasting models and observation models. These models are termed as the model operator and the observation operator in data assimilation systems.

In the nonlinear forecasting model of state space, the model operator is defined as follows:

$$\mathbf{X}^t(t_k) = M_k(\mathbf{X}^t(t_{k-1})) + \eta(t_{k-1}), \quad \mathbf{X} \in \Re^{n\times 1}, \ \eta \in \Re^{n\times 1} \tag{5}$$

where $\mathbf{X}$ is the state vector of the system, t represents the true value, η is the model error with independent identical distribution, n is the dimension of the system, k is the time, and M is the nonlinear model $M(\cdot) : \Re^{n\times 1} \times \Re^{n\times 1} \to \Re^{n\times 1}$. Because Eq. (5) may depend on time, we use the time subscript k for M, although in data assimilation systems, M will generally not change with time.

The observation model is given as follows:

$$\mathbf{Y}^o(t_k) = H_k(\mathbf{X}^t(t_k)) + \varepsilon(t_k) \quad \mathbf{Y} \in \Re^{m\times 1}, \ \varepsilon \in \Re^{m\times 1} \tag{6}$$

where the observation at time t_k is $\mathbf{Y}^o$, ε is the observation error with independent identical distribution, and m is the dimension of the observation vector $H(\cdot) : \Re^{n\times 1} \times \Re^{m\times 1} \to \Re^{m\times 1}$.

In the above state-forecasting and observation models, the distribution of the error can take any form.

The following assumptions are made accordingly (Han *et al.*, 2006):

(1) $p(\mathbf{X}^t(t_0))$ is available, and the probability distribution function (PDF) of the background is available with background expectations $E(\mathbf{X}^t(t_0)) = \mathbf{X}^b$. The error covariance matrix of the background is $\mathrm{Var}(\mathbf{X}^t(t_0)) = \mathbf{P}(t_0)$, or the error covariance matrix at time t_0 can be denoted as background error B. In the equation, $p(\mathrm{X})$ is the *a priori* PDF of the model state, and can be considered as the knowledge accumulated from past observations.

(2) Model error and observation error are both independent processes, independent from each other and from the initial state. Their PDFs are defined as follows:

$$p(\eta(t_k)), \quad p(\varepsilon(t_k))$$

where the mean of the PDFs is zero and the covariance matrix is expressed as $\mathbf{P}(t_k)$ and $\mathbf{R}(t_k)$.

(3) All PDFs can be calculated and defined as follows:

(a) All observations from time 1 to k are defined as $\mathbf{Y}^o(t_{1:k}) \equiv \{\mathbf{Y}^o(t_i), i = 1, \ldots, k\}$.

(b) The true value at time t_k is $\mathbf{X}^t(t_k) \equiv \mathbf{X}(t_k)$.

(c) The state transition function described in possibility form can be represented by the conditional possibility $p(\mathbf{X}(t_k)|\mathbf{X}(t_{k-1}))$.

(d) The observation function $p(\mathbf{Y}^o(t_k)|\mathbf{X}(t_k))$, denoted in probability form, is termed as the likelihood function. It can be considered as the most likely appearance of the model state X when the observation $\mathbf{Y}^o$ is presented. The likelihood function sets up the dependency relationship between $\mathbf{Y}^o$ and $\mathbf{X}$, which is a probability representation of the observation operator.

Regarding the Bayesian filter problem, the objective of the Bayes' filter is to obtain the posterior PDF $p(\mathbf{X}(t_k)|\mathbf{Y}^o(t_{1:k}))$ of the system state at time k based on all observation information available before time k.

6.2.2.2 *Data assimilation and Bayes estimation*

We begin by considering a simplified case to demonstrate the relationship between Bayes' estimation and data assimilation.

Without considering the propagation of the model state with time, the posterior distribution $p(\mathbf{X}|\mathbf{Y}^o)$ can be denoted as follows:

$$p(\mathbf{X}|\mathbf{Y}^o) = \frac{p(\mathbf{Y}^o|\mathbf{X})P(\mathbf{X})}{p(\mathbf{Y}^o)} = \frac{p(\mathbf{Y}^o|\mathbf{X})P(\mathbf{X})}{\int p(\mathbf{Y}^o|\mathbf{X})p(\mathbf{X})d\mathbf{X}} \quad (7)$$

assuming that $p(\mathbf{X})$ and $p(\mathbf{Y}^o|\mathbf{X})$ are multi-dimensional normal distributions with zero mean and error covariance matrix denoted by B and R, respectively. The following objective function can be obtained by taking a logarithm (7) of both sides and maximizing the posterior possibility density function (Lorenc, 1995) as follows:

$$J(\mathbf{X}) = \frac{1}{2}(\mathbf{X} - \mathbf{X}^b)^T\mathbf{B}^{-1}(\mathbf{X} - \mathbf{X}^b) + \frac{1}{2}[\mathbf{Y}^o - H(\mathbf{X})]^T\mathbf{R}^{-1}[\mathbf{Y}^o - H(\mathbf{X})] \quad (8)$$

The result is identical to the 3DVar objective function derived from the optimal linear unbiased estimation (Talagrand, 1997). Therefore, it has been proven that Bayes' theorem can provide a more general framework for data assimilation. Meanwhile, the fundamental principle of data assimilation is revealed under a more basic mathematical theory, i.e., the reliability of the estimation can be improved and the uncertainty can be reduced by merging model information with observations.

6.2.2.3 *Data assimilation and Bayesian filtering*

In this section, by considering the propagation of model state with time, the state estimation of nonlinear systems is introduced in the following two steps.

(1) The forecasting step

Assuming that the posterior possibility density function of the system state at a previous time is available, based on the Chapman–Kolmogorov equation of the Markov process transition density, the PDF of state forecasting at the next time can be given as follows:

$$\begin{aligned} &p(\mathbf{X}(t_k)|\mathbf{Y}^o(t_{1:k-1})) \\ &\quad = \int p(\mathbf{X}(t_k)|\mathbf{X}(t_{k-1}))p(\mathbf{X}(t_{k-1})|\mathbf{Y}^o(t_{1:k-1}))d\mathbf{X}(t_{k-1}) \end{aligned} \quad (9)$$

where

$$p(\mathbf{X}(t_k)|\mathbf{X}(t_{k-1})) = p[\mathbf{X}(t_k) - M_k(\mathbf{X}(t_{k-1}))] \quad (10)$$

Therefore, formula (9) can be derived as

$$p(\mathbf{X}(t_k)|\mathbf{Y}^o(t_{1:k-1})) \\ = \int p[\mathbf{X}(t_k) - M_k(\mathbf{X}(t_{k-1}))]p(\mathbf{X}(t_{k-1})|\mathbf{Y}^o(t_{1:k-1}))d\mathbf{X}(t_{k-1}) \quad (11)$$

The result is identical to that described by Pham (2001).

(2) The updating step

According to Bayes' theorem, the posterior PDF of the system state is written as follows (Han *et al.*, 2006):

$$p(\mathbf{X}(t_k)|\mathbf{Y}^o(t_{1:k})) = \frac{p(\mathbf{Y}^o(t_k)|\mathbf{X}(t_k))p(\mathbf{X}(t_k)|\mathbf{Y}^o(t_{1:k-1}))}{\int p(\mathbf{Y}^o(t_k)|\mathbf{X}(t_k))p(\mathbf{X}(t_k)|\mathbf{Y}^o(t_{1:k-1}))d\mathbf{X}(t_k)} \quad (12)$$

where the likelihood function is

$$p(\mathbf{Y}^o(t_k)|\mathbf{X}(t_k)) = p[\mathbf{Y}^o(t_k) - H_k(\mathbf{X}(t_k))] \quad (13)$$

Then, formula (12) can be derived as follows:

$$p(\mathbf{X}(t_k)|\mathbf{Y}^o(t_{1:k})) = \frac{p[\mathbf{Y}^o(t_k) - H_k(\mathbf{X}(t_k))]p(\mathbf{X}(t_k)|\mathbf{Y}^o(t_{1:k-1}))}{\int p[\mathbf{Y}^o(t_k) - H_k(\mathbf{X}(t_k))]p(\mathbf{X}(t_k)|\mathbf{Y}^o(t_{1:k-1}))d\mathbf{X}(t_k)} \quad (14)$$

The above result is identical to the description by Pham (2001) and Moradkhani *et al.* (2005).

By combining Eqs. (11) and (14), we obtain

$$p(\mathbf{X}(t_k)|\mathbf{Y}^o(t_{1:k})) = \frac{\begin{matrix} p[\mathbf{Y}^o(t_k) - H_k(\mathbf{X}(t_k))]\int p[\mathbf{X}(t_k) - M_k(\mathbf{X}(t_{k-1}))] \\ \times\, p(\mathbf{X}(t_{k-1})|\mathbf{Y}^o(t_{1:k-1}))d\mathbf{X}(t_{k-1}) \end{matrix}}{\begin{matrix} \int\{p[\mathbf{Y}^o(t_k) - H_k(\mathbf{X}(t_k))]\int p[\mathbf{X}(t_k) - M_k(\mathbf{X}(t_{k-1}))] \\ \times\, p(\mathbf{X}(t_{k-1})|\mathbf{Y}^o(t_{1:k-1}))d\mathbf{X}(t_{k-1})\}d\mathbf{X}(t_k) \end{matrix}}. \quad (15)$$

Thus far, recursive Bayesian filtering forms of sequential data assimilation have been completely obtained as Eq. (15), which contains three types of information: (1) the information to represent system dynamics, i.e., the model operator M_k; (2) the observation information, including all observations $\mathbf{Y}^o(t_k)$ before time k and observation operator H_k; and (3) error information, including the model error $p[\mathbf{X}(t_k) - M_k(\mathbf{X}(t_{k-1}))]$ and observation error $p[\mathbf{Y}^o(t_k) - H_k(\mathbf{X}(t_k))]$. Equation (15) fully reflects the

Houser, P. R., Shuttleworth, W. J., Famiglietti, J. S., Gupta, H. V., Syed, K. H. and Goodrich, D. C. (1998). Integration of soil moisture remote sensing and hydrologic modeling using data assimilation. *Water Resources Research*, **34**(12), 3405–3420.

Huang, C. L., Li, X., Lu, L. and Gu, J. (2008a). Experiments of one-dimensional soil moisture assimilation system based on ensemble Kalman filter. *Remote Sensing of Environment*, **112**(3), 888–900.

Huang, C. L., Li, X. and Lu, L. (2008b). Retrieving soil temperature profile by assimilating MODIS LST products with ensemble Kalman filter. *Remote Sensing of Environment*, **112**(4), 1320–1336.

Ide, K., Courtier, P., Ghil, M. and Lorenc, A. C. (1997). Unified notation for data assimilation: Operational, sequential and variational. *Journal of the Meteorological Society of Japan*, **75**(1B), 181–189.

Jin, R. and Li, X. (2009). Improve the estimation of hydrothermal state variables in the active layer of frozen ground by assimilating *in situ* observations and SSM/I data. *Science in China, Series D, Earth Sciences*, **52**(11), 1732–1745, doi: 10.1007/s11430-009-0174-0.

Kalnay, E., Li, H., Miyoshi, T., Yang, S.-C. and Ballabrera-Poy, J. (2007). 4D-Var or ensemble Kalman filter? *Tellus A*, **59A**(5), 758–773.

Kalnay, E. (2002). *Atmospheric Modeling, Data Assimilation and Predictability*, 512pp. Cambridge: Cambridge University Press.

Koike, T. (2004). Coordinated Enhanced Observing Period (CEOP) — An initial step for integrated global water cycle observation. *World Meteorological Organization Bulletin*, **53**(2), 115–121.

Li, X., Koike, T. and Mahadevan, P. (2004). A very fast simulated re-annealing (VFSA) approach for land data assimilation. *Computers and Geosciences*, **30**(3), 239–248.

Li, X., Huang, C. L., Che, T., Jin, R., Wang, S. G., Wang, J. M., Gao, F., Zhang, S. W., Qiu, C. J. and Wang, C. H. (2007). Development of a Chinese land data assimilation system: Its progress and prospects. *Progress in Natural Science*, **17**(8), 881–892.

Li, X. and Bai, Y. L. (2010). A Bayesian filter framework for sequential data assimilation. *Advances in Earth Science*, **25**(5), 515–523 (in Chinese).

Lorenc, A. C. (1995). Atmospheric data assimilation [R]. Scientific Paper No. 34, Forecasting Research, Meteorological Office, Bracknell.

McLaughlin, D. (1995). Recent development in hydrologic data assimilation. *Reviews of Geophysics*, **33**(suppl), 977–984.

Mitchell, K. E. *et al.* (2004). The multi-institution North American Land Data Assimilation System (NLDAS): Utilizing multiple GCIP products and partners in a continental distributed hydrological modeling system. *Journal of Geophysical Research*, **109**(D07). doi: 10.1029/2003JD003823.

Moradkhani, H., Hsu, K. L., Gupta, H. *et. al.* (2005). Uncertainty assessment of hydrologic model states and parameters: Sequential data assimilation using the particle filter. *Water Resources Research*, **41**, doi:10.1029/2004WR003604.

Nakano, S., Ueno, G. and Higuchi, T. (2007). Merging particle filter for sequential data assimilation. *Nonlinear Processes in Geophysics*, **14**(4), 395–408.

National Research Council, Panel on Model-Assimilated Data Sets for Atmospheric and Oceanic Research (1991). *Four-Dimensional Model Assimilation of Data*, 78pp. Washington, DC: National Academy Press.

Pham, D. T. (2001). Stochastic methods for sequential data assimilation in strongly nonlinear systems. *Monthly Weather Review*, **129**(5), 1194–1207.

Reichle, R. H., McLaughlin, D. B. and Entekhabi, D. (2001). Variational data assimilation of microwave radio brightness observations for land surface hydrology applications. *IEEE Transactions on Geoscience and Remote Sensing*, **39**(8), 1708–1718.

Rodell, M., Houser, P. R., Jambor, U. *et al.* (2004). The global land data assimilation system. *Bulletin of the American Meteorological Society*, **85**(3), 381–394.

Sellers, P. J., Randall, D. A., Collatz, G. J., Berry, J. A., Field, C. B., Dazlich, D. A., Zhang, C., Collelo, G. D. and Bounoua, L. (1996). A revised land surface parameterization (SiB2) for atmospheric GCMs, Part I: Model formulation. *Journal of Climate*, **9**(4), 676–705.

Talagrand, O. (1997). Assimilation of observations, an introduction. *Journal of the Meteorological Society of Japan*, **75**(1B), 191–209.

Tsien, H. S. (1954). *Engineering Cybernetics*. New York: McGraw-Hill.

van den Hurk, B. (2005). ELDAS Final Report, KNMI and ECMWF.

van Leeuwen, P. J. (2003) A variance-minimizing filter for large-scale applications. *Monthly Weather Review*, **131**(9), 2071–2084.

Walker, J. P., Willgoose, G. R. and Kalma, J. D. (2001). One-dimensional soil moisture profile retrieval by assimilation of near-surface observations: A comparision of retrival algorithms. *Advances in Water Resources*, **24**(6), 631–650.

Weerts, A. H. and El Serafy, G. (2006). Particle filtering and ensemble Kalman filtering for state updating with hydrological conceptual rainfall-runoff models. *Water Resources Research*, **42**(W0940), doi:10.1029/2005WR004093.

Xue, Y., Sellers, P. J., Kinter, J. L. and Shukla, J. (1991). A simplified biosphere model for global climate studies. *Journal of Climate*, **4**(3), 345–364.

Yang, K., Watanabe, T., Koike, T., Li., X., Fujii, H., Tamagawa, K., Ma, Y. and Ishikawa, H. (2007). Auto-calibration system developed to assimilate AMSR-E data into a land surface model for estimating soil moisture and the surface energy budget. *Journal of the Meteorological Society of Japan*, **85A**, 229–242.

Chapter 7

Estimating Model and Observation Error Covariance Information for Land Data Assimilation Systems

Wade T. Crow

USDA ARS Hydrology and Remote Sensing Laboratory,
Beltsville, MD, 20705, USA
wade.crow@ars.usda.gov

In order to operate efficiently, data assimilation systems require accurate assumptions concerning the statistical magnitude and cross-correlation structure of error in model forecasts and assimilated observations. Such information is seldom available for the operational implementation of land data assimilation systems designed to ingest remotely sensed observations. As a consequence, it is important to understand the impact of poor error assumptions on the performance of land data assimilation systems and strive to develop new tools to obtain required statistical error information. After presenting a simplified theoretical background for the problem, this chapter will describe recent research results which document the impact of incorrect model and observation error assumptions on the assimilation of surface soil moisture fields into a land surface model and review the recent development of adaptive filtering systems which attempt to iteratively estimate modeling and observation error covariance information required by a sequential data assimilation filter.

7.1 Introduction

In the geophysical sciences, data assimilation is the dynamic updating of prognostic model states using incomplete and uncertain observations of model state variables. The past 15 years has seen a rapid increase in the application of data assimilation techniques to land surface models (LSMs) aimed at estimating components of the terrestrial water and energy cycles (see, e.g., Reichle *et al.*, 2008b). In order to function optimally, these techniques require information concerning the statistical properties of both

---

S. Liang, X. Li & X. Xie (eds.), *Land Surface Observation, Modeling and Data Assimilation*, 169–204.

model and observation errors (see, e.g., Crow and Van Loon, 2006). This information is vital for determining the appropriate response of the model to observations. Take the common case of using a remotely sensed surface soil moisture retrieval to update a soil-water balance model. If the modeling error is high and observation error is low, then model-predicted soil moisture (acquired via rainfall observations and a water balance approach) should be updated vigorously in response to a soil moisture observation. Conversely, if the model error is low and soil moisture retrievals are known to be of poor quality, then retrievals should be given little weight and the best predictions of soil moisture will be obtained by simply allowing the model to run with little or no attention made to observations. In this way, the functioning of the filter is directly tied to the relative magnitude of assumed modeling and observation errors.

In addition to the magnitude (i.e., variance) of error, accurate information regarding the cross-covariance of error in various model states is necessary in order to update model states not directly linked to observations. For instance, remotely sensed surface soil moisture retrievals are sensitive only to the amount of water in the top 2–5 cm of the vertical soil column. However, these observations can be used to update deeper soil moisture (and temperature) states if model errors at the surface are known to be vertically correlated with error in adjoining soil layers (see, e.g., Entekhabi *et al.*, 1994). Analogously, updates can be laterally translated from observed to unobserved spatial locations if modeling errors are known to be horizontally auto-correlated (see, e.g., De Lannoy *et al.*, 2009).

Modern data assimilation techniques like the Ensemble Kalman filter (EnKF) provide great flexibility in capturing different types of modeling error (Reichle *et al.*, 2002). Any error source that can be represented by introducing random perturbations to create a Monte Carlo model ensemble can be implemented via an EnKF. However, very little guidance is currently available to constrain either the source or magnitude of such error. This is a particular problem for LSMs, given their diverse range of possible errors sources (e.g., poorly specified initial conditions, errors in micrometeorological forcing variables, inappropriate model parameter choices, neglect of sub-grid land surface heterogeneity, and inaccurate model physics) (Crow and Van Loon, 2006). Likewise, observation errors in many remote sensing retrieval products are difficult to characterize and may vary strongly in space and/or time (Scipal *et al.*, 2008). Consequently, developers of land data assimilation are inevitably forced to make oversimplified and arbitrary decisions regarding the source, magnitude, and correlation structure of both model and observation errors.

Using examples based on recent efforts to assimilate remotely sensed surface soil moisture retrievals into LSMs, this chapter addresses two specific questions related to the characterization of error in a land data assimilation system:

(1) How sensitive are land data assimilation results to the misspecification of modeling and/or observation error parameters?
(2) What steps can be taken to reduce the negative impact of such misspecification on the accuracy of land data assimilation predictions?

Section 7.2 will introduce required data assimilation concepts using a simplified modeling and assimilation framework. These concepts will include the idea of an "adaptive" filtering system in which model and observation error parameters are defined during the analysis cycle of a sequential data assimilation system (Dee, 1995; Mitchell and Houtekamer, 1999; Reichle *et al.*, 2008a). Section 7.3 describes the application of adaptive filtering concepts in Sec. 7.2 to a more realistically complex land surface modeling system, and Sec. 7.4 presents a list of remaining challenges facing the widespread implementation of such systems. Finally, Sec. 7.5 describes recent advances in addressing these challenges.

Here we will focus exclusively on examples derived from the assimilation of surface soil moisture retrievals into LSMs which make prognostic surface soil moisture predictions using water and energy balance considerations. This type of land surface data assimilation system is (arguably) the best-studied and most well-known land surface data assimilation problem. Nevertheless, results presented here should be broadly applicable to other land data assimilation problems.

7.2 Background

Here we present the necessary theoretical background to examine the issue of error parameterization uncertainty in a simple land data assimilation system. In this basic system, land surface modeling is based on using a daily precipitation accumulation product (I) to force the so-called antecedent precipitation index (API):

$$x_i = \beta x_{i-1} + I_i \tag{1}$$

where x is the API-based soil moisture proxy, i is a daily time index, and β a constant loss coefficient. In sequential data assimilation, (1) is typically referred to as the "forecast step" in which a prognostic model advances the prediction of state variables in time until the next available observation. When available, remotely sensed soil moisture estimates θ^{RS} are used to update (1) using the Kalman filter (KF) update equation:

$$x_i^a = x_i^f + K_i(\theta_i^{RS} - x_i^f). \tag{2}$$

In sequential data assimilation, (2) is usually referred to as the "update" step, where model forecasts are modified based on information contained in θ^{RS}, and forecast (f) and analysis (a) superscripts denote state variables before and after KF updating, respectively. For simplicity, we neglect an observation operator in (2) and instead assume that daily θ^{RS} are obtained by linearly rescaling a time series of raw soil moisture retrievals such that their long-term mean and variance match those derived from a multi-year soil moisture time series derived from (1) and no data assimilation. The magnitude of the update (and the relative weighting applied to the forecast relative to the observations) is given by the Kalman gain K in (2):

$$K_i = P_i^f(P_i^f + R)^{-1} \tag{3}$$

where P^f is the error variance in the forecasts of x and R the error variance in θ^{RS} retrievals. Consequently, the adjustment of model forecasts via observations is determined by the relative magnitude of P (the uncertainty in the model forecast) versus R (the uncertainty in the assimilated observation). Like x, P has a forecast step:

$$P_i^f = \beta^2 P_{i-1}^a + Q \tag{4}$$

which describes its temporal evolution between observations. This evolution is a function of both the structural form of the model (here captured by β) and the random uncertainty of the model represented by the forecast noise variance Q. Here, Q represents the magnitude of random error incurred by (1) as it propagates soil moisture forecasts forward in time. Some of the aggregated impact of this uncertainty is reduced at measurement times, where P is updated by:

$$P_i^a = (1 - K_i)P_i^f. \tag{5}$$

Note that P is therefore a function of the (user-specified) forecast noise Q, the structure of the model (β) and the proximity/strength of recent updates.

The overall cycling of the system is as follows: (1) initialize P and x at $i = 0$, (2) forecast P and x using (1) and (4) until the first/next observation time, (3) update P and x using (2) and (5), and (4) return to step #2 and repeat.

Assuming the model operator β is known, the two vital parameter inputs into the KF system are Q in (4) and R in (3). Setting $Q = 0$ (i.e., assuming a perfect model forecast) eventually leads to $P = 0$, $K = 0$ and no weight applied to observations. Conversely, setting $R = 0$ (i.e., assuming perfect observations), reduces (3) to $K = 1$ and (2) to $x_i^f = \theta_i^{RS}$ and places all updating weight on the assimilated observations. Therefore, the functioning of the filter is effectively determined by the relationship between Q and R. In fact, it can be shown that, under steady state conditions, the magnitude of K is fixed by the ratio Q/R (Drecourt *et al.*, 2006).

Figure 7.1 numerically illustrates this point for a simple synthetic experiment in which (1) is run without data assimilation to construct a "truth" soil moisture times. This time series is then artificially perturbed using uncorrelated random noise of magnitude $R = 24\,\text{mm}^2$ to obtain a time series of synthetic API observations. These observations are then re-assimilated into (1) using (2)–(5) for the case where I in (1) has been artificially perturbed using additive random noise of variance $Q = 16\,\text{mm}^2$. Since all model and observation error has been artificially generated, both Q and R are perfectly known in this case. However, we also attempt assimilation for the range of incorrect assumed values for Q and R (Q_o and R_o).

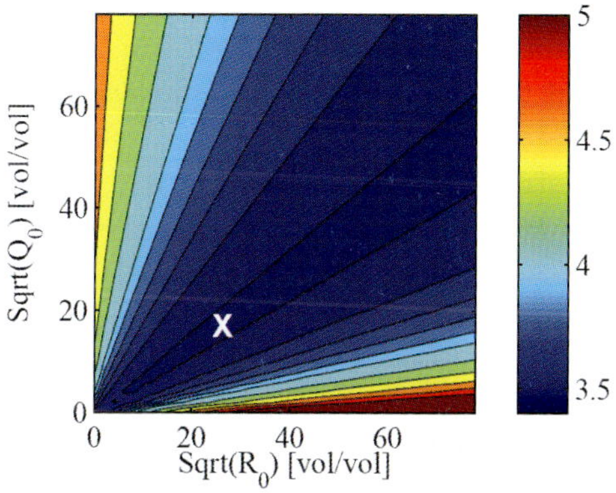

Fig. 7.1. Impact of incorrect Q and R assumptions (Q_o and R_o), on the RMSE accuracy of surface and root-zone soil moisture predictions derived from the assimilation of surface soil moisture retrievals into the API model using a KF. The white "X" indicates true Q and R values.

Hereinafter, we will withhold the use of Q and R notation to refer only to true values of these variables (i.e., those associated with optimal filtering performance) and instead use Q_o and R_o to refer to incorrect or derived estimates of the same variables.

The contour plot in Fig. 7.1 shows how error in the subsequent KF soil moisture analysis (relative to the known truth) is impacted by various choices for Q_o and R_o. True Q and R values are indicated by the white "X". Note how the use of the true Q and R (or, more precisely, the true ratio Q/R) is associated with optimal filter performance. However, the performance of the KF degrades when assumed Q_o and R_o values stray from the correct Q/R ratio.

The classical solution to the Q and R specification problem described above is based on a statistical analysis of filtering innovations. Here, we define these innovations as:

$$v_i = (\theta_i^{RS} - x_i^f)(P_i^f + R_o)^{-1/2} \tag{6}$$

A properly constructed linear filter produces a ν time series that is serially uncorrelated ($E[v_i v_{i-k}^T] = \rho_v(k) = 0$) and has a temporal variance of one ($\text{var}[\nu] = 1$) (Gelb, 1974). Hereinafter, these two constraints will be referred to as the innovation "whiteness" and "variance" constraints, respectively. In addition, ν should be mean-zero ($E[\nu] = 0$); however, this constraint is trivially met by our original rescaling of θ^{RS} (see above). See Appendix A for an analytical derivation of these statistical constraints in the optimal application of a linear filter to a multi-variable, linear, discrete, dynamic system.

For the same synthetic data assimilation case, Fig. 7.2 illustrates the relationship between Q_o and R_o and these innovation-based constraints. True Q and R values are indicated by a black dashed line, but KF is also implemented for a stratified set of other Q_o and R_o combinations. The set of scalar Q_o and R_o combinations satisfying the ν-whiteness constraint are indicated by the solid black line and the set satisfying the ν-variance constraints by the solid red line. Circles indicate Q_o and R_o combinations associated with optimal filter performance (i.e., the RMSE minimum in Fig. 7.1). Note how the two constraints intersect at the true values of Q and R. However, the whiteness constraint is relatively more important since it represents the necessary and sufficient condition for the filter to function optimally (see Appendix A).

Crow and Bolten (2007) provide an example of using ν statistical constraints to infer useful model error information. Their particular strategy

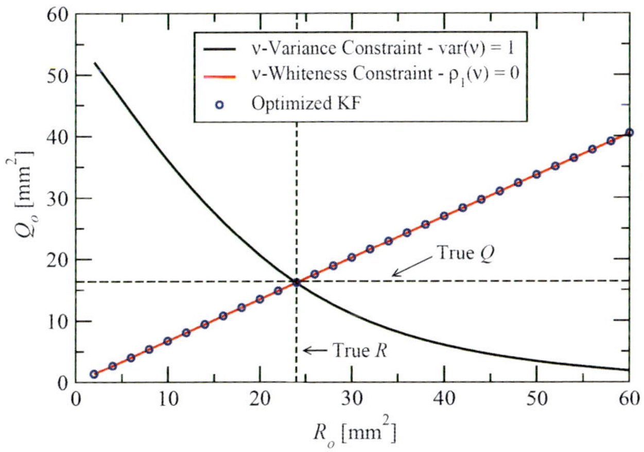

Fig. 7.2. Combinations of Q_o and R_o: satisfying the ν-variance constraint (solid black line), satisfying the ν-whitening constraint (red line), and providing optimal filter performance (blue circles).

is based on a "batch" processing approach in which a retrospective KF analysis over a long-term period is repeated multiple times until calibrated Q_o and R_o values are obtained which satisfy both the ν-whiteness and ν-variance constraints. In a batch approach, iterative modifications are made to Q_o and R_o only after the end of a single retrospective analysis, and a single retrospective period must be repeatedly rerun in order to obtain a stable calibration. Note that this type of retrospective batch strategy differs from an "adaptive" strategy (discussed later) in which Q_o and R_o are continuously updated as the KF makes a single pass through the analysis. In operational terms, the practical difference is that an adaptive approach can be implemented in a real-time (or quasi-real time) forecasting system while a batch approach is only applicable to a retrospective reanalysis.

For the case examined in Crow and Bolten (2007), a range of precipitation products are used for I in (1), and the uncertainty in each product is assumed to be the dominant error source of random modeling error. Using a batch calibration approach, Crow and Bolten (2007) calculate the set of Q_o and R_o which satisfy the ν-whiteness and ν-variance constraints. They then compared derived sqrt(Q_o) to the root-mean-square error (RMSE) of various I over the Southern Great Plains (SGP) portion of the USA (Fig. 7.3). Each grey circle represents a single precipitation product within a single 1° box in the SGP region. Black symbols represent the SGP-wide average of results for each individual precipitation product.

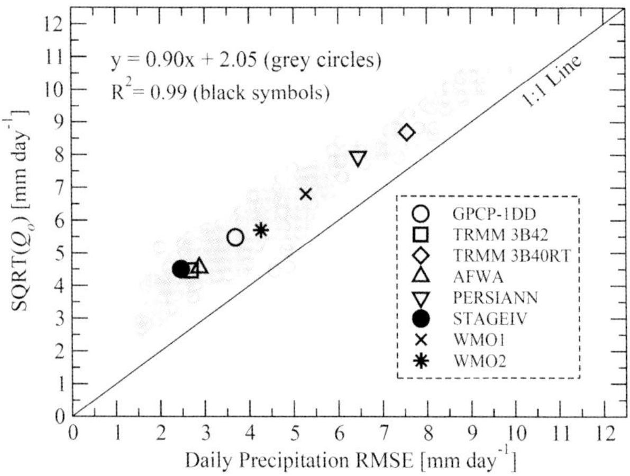

Fig. 7.3. For a range of different precipitation products in the US Southern Great Plains region, the relationship between Q_o, derived from tuning to satisfy the ν-whitening and ν-variance constraints during assimilation into (1) and RMSE in daily precipitation products used to force (1). Adapted from Crow and Bolten (2007). Copyright 2007, John Wiley & Sons, Inc.

The strong linear relationship in Fig. 7.3 between derived sqrt(Q_o) and RMSE in I suggests that imposing ν statistical constraints on soil moisture data assimilation provides a robust method for determining error in daily precipitation accumulation products used to force (1). This is an indication that constraints on temporal ν-statistics can be used to derive useful information concerning the magnitude of stochastic error in LSM inputs.

Based on Fig. 7.3, Crow and Bolten (2007) propose using remotely sensed surface soil moisture retrievals and batch adaptive filtering to validate remotely sensed precipitation products in land areas lacking adequate ground rain gauge coverage. However, it is interesting to note that the relationship between sqrt(Q_o) and rainfall RMSE in Fig. 7.3 does not fall along a one-to-one line. In particular, KF-derived sqrt(Q_o) tends to uniformly overestimate rainfall RMSE. Crow and Bolten (2007) attributed this offset to the impact of non-rainfall-based error sources on API-based soil moisture predictions; however, recent work (summarized later in Sec. 7.4.1) indicates that additional factors might also contribute to this bias.

7.3 Application to a Modern LSM

All results in Sec. 7.2 are based on the application of a linear LSM which does not fully reflect the range of existing complexity in modern LSMs.

Differences between such simple modeling and commonly applied modern LSMs must therefore be considered when evaluating the relevance of results in Sec. 7.2. The first key difference is that, unlike (1), modern LSMs make predictions about multiple, vertically discretized soil moisture and temperature states and therefore require updating of an n-element state vector during data assimilation. A second key difference is that the dynamics of the LSM are typically nonlinear such that the forecast model cannot be expressed in matrix form. The most important nonlinear threshold in such LSMs is often the basic requirement that estimated soil moisture levels be bounded below by residual soil moisture and above by saturation soil moisture (Reichle *et al.*, 2002).

As a consequence, the KF-based data assimilation approach described in (2)–(5) is not sufficient for assimilation into a modern LSM. The most common alternative for such models is currently the application of an EnKF. The EnKF is based on generating an ensemble of state predictions from the Monte Carlo propagation of random error through a (potentially nonlinear) LSM. When an m-dimensional surface soil moisture measurement vector θ^{RS} is available, the n-dimensional state vector x associated with ensemble member j is updated following:

$$x_i^{j,a} = x_i^{j,f} + K_i\left[\theta_i^{RS} + \varepsilon_i^j - h(x_i^{j,f})\right] \tag{7}$$

where ε is an m-dimensional random variable sampled from mean-zero, Gaussian distribution with covariance properties derived from the $m \times m$ observations error covariance matrix R (Burgers *et al.*, 1998) and h is a (potentially nonlinear) function which maps derives an m-dimensional predicted observation vector based on an n-dimension vector forecasted state predictions. In an EnKF, the $n \times m$ matrix K is calculated as:

$$K_i = C_{YM,i}(C_{M,i} + R)^{-1} \tag{8}$$

where C_{YM} is the ensemble-sampled, $n \times m$ cross-covariance matrix between all model states and forecasted soil moisture observations and the C_M, the ensemble-sampled, $m \times m$ covariance matrix of the forecasted soil moisture observations. In this way, the EnKF uses an ensemble of nonlinear model forecasts to acquire the covariance information required by (8). This strategy is in contrast with the analytical KF forecast provided by (4), which requires that the LSM be represented in matrix form. Consequently, the EnKF provides a much more flexible approach for forecasting error information using nonlinear LSMs. Following the update of each ensemble member via (7), the resulting ensemble of x is repropagated until the next available observation. Deterministic EnKF state predictions are then obtained at any point in time by averaging across the ensemble. The EnKF examples shown below follow the typical land data assimilation tendency of assimilating only a single observation of surface soil moisture (i.e., the $m = 1$ case).

Note that Q (or its matrix equivalent) does not appear explicitly in (7) or (8). However, EnKF users must still make an assumption regarding model error characteristics when they select a particular statistical magnitude and form of perturbations to generate the forecast ensemble. This selection of perturbation statistics for the EnKF is equivalent to the selection of Q for the KF in that poor choices for either will degrade the performance of the filter. For example, when updating multi-state models, a critical aspect of (8) is the assumption that the forecast ensemble accurately captures both the error variances of observations predicted by model (C_M) as well as the error covariance between such observations and all (other) model states (C_{YM}). Since elements in the C_{YM} matrix are largely determined by the type of perturbations used to generate the model ensemble, the model perturbation scheme must accurately capture the statistical relationship between errors in observed and unobserved modeling states. This requirement is especially important in land data assimilation cases where remotely sensed observations of the near-surface are frequently used to update profile soil temperature and moisture states far below the vertical penetration depth of the microwave observations (Entekhabi *et al.*, 1994). Note that the vertical relationship between error in (observed) surface soil moisture and (unobserved) root-zone soil, and thus the functioning of the EnKF, will differ if model error used to generate the ensemble is assumed to be concentrated in precipitation errors propagating downward from the surface versus evapotranspiration error propagating upward from the root-zone or water-table. In such cases, the developer of a land data assimilation system is responsible for making appropriate assumptions concerning both the magnitude and the cross-covariance of vertical modeling errors. We will return to this concept later in Sec. 7.4.2.

Recent work has also started the process of duplicating KF/API results in Figs. 7.1 and 7.2 (generated using a simplistic API/KF system) with a fully complex LSM and EnKF data assimilation system. Figure 7.4, taken from Crow and Reichle (2008), duplicates synthetic data assimilation results in Fig. 7.1, but for the multi-layer, nonlinear Water and Energy Balance Surface/Vegetation/Atmosphere Transfer Scheme (WEB-SVAT;

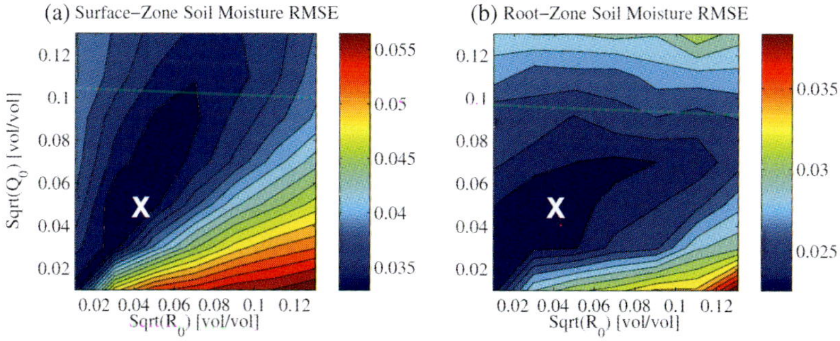

Fig. 7.4. Impact of incorrect Q and R assumptions (Q_o and R_o), on the RMSE accuracy of WEB-SVAT (a) surface and (b) root-zone soil moisture predictions derived from the assimilation of surface soil moisture retrievals into the WEB-SVAT model using an EnKF. The white "X" indicates true Q and R values. Adapted from Crow and Reichle (2008). Copyright 2008, John Wiley & Sons, Inc.

Crow *et al.*, 2008). As before, R_o is the assumed variance of error in assimilated surface soil moisture observations and Q_o relates the variance of random, additive Gaussian error perturbations applied to WEB-SVAT soil moisture states in order to generate the EnKF forecast ensemble. Since WEB-SVAT is a 2-layer model, the full covariance matrix of the applied perturbations is now a 2×2 matrix of the form:

$$\mathbf{Q}_o = \begin{bmatrix} Q_o & \alpha Q_o \rho \\ \alpha Q_o \rho & \alpha^2 Q_o \end{bmatrix}. \tag{9}$$

The cross-correlation of perturbations applied to each layer (ρ) and the ratio of perturbation standard deviations applied to each layer (α) are both assumed known. Note that this greatly simplifies the adaptive filtering problem by requiring it to constrain only the scalar value Q_o.

As in Fig. 7.1, true (as opposed to assumed) values for R and Q in Fig. 7.4 are represented by a white "X" and contour lines capture value of surface (Fig. 7.4(a)) and root-zone (Fig. 7.4(b)) RMSE associated with various incorrect choices for Q_o and R_o. For sufficiently low Q_o and R_o, surface and root-zone soil moisture results in Fig. 7.4 mirror Fig. 7.1 in that a clear minimum is evident when assumed Q_o and R_o combinations fall along a line corresponding to the correct Q/R ratio. However, unlike Fig. 7.1, larger values of Q_o and R_o in Fig. 7.4 are associated with degraded filter performance — even for the case where the optimal Q/R ratio is preserved. This degradation occurs because large assumed values of Q_o cause the EnKF ensemble to encounter nonlinear dynamic thresholds (e.g., soil moisture saturation levels) existing within the

WEB-SVAT model. Such thresholds impose non-desirable characteristics on the EnKF forecast ensemble, including forecast bias and a non-Gaussian distribution shape, which degrade the subsequent performance of an EnKF. Note that these negative impacts are absent for the linear modeling case examined in Fig. 7.1. See Ryu *et al.* (2009) for a detailed description of nonlinear modeling effects on EnKF performance.

An important question is whether or not such nonlinear effects will preclude the application of adaptive filtering techniques (designed originally for linear modeling systems) to nonlinear LSMs and EnKF-based data assimilation systems. EnKF innovations are defined as:

$$\nu_i = \left\langle \left(\theta_i^{RS} - h(x_i^{j,f})\right)(C_{M,i} + R)^{-1/2} \right\rangle \tag{10}$$

where angled bracketing indicates ensemble averaging. Crow and Reichle (2008) and Reichle *et al.* (2008a) test a number of different adaptive filtering techniques designed to iteratively estimate Q_o and R_o based on the temporal statistics of ν. Relative to earlier attempts presented in Crow and Bolten (2007), they attempt the more ambitious case of assimilation into a nonlinear, multi-state LSM where Q_o and R_o are adaptively modified within the analysis cycle.

Crow and Reichle (2008) examined EnKF/WEB-SVAT adaptive filtering results for four different adaptive filters approaches which they referred to as the: "Mehra", "Secant Whitening", "Nudge Whitening" and "Desrozier" algorithms. The first three of these algorithms (Mehra, Secant Whitening, and Nudge Whitening) are based on the following sequential steps: (1) start with an initial guess Q_o and R_o, (2) run the EnKF for N-days using these initial values, (3) use the last N-days to sample temporal values of $\rho_\nu(1)$ and var(ν), (4) adjust the ratio between Q_o and R_o based on these sampled values, and (5) go to step #2 and repeat. In this way, all three approaches attempt to iteratively evolve Q_o and R_o during via consideration of temporal ν statistics. The approaches vary only in there particular strategy for updating non-optimal Q_o and R_o in response to sampled $\rho_\nu(1)$ and var(ν).

For example, the Nudge Whitening approach is based on a simple set of updating rules at the end of each N-day period. Defining $\Delta = Q_o/R_o$ and $\rho_\nu(1)_j$ to be the sampled lag-1 auto-correlation of ν in N-day time window j, the update at the end of time-window j is given by

$$\Delta_{j+1} = \left\{ \begin{matrix} 2\Delta_j & \rho_\nu(1)_j & > 0 \\ \Delta_j & \rho_\nu(1)_j & = 0 \\ 0.5\Delta_j & \rho_\nu(1)_j & < 0 \end{matrix} \right\} \tag{11}$$

Note the focus on the sampled $\rho_\nu(1)$ diagnostic and direct tuning of the Q/R ratio. This is due to the fundamental role both play in diagnosing and/or ensuring optimal filter performance (see Figs. 7.1 and 7.2). Sampled var(ν) is used only as a secondary constraint to obtain R_o^{j+1} and Q_o^{j+1} given Δ_{j+1} (Crow and Reichle, 2008). The Secant Whitening and Mehra approaches are conceptually similar to (11) except based on a more sophisticated approach for updating Δ according to sampled $\rho_\nu(1)$. See Crow and Reichle (2008) and Mehra (1970) for further details. Also, note that the Desrozier approach is based on a slightly different adaptive strategy outlined in Reichle *et al.* (2008a).

For the case of $N = 50$ days, Fig. 7.5 demonstrates the ability of these adaptive techniques to iteratively converge toward true Q and R values (see black "X"). Assuming a stratified grid of initial guesses for Q_o and R_o (see black "+"), each approach is able to gradually converge toward the true error parameters. As Q_o and R_o converge toward their correct values, the performance of the EnKF gradually improves. For various initial choices of Q_o and R_o, Fig. 7.6 relates the long-term improvement in filter surface and root-zone soil moisture RMSE results associated with the implementation of various adaptive filtering strategies relative to a baseline of not modifying Q_o and R_o relative to their initial stratified values. Therefore, contour results for each Q_o and R_o combinations relates the improvement in adaptive filtering results relative to a baseline of simply maintaining the incorrect initial specification of Q_o and R_o. Results in Fig. 7.6 show a general trend of improvement (i.e., negative changes in RMSE) for all adaptive filtering approaches. The magnitude of this improvement varies as a function of accuracy in initial Q_o and R_o guesses. In fact, for all techniques, a small positive increase in RMSE performance (see red shading within solid black lines) is evident for the case of relatively good initial guesses for Q_o and R_o. However, the magnitude of this degradation is small relative to potential improvements, and overall results demonstrate that adaptive filtering approaches are generally viable for complex LSMs.

7.4 Remaining Challenges

Despite encouraging results in Figs. 7.5 and 7.6, several remaining challenges must be addressed before adaptive filters can be successfully applied to real data assimilation problems. This section will identify and discuss a number of such challenges. To begin, it should be stressed that all results presented to this point have been based on synthetic twin data assimilation experiments in which the type and/or vertical source of LSM modeling error

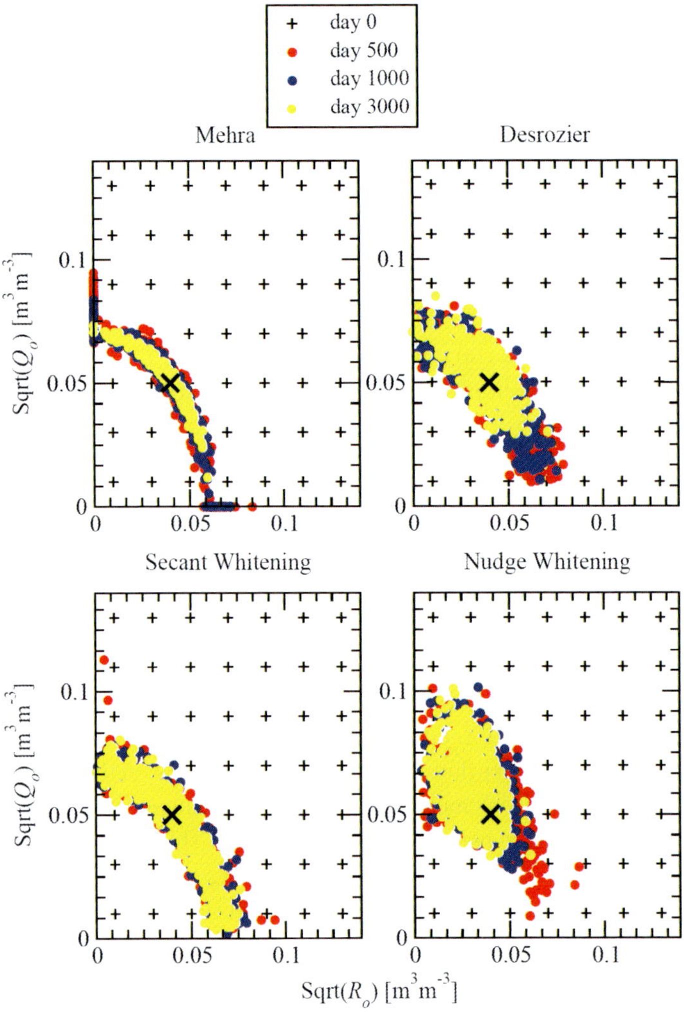

Fig. 7.5. Starting with a stratified grid of initial R_o and Q_o values (small black "+"), the temporal convergence of all four adaptive filtering strategies toward the correct Q and R combinations (large black "X"). Adapted from Crow and Reichle (2008).

were assumed to be known. Furthermore, synthetically-induced modeling and observation error in these experiments was generated to conform perfectly to the statistical assumptions underlying the optimality of the EnKF. The first two challenges described below (Secs. 7.4.1. and 7.4.2.) emerge when attempting to graduate from this type of highly controlled

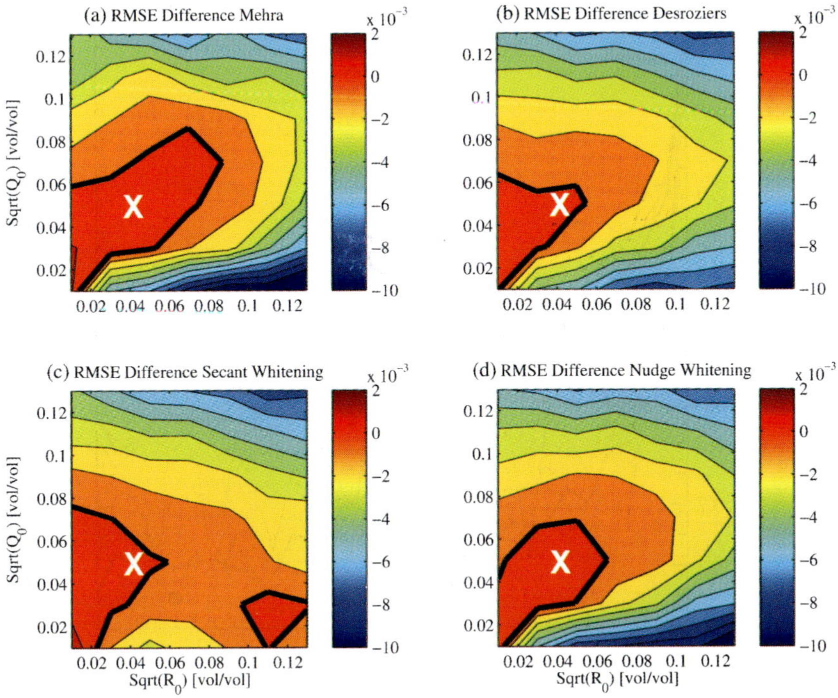

Fig. 7.6. Impact of incorrect Q and R assumptions (Q_o and R_o) on the reduction of RMSE in WEB-SVAT root-zone soil moisture predictions associated with the application of all four adaptive filtering systems. The large white "X" indicates true Q and R values. Adapted from Crow and Reichle (2008). Copyright 2008, John Wiley & Sons, Inc.

synthetic experiment to the more challenging case of assimilating real remote sensing data into an LSM with uncertain error characteristics. In addition, a close examination of Fig. 7.4 suggests that, while available adaptive filtering approaches reliably converge toward correct error values, their rate of convergence tends to be very slow. The third challenge described below addresses this rate-of-convergence issue (see Sec. 7.4.3).

7.4.1 *Auto-correlated observation errors*

A key assumption underlying the optimality of a KF (and/or EnKF) is that observation errors are serially uncorrelated (i.e., "white"). Figure 7.7 directly examines this issue by calculating the autocorrelation function for errors found in three surface soil moisture retrieval products derived from Advanced Microwave Scanning Radiometer (AMSR-E) brightness

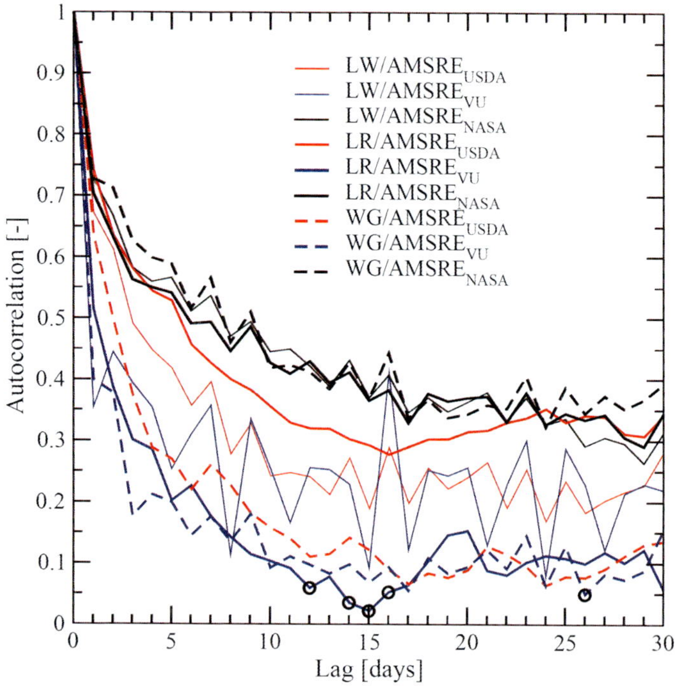

Fig. 7.7. Autocorrelation functions for three different AMSR-E soil moisture products (USDA, VU and NASA) at three different watershed sites (LW, LR and WG) in the USA. Open circles indicate lagged autocorrelation values which are not statistically significant at a 95% confidence level. Adapted from Crow and Van den Berg (2010). Copyright 2010, John Wiley & Sons, Inc.

temperature observations ($AMSRE_{USDA}$, $AMSRE_{VU}$, and $AMSRE_{NASA}$). For details and further references, see Crow and Van den Berg (2010).

Here error is calculated as the difference between ground-based observations (acquired at three separate highly instrumented ground data sites located in the Little Washita (LW), Walnut Gulch (WG), and Little River (LR) US Department of Agricultural (USDA) Experimental Watersheds — see Jackson *et al.* (2010) for details) after a separate 31-day moving average climatology has been removed from both the retrievals and the ground observations. Even after the removal of such seasonality, results in Fig. 7.7 demonstrate the presence of significant lagged autocorrelation in retrieval errors for all three AMSR-E surface soil moisture products at all sites.

Such autocorrelation has a demonstrable impact on the appropriateness of R_o estimates derived from the ν-whiteness constraint. Returning briefly

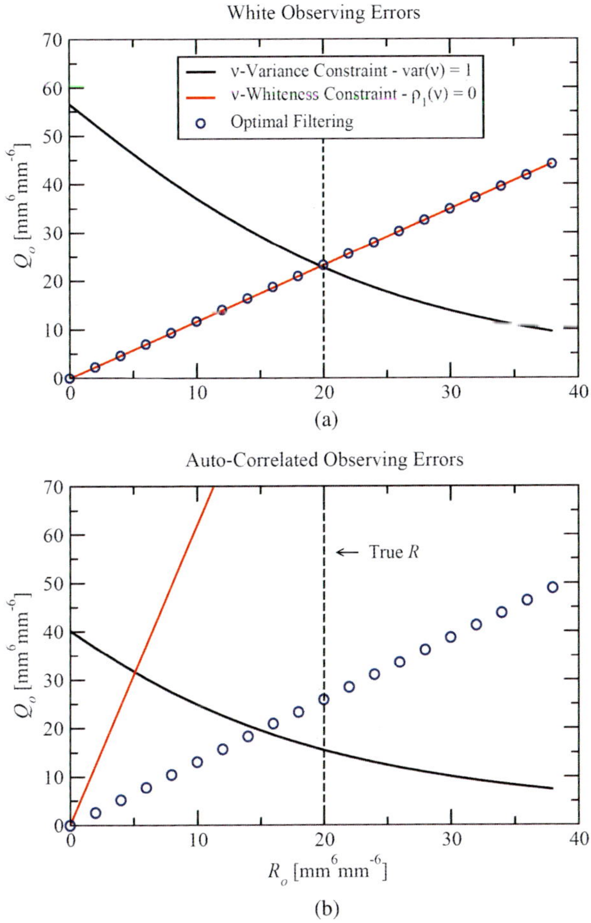

Fig. 7.8. Combinations of Q_o and R_o satisfying the ν-variance constraint (black line), satisfying the ν-whitening constraint (red line) and providing optimal filter performance (blue circles) for the case of (a) white observing error and (b) auto-correlated observing errors. Adapted from Crow and Van den Berg (2010). Copyright 2010, John Wiley & Sons, Inc.

to the simple, synthetic data assimilation system introduced in Sec. 7.2, Fig. 7.8 demonstrates the theoretical impact of auto-correlated observation error on LSM adaptive filtering results. For the case of serially white observing error, the ν-whiteness constraint is aligned with the Q_o/R_o ratio associated with optimal filter performance (Fig. 7.8(a)). However, this alignment is disrupted when assimilated observations are artificially perturbed

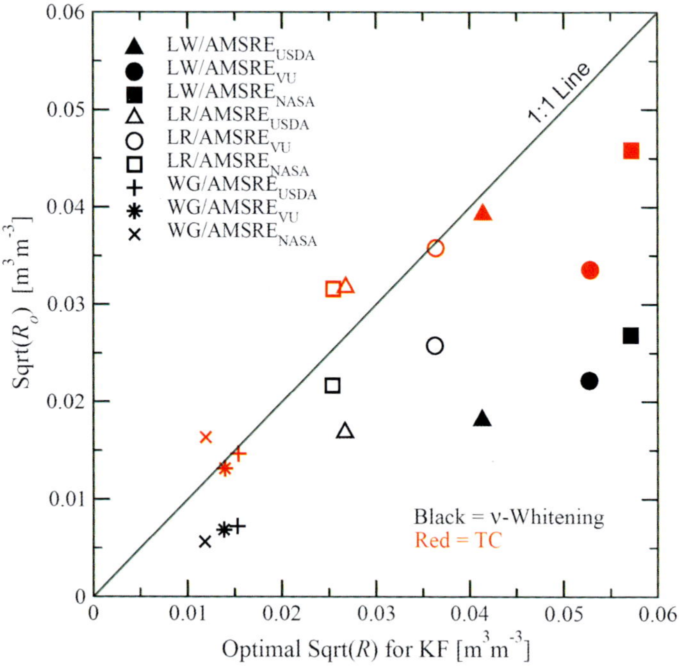

Fig. 7.9. Comparison of R_o estimates obtained from both TC (in red) and v-whitening (in black) with R values associated with the optimal performance of the KF. Adapted from Crow and Van den Berg (2010).

with serially correlated error (Fig. 7.8(b)). In this case, the Q_o/R_o ratio associated with serially white ν increases such that serially uncorrelated ν are obtained only when the true Q/R ratio is *overestimated*. When the ν-variance constraint is also enforced, this leads to a low bias in recovered values of R_o and, consequently, excessive weight placed on assimilated observations. Note also that a low bias in R_o is accompanied by a high bias in Q_o in Fig. 7.8. Therefore, part of the overestimation of sqrt(Q_o) in Fig. 7.3 (relative to rainfall RMSE) is likely attributable to the presence of auto-correlated error in assimilated surface soil moisture retrievals.

This bias is also apparent in real data KF and EnKF results. For the same set of watershed sites and AMSR-E soil moisture products examined in Fig. 7.7, Fig. 7.9 plots the relationship between estimated R_o values (via a simple batch ν-whitening procedure) and optimal R values which minimize the RMSE difference between KF/API results. Results demonstrate that R_o estimates obtained via ν-whitening consistently underestimate values of R associated with the optimal performance of the filter. These real data assimilation results are therefore consistent with earlier theoretical predictions made in Fig. 7.8. Innovation whitening in the presence of observations containing auto-correlated errors leads to excessively low R_o estimates and, consequently, overconfidence concerning the quality of assimilated observations. Therefore, any operationally viable adaptive filtering technique needs to account for the possibility of auto-correlated errors in surface soil moisture retrieval products. In addition to the KF/API results shown here, comparable results were also obtained for conducting the same data assimilation analysis using an EnKF to assimilate surface soil moisture retrievals into a fully complex multi-layer LSM (Crow and Van den Berg, 2010).

7.4.2 *Uncertainty in the source and structure of model error*

Another substantial simplification for synthetic data assimilation results in Sec. 7.3 is that the relationship between modeling error in various soil moisture states is assumed to be perfectly known. That is, the parameters α and ρ in (9) are assumed known and the adaptive filter can therefore focus exclusively on estimating the overall magnitude of errors (i.e., Q_o).

In reality, relatively little is known about the relationship between LSM errors in various vertical soil moisture states. Crow and Van Loon (2006) illustrate a case where such ignorance has serious impacts on the functioning of an EnKF. Their analysis is based on a synthetic data assimilation strategy in which the errors used to synthetically perturb an LSM (to create the open loop simulation) vary relative to the assumed errors used to create the Monte Carlo EnKF forecast ensemble. In this way, they analyze cases where the type of model error captured in the EnKF forecast ensemble is *not* an accurate representation of the true error actually degrading the LSM.

The LSM they use, the Topographically Land Atmosphere Transfer Scheme (TOPLATS) (Peters-Lidard *et al.*, 1997), tries to capture the direct impact of a dynamic water table on surface soil moisture levels. Therefore the TOPLATS surface soil moisture state θ_{sz} is calculated as the weighed average of soil moisture in non-saturated areas (θ_{unsat}) and saturation soil moisture values (θ_{sat}) within areas of a catchment where topographic influences lead to the intersection of the water table with the surface:

$$\theta_{sz} = [1 - f_{\text{sat}}(\bar{z})]\theta_{\text{unsat}} + f_{\text{sat}}(\bar{z})\theta_{\text{sat}}. \tag{12}$$

Here, f_{sat} is the dynamic areal fraction of a catchment which is completely saturated by a shallow water table. Note that this fraction varies as a function of the spatially averaged water-table depth in the catchment ($\bar{z}$).

When attempting to characterize error in θ_{sz} predictions, derived from (12), a decision must be made concerning the fraction of θ_{sz} error arising from surface processes (e.g., rainfall and/or soil evaporation) directly impacting θ_{unsat} or sub-surface processes (e.g., root extraction from the water table, recharge or baseflow) which affect $\bar{z}$ and thus f_{sat}. For an EnKF application, this decision dictates the relative strength of random perturbations applied to either TOPLATS θ_{unsat} or $\bar{z}$ state estimates in order to generate the EnKF forecast ensemble. In choosing a particular TOPLATS state prediction to perturb, the developer of an EnKF-based data assimilation system is guessing about the type of error they believe is actually degrading TOPLATS predictions. Unfortunately, these decisions are frequently arbitrary and have a strong impact on subsequent EnKF results.

For example, Crow and Van Loon (2006) illustrate cases where poor decisions concerning the vertical source of TOPLATS modeling error leads to sub-optimal EnKF performance. Their approach is based on conducting a series of synthetic twin data assimilation experiments in which the source and magnitude of modeling errors are varied for both the actual error used to perturb the model and the Monte Carlo perturbations (i.e., assumed error) used to generate the EnKF. For this particular synthetic analysis, R is assumed to be perfectly known and observation errors are generated to be serially white. As a result, optimal filter performance is ensured if the ν-variance constraint is met (see Fig. 7.2).

Figure 7.10 shows an example of this analysis and illustrates the danger of poorly parameterizing the vertical source of TOPLATS modeling errors. The three vertical panels in the figure each refer to a synthetic EnKF analysis where actual modeling error is created via three different types of modeling error: uncertainty in θ_{unsat} water balance calculations (Fig. 7.10(a)), uncertainty in precipitation inputs (Fig. 7.10(b)), and error in $\bar{z}$ water balance calculations (Fig. 7.10(c)). The three lines on each panel in Fig. 7.10 reflect attempts to correct each of these error types using an EnKF forecast ensemble generated via random perturbations consistent with each of these three potential error sources. Therefore, for each panel in Fig. 7.10, one line corresponds to a correct assumption concerning the source of modeling error while the other two lines capture the impact of making an incorrect assumption concerning the type of

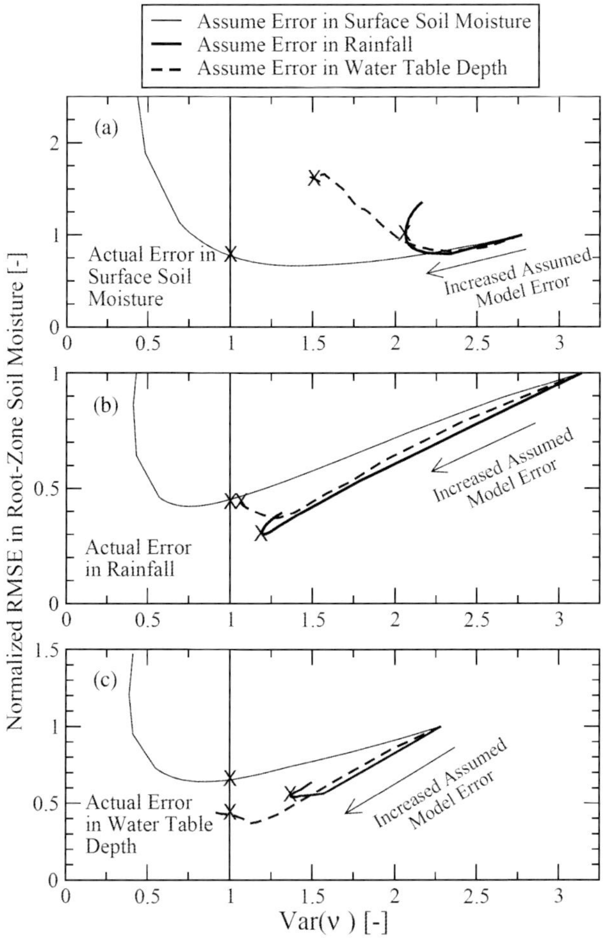

Fig. 7.10. Relationship the variance of ν and the normalized RMSE of EnKF-based root-zone soil moisture predictions for the case of actual error in model (a) surface soil moisture predictions, (b) rainfall forcing, and (c) water-table depth predictions for the case of assimilating surface soil moisture. Assumed model error for each panel is varied according to both its source and magnitude. Normalized RMSE is defined as the RMSE of the EnKF results divided by RMSE of the open loop simulation. The large black "X" indicates the closest approach to the optimal condition of $\text{var}(\nu) = 1$. Adapted from Crow and Van Loon (2006).

modeling error actually present in TOPLATS predictions. Each point on these lines represents a different variance for assumed modeling error (i.e., a different variance of perturbations used to generate the EnKF ensemble).

The x-axis of Fig. 7.10 lists the variance of ν associated with each assumption. Since R is assumed to be perfectly known in this synthetic experiment, EnKF root-zone soil moisture results (zero to 40-cm depth) *should* be optimized (i.e., exhibit minimum RMSE) along the vertical line corresponding to $\mathrm{var}(v) = 1$. This generalization is supported by results in Fig. 7.10 but only for cases in which the random error source used to generate the EnKF ensemble corresponds to the true model error type. For instance, in Fig. 7.10(a) (where actual error is due to random noise in θ_{unsat} water balance predictions), the best EnKF results are found when generating the EnKF via random perturbations applied continuously to TOPLATS θ_{unsat} predictions and then tuning the variance of these perturbations until $\mathrm{var}(v) = 1$. Analogous results are seen in Figs. 7.10(b) and 7.10(c), where the best root-zone EnKF results (i.e., lowest RMSE along the y-axis) are obtained by making a correct assumption concerning the source of modeling errors (i.e., uncertain precipitation inputs in Fig. 7.10(b) and uncertain $\bar{z}$ dynamics impacting f_{sat} in Fig. 7.10(c)) and then tuning the magnitude of random perturbations until $\mathrm{var}(v)$ is as close to one as possible.

However, results in Fig. 7.10 become markedly worse in cases where the assumed source of the error differs from that actually applied. For instance, in Fig. 7.10(a) (where actual error is due to random perturbations applied continuously to TOPLATS θ_{unsat} predictions), falsely assuming that modeling error arises from random errors in $\bar{z}$ leads to an unfortunate case where increasing the variance of random perturbations applied to TOPLATS $\bar{z}$ predictions produces progressively better innovation results (i.e., $\mathrm{var}(v)$ moves toward one) while simultaneously *degrading* the performance of the filter. That is, in the presence of a poor assumption regarding the type of modeling error present, using adaptive filtering to calibrate an EnKF can actually degrade the accuracy of data assimilation results. Similar problems are seen in Figs. 7.10(a) and 7.10(b), when error is falsely attributed to either precipitation error or $\bar{z}$ dynamics, respectively.

Earlier adaptive filtering results presented were based on synthetic data assimilation experiments in which both the source and relative vertical distribution of modeling errors was assumed known. Consequently, adaptive filtering was tasked solely with estimating the *magnitude* of modeling errors. Results from Crow and Van Loon (2006) summarized here stress the importance of characterizing the source and type of these errors as well.

7.4.3 *Speed of adaptive filter convergence*

A final problem is that, even in well-behaved cases in which observing errors are serially uncorrelated and the source of modeling errors is known perfectly, existing adaptive filters have been observed to converge at very slow rates when applied to the assimilation of surface soil moisture retrievals into an LSM. For instance, Fig. 7.11, adapted from Crow and Reichle (2008), illustrates the rate at which the Mehra adaptive filter converges to the correct value of Q. Because it updates Q_o and R_o based on an analytical formulation, the Mehra approach can be assumed to converge as fast as any competing adaptive filtering approach. Nevertheless, it demonstrates an extremely slow rate of convergence. Even assuming the availability of one observation per day, the Mehra adaptive filter is still converging after ~3000 days of operation (Fig. 7.11).

Crow and Reichle (2008) examine the source of this convergence problem and find it to be related to the physical characteristics of the surface soil moisture retrievals themselves. Most adaptive filtering

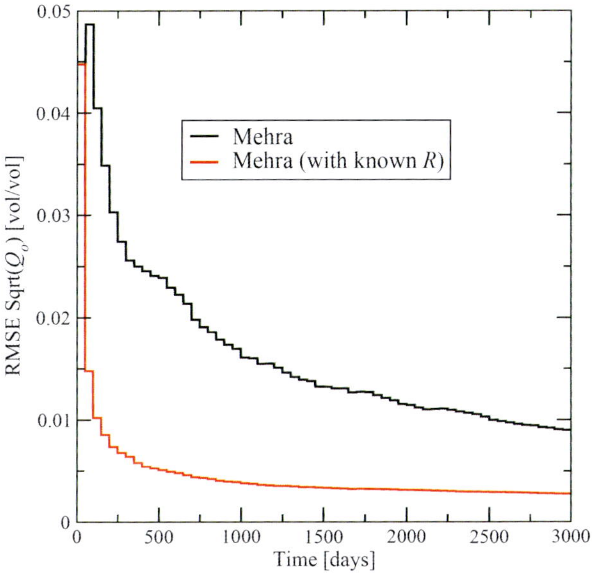

Fig. 7.11. Convergence of the Mehra adaptive filter toward true Q for the case of simultaneously estimating both Q_o and R_o (black line) versus adaptively estimating only Q_o assuming that true R is known (red line). Adapted from Crow and Reichle (2008). Copyright 2008, John Wiley & Sons, Inc.

approaches (including the Mehra approach in Fig. 7.11) are based on the detection — and subsequent removal — of serial ν-autocorrelation. As a result, observation traits that hamper such detection will tend to degrade the performance of the adaptive filter. For instance, many soil moisture observations assimilated into LSMs are, in actuality, remote sensing retrievals based on complex inverse radiative transfer modeling. Consequently, they are prone to high levels of error (particularly in highly vegetated areas) which can mask weak autocorrelation signals associated with a poorly calibrated filter. Remotely sensed surface soil moisture retrievals are also negatively impacted by restrictions in the vertical penetration depth of microwave remote sensing observations (typically less than 5 cm). This lack of vertical support limits the temporal memory present in remotely sensed surface soil moisture retrievals. This is critical since it is the contrast between auto-correlated errors in background WEB-SVAT soil moisture predictions (due to soil moisture memory) and the non-correlated temporal error assumed for surface soil moisture retrievals that forms the basis on which the Mehra algorithm simultaneously estimates Q_o and R_o. This basis for estimation is weakened if the amount of memory present in the observed surface soil moisture state is reduced. Unfortunately, both of these characteristics (high levels of retrieval noise and shallow observation depths) are typical of many land remote sensing observations. Consequently, an inadequate rate of convergence is likely to be a general challenge for the direct application of adaptive filtering techniques to the assimilation of remotely sensed observations into LSMs.

7.5 Potential Solutions

Issues identified in Sec. 7.4 represent serious challenges to the successful implementation of adaptive filtering techniques in operational settings. In this section, we will review potential strategies for addressing some of these problems.

7.5.1 *Use of triple collocation to estimate R*

The speed of convergence issue (examined in Sec. 7.4.3) has a simple solution if reliable estimates of the true observation error covariance R can be obtained independently (Crow and Reichle, 2008). In this case, the adaptive filtering problem reduces to the estimation of Q_o only and demonstrates much faster convergence than the case of simultaneously estimating both Q_o and R_o (see Fig. 7.11). However, for many remotely sensed variables, including soil moisture, the wide-scale estimation of R is not possible using existing ground-based instrumentation.

A potential solution to this problem is the application of so-called Triple Collocation (TC) error estimation techniques to independently acquire R_o for remotely sensed surface soil moisture retrievals (Scipal *et al.*, 2008). TC is based on the acquisition of three independent measurements (or predictions) of a given geophysical quantity and the estimation of error in all three measurements using cross-multiplication and temporal averaging. While the approach requires strict mutual independence of errors in each product, it is not based on an assumption that any of the three products are error-free. Crow and Van den Berg (2010) describe the application of a TC approach to estimate R for surface soil moisture retrievals assimilated into an LSM.

Their approach follows Scipal *et al.* (2008) in acquiring independent surface soil moisture products (θ) acquired from: (1) European Space Radar (ERS) scatterometer measurements (θ_{ERS}), (2) AMSR-E radiometer measurements (θ_{AMSRE}) and (3) predictions from the water balance portion of an LSM (θ_{LSM}). To begin, a 31-day mean climatology is removed from each product to create a time series of soil moisture anomalies (θ'). The resulting anomalies products are then rescaled to have the same temporal variance as the θ'_{LSM} anomaly product. If error in each of the three resulting anomaly products (θ'_{LSM}, θ'_{AMSRE}, and θ'_{ERS}) is assumed to be mutually independent, then R_o for θ'_{AMSRE} can be obtained from:

$$R_o = \overline{(\theta'_{\mathrm{AMSRE}} - \theta'_{\mathrm{LSM}})(\theta'_{\mathrm{AMSRE}} - \theta'_{\mathrm{ERS}})} \qquad (13)$$

where the over-bar signifies temporal averaging over a multi-year time period. The advantage of (13) is that the error variance of a remotely sensed product (θ'_{AMSRE}) can be estimated without resorting to the use of ground-based soil moisture observations. Such observations are difficult to obtain and seldom available at a sufficient spatial density to allow for direct comparisons with footprint-scale (10–40 km) remotely sensed retrievals. Therefore the application of (13) expands the geographic area over which reliable observation error information is available for data assimilation applications. As indicated in Fig. 7.11, obtaining an independent estimate of R prior to attempting adaptive filtering greatly increases the rate at which such filters converge.

However, the benefits of (13) go beyond improved convergence for adaptive filters. Crow and Van den Berg (2010) evaluate R_o derived from (13) by comparing them to calibrated values of R associated with optimal KF performance (Fig. 7.9). To do this, they assimilated a variety of AMSR-E-based surface soil moisture products into (1) over the same set of highly instrumented watersheds used to construct Fig. 7.7. In all cases, R_o was derived from (13) and Q_o was fixed via tuning until the ν-variance constraint was met. Figure 7.9 shows comparisons between R_o values derived from (13) and optimized R which were explicitly calibrated to minimize the RMSE difference between KF soil moisture predictions and ground-based surface soil moisture observations. For reference, Fig. 7.9 also shows a similar comparison for R_o values derived from ν-whitening. As discussed in Sec. 7.4.1 and demonstrated in Fig. 7.8(b), the presence of auto-correlated errors in AMSR-E surface soil moisture retrievals causes ν-whitening to return R_o which are biased low relative to optimum values. However, the TC procedure in (13) is robust to the presence of auto-correlated error in any of its three input soil moisture datasets. Consequently, it provides a much better estimate of optimal R in the presence of auto-correlated retrieval errors (Fig. 7.9).

Results in Figs. 7.8 and 7.9 suggest an alternative filtering calibration strategy based on estimating R via (13) and then calibrating Q_o using the ν-variance constraint. Using both the simple API model in (1) and the more complex Noah model (Mitchell, 2005) as the LSM, Fig. 7.12 compares surface soil moisture data assimilation results using this new TC-based methodology to a range of existing strategies for estimating R_o. Results represent mean KF (for API) or EnKF (for Noah) surface soil moisture results over the same set of heavily-instrumented watershed sites examined in Figs. 7.7 and 7.9. Note, for both the KF/API system and (more complex) EnKF/Noah system, the TC-based approach consistency improves upon a ν-whitening approach. In addition, because of the improved convergence of adaptive filtering systems tasked with retrieving just Q_o (versus *both* Q_o and R_o simultaneously), the new TC-based approach performs well within an adaptive filtering systems based on nudging rules similar to (11). Note the lack of negative impact in Fig. 7.12 when changing from a batch calibration strategy (for tuning Q_o until $\text{var}(\nu) = 1$) to a more difficult adaptive filtering strategy for iteratively updating Q_o. This suggests that providing the adaptive filtering system with outside R_o information obtained from (13) makes the adaptive calculation converge quickly enough to function in a real-time forecasting framework. As a consequence, the application of TC in (13) appears to solve both the autocorrelation error problem introduced in Sec. 7.4.1 and the temporal convergence problem discussed in Sec. 7.4.3.

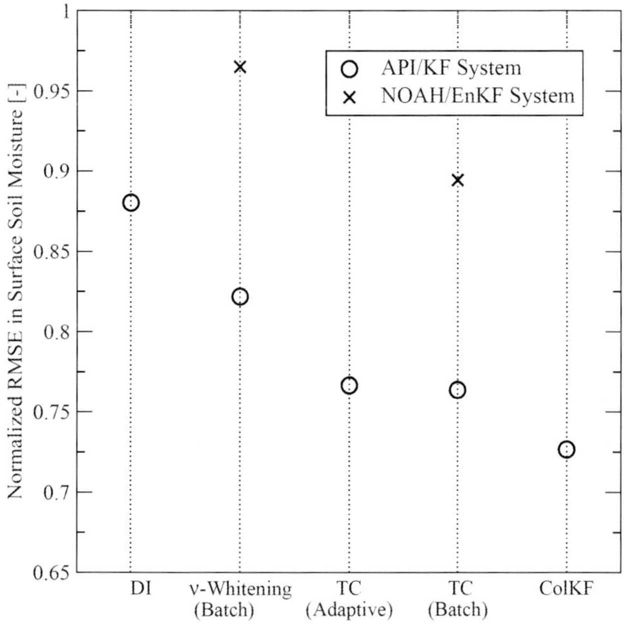

Fig. 7.12. For both the API/KF and Noah/EnKF modeling systems, the impact of different filter calibration strategies on the normalized RMSE (filtering RMSE divided by open loop RMSE) of surface soil moisture predictions. "DI" refers to direct insertion (i.e., the assumption $R_o = 0$) and "ColKF" to the implementation of a fully-calibrated colored Kalman filter to optimally account for auto-correlated soil moisture retrieval errors. Adapted from Crow and Van den Berg (2010). Copyright 2010, John Wiley & Sons, Inc.

Despite this improvement, it should be stressed that in the presence of auto-correlated observing errors, no KF or EnKF approach will be optimal. For the KF/API results, the Colored Kalman filtering (ColKF) approach introduced in Fig. 7.12 represents an optimal filtering approach for the case of observation errors following a first order auto-regressive process in time (Chui and Chen, 1991). Therefore, unlike the standard KF, the ColKF is designed to explicitly account for the presence of autocorrelations in observation errors. Unfortunately, the optimality of the ColKF in this case comes at a cost of specifying even more model error parameters. Here, we avoid this practical consideration by calibrating ColKF parameters to minimize the RMSE of ColKF predictions versus ground-based surface soil moisture observations available within heavily instrumented watershed sites. As a consequence, calibrated ColKF results provide a theoretical baseline representing the optimal treatment of auto-correlated errors in a sequential data assimilation system. However, despite the theoretical limitations of a KF in the presence of auto-correlated observing errors, and its lack of reliance on any ground-based observations, application of (13) to obtain R_o and the ν-variance constraint to obtain Q_o leads to assimilation results which are only slightly worse than the theoretical minimum defined by a calibrated ColKF case (Fig. 7.12).

7.5.2 *Robust filtering strategies*

Results in Secs. 7.3 and 7.5 are based on the application of an adaptive filtering strategy to estimate only the *magnitude* of modeling errors. For instance, during the application of the EnKF to the Noah model described in Sec. 7.5.3, a fixed cross-covariance structure, analogous to the assumption of a known ρ in (9), is assumed for perturbations applied to various Noah vertical soil moisture layers. In addition, subsequent data assimilation results are evaluated only for their surface soil moisture predictions and not deeper soil moisture states whose accurate updating requires credible information concerning the vertical cross-correlation of model error. Consequently, these results do not address challenges surrounding uncertainties in the vertical structure of modeling errors (see Sec. 7.4.2).

Theoretically, an adaptive filtering strategy could be expanded to recover both a single parameter describing the magnitude of modeling error (i.e., Q_o) and second parameter to characterize the cross-correlation structure of error (i.e., ρ). See Mehra *et al.* (1970) for an example of this strategy. However, the implementation of such strategies is likely to be plagued by identifiability issues and severe convergence problems. An alternative approach is defining robust data assimilation and modeling approaches which either help identify poor model error assumptions or help minimize their eventual impact on data assimilation results.

Kumar *et al.* (2009) provides an example of such a strategy. Their analysis is based on a set of fraternal twin EnKF data assimilation experiments in which three different LSMs are used to generate three different time series of soil moisture "truth" and synthetic observations (generated by randomly perturbing surface soil moisture prediction made by the truth LSM simulation). Each LSM employs different parameterizations of sub-surface soil moisture dynamics and thus demonstrates different degrees of vertical coupling between errors in their surface and root-zone soil moisture predictions. Simulated synthetic observations from each of the three LSMs are then cross-assimilated into each of the other two LSMs.

They do this with the goal of investigating the impact of assimilating surface soil moisture retrievals into LSMs with vertical coupling characteristics which differ from the source LSM simulations used to create the assimilated synthetic observations.

Results from these synthetic data assimilation experiments indicate that assimilating surface soil moisture retrievals derived from a tightly coupled LSM into a loosely coupled LSM tends to produce worse results than the opposite case (i.e., assimilating observations from a loosely coupled LSM into a tightly coupled LSM). Or, put another way, it is better to overestimate vertical coupling strength in the assimilation model than to underestimate it. This asymmetry suggests that, in the presence of uncertainty regarding the proper model vertical coupling strength, and thus the proper vertical cross-correlation of soil moisture model errors, the least-risky choice is selecting an LSM (and $\mathbf{Q}$ matrix) which demonstrates strong vertical coupling. Therefore, these results point to the existence of a possible *ad hoc* rule which minimizes the sensitivity of EnKF results to uncertainties in the vertical structure of modeling errors.

Another potential solution to uncertainty concerning the vertical source of modeling errors is increasing the range of observation types assimilated into an LSM. As a follow-up to the surface soil moisture only data assimilation analysis is presented in Fig. 7.10. Crow and Van Loon (2006) examined the impact of simultaneously assimilating both surface soil moisture and stream flow observations into an LSM. They found that assimilating both soil moisture and stream flow (as opposed to surface soil moisture alone) helped the v-variance diagnostic correctly identify the true source of modeling error. That is, with more observations, it became progressively more difficult to get acceptable var(v) results using an erroneous assumption concerning the source of modeling errors. Therefore, more complex error attribution decisions will likely require a wider range of observation types in order to successfully apply an adaptive filter.

7.6 Summary

Within the next decade, advances in remote sensing (see, e.g., Entekhabi *et al.*, 2010) and computational resources should lead to a new era in the assimilation of land observations into LSMs used for operational forecasting systems. Data assimilation techniques like the EnKF allow for a sophisticated treatment of modeling error in land data assimilation. However, the flexibility of the EnKF with regards to error significantly outpaces our current knowledge concerning the statistical properties of error in LSMs or the observations we wish to assimilate them into. Consequently, the next decade could see a struggle for operational users to obtain — currently unavailable — statistical error parameters on which to base an EnKF implementation.

In anticipation of this possibility, this chapter reviews recent advances in the development of adaptive filtering strategies capable of recovering model and observing error statistical information during the cycling of an EnKF or KF. These strategies are based on the statistical evaluation of filtering innovations calculated during the analysis cycle of a sequential data assimilation system. Preliminary synthetic twin data assimilation results indicate that existing adaptive filtering techniques can be applied to modern LSMs (Sec. 7.3) but a number of specific problems still limit their utility when applied in real data assimilation cases (Sec. 7.4). Ongoing research should be directed toward these problems and the application of new statistical techniques to characterize observational errors has shown particular promise (Sec. 7.5).

Finally, it should be noted that while results here have focused on the direct assimilation of soil moisture observations into a LSM, they are also directly to other data assimilation cases — in particular, the closely related case of assimilating radiances and using a radiance transfer model as an observation operator. Here, geophysical retrieval errors (i.e., uncertainty in the linkage of geophysical variables like soil moisture to radiance observations) are simply removed from the observation covariance R and added to the model error covariance Q (Crow and Van den Berg, 2010). It should be stressed that this change in error accounting does not fundamentally change the adaptive filtering problem.

Appendix A. Innovation Properties in an Optimal KF

Material in this appendix — based on Mehra (1970) — develops the theoretical constraints on innovation statistics for an optimal linear filter. Consider an n-dimension, linear, discrete system evolving in time according to:

$$x_{i+1}^t = \Phi x_i^t + w_i \tag{A.1}$$

where x^t is the true $n \times 1$ state vector, Φ is a (known) $n \times n$ model operator matrix, and w is an uncorrelated Gaussian white noise sequence capturing error in the model forecast. In addition, assume that an observation vector y of size $m \times 1$ is periodically available and can be related to x^t by:

$$y_i = Hx_i^t + \varepsilon_i \tag{A.2}$$

where H is a (known) $m \times n$ matrix and ε is an uncorrelated Gaussian white noise sequence capturing error in the observations.

The iterative filtering equations required to estimate x are then:

$$x_{i+1}^f = \Phi x_i^a \tag{A.3}$$

and

$$x_i^a = x_i^f + K(y_i - Hx_i^f) \tag{A.4}$$

where K is the filtering gain and the superscript a and f refers to analysis (i.e., post-updating) and forecast (i.e., prior to updating) state estimates respectively. The novel information provided to the filter by observations is captured by the filter innovations:

$$d_i = y_i - Hx_i^f, \tag{A.5}$$

and filtering error e is defined as the difference between the true state and the forecast state estimate:

$$e_i = x_i^t - x_i^f. \tag{A.6}$$

Substituting (A.2) into (A.6) and applying the definition in (A.5) yields as alternative definition for filtering innovations:

$$d_i = He_i + \varepsilon_i \tag{A.7}$$

The variance of d can be calculated as:

$$E[d_i d_i^T] = E[(He_i + \varepsilon_i)(He_i + \varepsilon_i)^T] = HE[e_i e_i^T]H^T + E[\varepsilon_i \varepsilon_i^T] \tag{A.8}$$

since forecast error at time i (e_i) is independent of error in the subsequently assimilated observation (ε_i). Assigning:

$$P_i^f = E[e_i e_i^T] \tag{A.9}$$

$$R_i = E[\varepsilon_i \varepsilon_i^T] \tag{A.10}$$

allows (A.8) to be rewritten as

$$E[d_i d_i^T] = HP_i^f H^T + R_i \tag{A.11}$$

A similar procedure can be applied to calculate the entire autocovariance function for d. Using (A.7), the covariance of d obtained at times i and $i-k$ (where $k>0$) is:

$$\begin{aligned} E[d_i d_{i-k}^T] &= E[(He_i + \varepsilon_i)(He_{i-k} + \varepsilon_{i-k})^T] \\ &= HE[e_i e_{i-k}^T]H^T + HE[e_i \varepsilon_{i-k}^T] \end{aligned} \tag{A.12}$$

since $E[\varepsilon_i \varepsilon_{i-k}^T] = 0$ and $E[\varepsilon_i e_{i-k}^T] = 0$ for $k > 0$ (i.e., *future* observation errors will be independent of *past* filtering errors).

Combining (A.1–A.4), a recursive relationship can be obtained for e:

$$e_i = \Phi(I - KH)e_{i-1} - \Phi K \varepsilon_{i-1} + w_i \tag{A.13}$$

Applying (A.13) k steps back in time yields:

$$\begin{aligned} e_i = [\Phi(I - KH)]^k e_{i-1} - \sum_{j=1}^{k} [\Phi(I - KH)]^{j-1} \Phi K \varepsilon_{i-j} \\ + \sum_{j=1}^{k} [\Phi(I - KH)]^{j-1} \omega_{i-j} \end{aligned} \tag{A.14}$$

Multiplying (A.14) by e_{i-k}^T and taking the expectation allows the first term in (A.12) to be calculated as:

$$E[e_i e_{i-k}^T] = [\Phi(I - KH)]^k E[e_{i-k}, e_{i-k}^T] = [\Phi(I - KH)]^k P_{i-k}^f \tag{A.15}$$

since $E[\varepsilon_i e_{i-k}^T] = 0$ and $E[\omega_i e_{i-k}^T] = 0$ for $k > 0$ (i.e., *future* observation and forecasting errors will be independent of *past* filtering errors). Likewise, multiplying (A.12) by ε_{i-k}^T provides the second term in (A.12):

$$\begin{aligned} E[e_i \varepsilon_{i-k}^T] &= -[\Phi(I - KH)]^{k-1} \Phi K E[\varepsilon_{i-k}, \varepsilon_{i-k}^T] \\ &= -[\Phi(I - KH)]^{k-1} \Phi K R_{i-k} \end{aligned} \tag{A.16}$$

Plugging (A.15) and (A.16) into (A.12) yields

$$E[d_i d_{i-k}^T] = H[\Phi(I - KH)]^k P_{i-k}^f H^T - H[\Phi(I - KH)]^{k-1} \Phi K R_{i-k} \tag{A.17}$$

Neglecting temporal variations in P and R, (A.17) can be rewritten as:

$$E[d_i d_{i-k}^T] = H[\Phi(I - KH)]^{k-1} \bullet \Phi([(P^f H^T - K(HP^f H^T + R)) \tag{A.18}$$

Note that inserting the optimal form of K or:

$$K = \frac{P^f H^T}{HP^f H^T + R} \tag{A.19}$$

into (A.18) leads to $E[d_i d_{i-k}^T] = 0$ for all $k > 0$. Therefore, the two key statistical traits of optimal filtering innovations are $E[d_i d_{i-k}^T] = HP^f H^T + R$ for the case $k = 0$ (from A.11 above) and $E[d_i d_{i-k}^T] = 0$ for the case $k > 0$.

For the simple case of $m = 1$, d can be normalized by the square root of the scalar $HP^f H^T + R$ to construct a time series of "normalized" innovations:

$$v_i = \frac{d_i}{(HP^f H^T + R)^{1/2}} \tag{A.20}$$

By the above development, if the filter is performing optimally (i.e., if P^f and R are being accurately estimated) then the resulting v time series should be sequentially white (i.e., $E[v_i v_{i-k}^T] = 0$ for $k > 0$) with second moment of one (i.e., $E[v_i v_{i-k}^T] = 1$ for $k = 0$). Note that in the simple case examined in the main text, H is a scalar equal to 1.

References

Burgers, G., van Leeuwen, P. J. and Evensen, G. (1998). Analysis scheme in the ensemble Kalman filter. *Monthly Weather Review*, **126**, 1719–1724.

Chui, C. K. and Chen, G. (1991). *Kalman Filtering with Real-Time Applications*, 195pp. New York: Springer-Verlag.

Crow, W. T. and Bolten, J. D. (2007). Estimating precipitation errors using spaceborne surface soil moisture retrievals. *Geophysical Research Letters*, **34**, L08403, doi:10.1029/2007GL029450.

Crow, W. T. and Reichle, R. H. (2008). Comparison of adaptive filtering techniques for land surface data assimilation. *Water Resources Research*, **44**, W08423, doi:10.1029/2008WR006883.

Crow, W. T. and Van den Berg, M. J. (2010). An improved approach for estimating observation and model error parameters for soil moisture data assimilation. *Water Resources Research*, **46**, W12519, doi:10.1029/2010WR009402.

Crow, W. T. and Van Loon, E. (2006). The impact of incorrect model error assumptions on the sequential assimilation of remotely-sensed surface soil moisture. *Journal of Hydrometeoroloy*, **8**(3), 421–431.

Crow, W. T., Kustas, W. P. and Prueger, J. (2008). Monitoring root-zone soil moisture through the assimilation of a thermal remote sensing-based soil moisture proxy into a water balance model. *Remote Sensing of Environment*, **112**, 1268–1281.

DeLannoy, G. J. M., Houser, P. R., Verhoest, N. E. C. and Pauwels, V. R. N. (2009). Adaptive soil moisture profile filtering for horizontal information propagation in the independent column-based CLM2.0. *Journal of Hydrometeorology*, **10**(3), 766–779.

Dee, D. P. (1995). On-line estimation of error covariance parameters for atmospheric data assimilation. *Monthly Weather Review*, **123**, 1128–1145.

Drecourt, J.-P., Madsen, H. and Rosbjerg, D. (2006). Calibration framework for a Kalman filter applied to a groundwater model. *Advances in Water Resources*, **29**(5), 719–734.

Entekhabi, D., Nakamura, H. and Njoku, E. G. (1994). Solving the inverse problem for soil moisture and temperature profiles by sequential assimilation of multifrequency remotely-sensed observations. *IEEE Transactions on Geoscience and Remote Sensing*, **32**, 438–447.

Entekhabi, D. *et al.* (2010). The Soil Moisture Active and Passive (SMAP) mission. *Proceedings of the IEEE*, **98**(5), 704–716.

Gelb, A. (1974). *Applied Optimal Estimation*, 374 pp. Cambridge, Massachusetts: The MIT Press.

Jackson, T. J. *et al.* (2010). Validation of advanced microwave scanning radiometer soil moisture products. *IEEE Transactions on Geoscience and Remote Sensing*, **48**(12), 4256–4272.

Kumar, S. V., Reichle, R. H., Koster, R. D., Crow, W. T. and Peters-Lidard, C. D. (2009). Role of subsurface physics in the assimilation of surface soil moisture observations. *Journal of Hydrometeorology*, **10**, 1534–1547, doi:10.1175/2009JHM1134.1.

Mehra, R. K. (1970). On identification of variances and adaptive Kalman filtering, *IEEE Transactions on Automatic Control*, **2**, 175–184.

Mitchell, K. (2005). The community Noah land-surface model: User guide public release, version 2.7.1. Available at: http://www.emc.ncep.noaa.gov/mmb/gcp/noahlsm/Noah LSM USER GUIDE2.7.1.html, data of April 1, 2003.

Mitchell, H. L. and Houtekamer, P. L. (1999). An adaptive ensemble Kalman filter. *Monthly Weather Review*, **128**, 416–433.

Peters-Lidard, C. D., Zion, M. S. and Wood, E. F. (1997). A soil–vegetation–atmosphere transfer scheme for modeling spatially variable water and energy balance processes. *Journal of Geophysical Research*, **102**, 4303–4324.

Reichle, R. H., McLaughlin, D. B. and Entekhabi, D. (2002). Hydrologic data assimilation with the ensemble Kalman filter. *Monthly Weather Review*, **130**, 103–114.

Reichle, R. H., Crow, W. T. and Keppenne, C. L. (2008a). An adaptive ensemble Kalman filter for soil moisture data assimilation. *Water Resources Research*, **44**, W03423, doi:10.1029/2007WR006357.

Reichle, R. H., Bosilovich, M. G., Crow, W. T., Koster, R. D., Kumar, S. V., Mahanama, S. P. P. and Zaitchik, B. F. (2008b). Recent advances in land data assimilation at the NASA global modeling and assimilation office. In *Data Assimilation for Atmospheric, Oceanic, and Hydrologic Applications*, Park, S. K. (ed.), pp. 407–428. New York: Springer-Verlag.

Ryu, D., Crow, W. T., Zhan, X. and Jackson, T. J. (2009). Correcting unintended perturbation biases in hydrologic data assimilation using Ensemble Kalman filter. *Journal of Hydrometeorology*, **10**(3), 734–750, doi:10.1175/2008JHM1038.1.

Scipal, K., Holmes, T., de Jeu, R., Naeimi, V. and Wagner, W. (2008). A possible solution for the problem of estimating the error structure of global soil moisture datasets. *Geophysical Research Letters*, **35**(24), L24403, doi:10.1029/2008GL035599.

Chapter 8

Inflation Adjustment on Error Covariance Matrices for Ensemble Kalman Filter Assimilation

Xiaogu Zheng†, Guocan Wu†, Xiao Liang‡ and Shupeng Zhang†

†*College of Global Change and Earth System Science, Beijing Normal University, China*

‡*National Meteorological Information Center, China Meteorological Adminstration, China*

In the ensemble Kalman filter assimilation, the estimated forecast error covariance matrix and prior observation error covariance matrix can be far from the true values, which is likely to significantly affect the assimilation results. This chapter deals with an adaptive inflation scheme for adjusting the forecast error covariance matrix and prior observation error covariance matrix by minimizing the -2log-likelihood of innovation (observation — forecast residuals) in the case of linear or tangent linear observation operator.

8.1 Introduction

Conceptually, data assimilation is an algorithm of assimilating observation data into the physical model based on the perception of physical model and observation. Data assimilation will produce an optimal combination (referred to as analysis state) of model forecast state and observation, which is closer to the true state than either the model forecast state or the observation. However, the optimization of the combination relies on the accuracy of the estimated forecast error covariance matrix and observation error covariance matrix. If an assimilation algorithm provides an inaccurate estimation of these error covariance matrices, the combination created by this algorithm may not be satisfactory. In other words, the resulting analysis state may be far from the true state. In practice, however, we usually

S. Liang, X. Li & X. Xie (eds.), *Land Surface Observation, Modeling and Data Assimilation*, 205–232.

know little about the statistical property of these error covariance matrices, mainly due to the fact that it is impossible to have a full knowledge of the true state to be estimated. As a result, it is difficult to calculate the errors of model forecast state and observation with respect to the true state (Sénégas *et al.*, 2001).

Ensemble Kalman filter (EnKF), which is a popular data assimilation method, was proposed by the Norwegian scientist Evensen (1994a, 1994b). Compared with the linear Kalman filter, EnKF is designed for nonlinear forecast operator and observation operator. Its major innovation is creating a forecast ensemble and then taking its sampling covariance matrix as the estimate of forecast error covariance matrix. Until now, there have been many researches concerning reviews, applications, and improvements of the EnKF. It is found that EnKF tends to significantly underestimate forecast error covariance matrix, possibly resulting in the degeneration of the filter (e.g., Anderson and Anderson, 1999; Constantinescu *et al.*, 2007). To compensate for this, the common approach is inflation adjustment on forecast error covariance matrix.

The early method of inflation adjustment is to multiply the sampling covariance matrix of ensemble forecast states by a constant larger than 1 and then empirically speculate this constant according to the assimilation effect of EnKF (e.g., Anderson and Anderson, 1999). Wang and Bishop (2003) first put forward an adaptive optimization algorithm for online estimation of inflation factor of forecast error covariance matrix, in which the inflation factor is obtained from the first-order moment estimation of innovation (i.e., observation — forecast residuals, Dee, 1995). Wang and Bishop's work is a significant progress in optimizing the estimation of inflation factor, while it needs improvement in the following two aspects: first, the observation operator is required to be linear; second, an exact observation error covariance matrix is required. To solve the first problem, Li *et al.* (2009) promoted Wang and Bishop's method by relaxing the linear requirement to be tangent linear requirement for observation operator. Directed at the second requirement, Li *et al.* (2009) suggested a method of estimating inflation factor of forecast error covariance matrix and observation error covariance matrix simultaneously. Recently, we tested this method and found that the effect was not so satisfactory when it was used to estimate the observation error covariance matrix. Therefore, it is necessary to look for a new method to solve the problem of simultaneous inflation on both forecast and observation error covariance matrices.

Back in the 1990s (Dee, 1995; Dee and Da Silva, 1999 and Dee *et al.*, 1999) proposed to use the likelihood of innovation to estimate forecast and observation error covariance matrices simultaneously. They suggested parameterizing these error covariance matrices and deducing likelihood of innovation with respect to these parameters; then, these parameters can be estimated using maximum likelihood estimation method. In practical estimation, however, due to the large number of parameters in error covariance matrices, it is difficult to obtain accurate parameters using maximum likelihood estimation method.

Recently, a team of researchers from Beijing Normal University combined the method proposed by Dee *et al.* and the method of inflation adjustment on error covariance matrices, to simultaneously optimize the inflation factors of forecast and observation error covariance matrices (Liang *et al.*, 2011). Our parameterization scheme is to multiply forecast and observation error covariance matrices of EnKF by inflation factors, respectively, as new error covariance matrices. Only two inflation factors are the parameters involved in this parameterization scheme, so the maximum likelihood method can be easily used to estimate the inflation factors. Compared with the estimation method proposed by Wang and Bishop (2003), the advantage of this method is the simultaneous estimation of two inflation factors. Taking two simplified ideal models as examples, we compare our method with that of Li *et al.* (2009) in this chapter. We found in these two experiments that our approach estimated the inflation factors more accurately, with better assimilation results.

We hereby state that inflation method is never the only adjustment method for error covariance matrices. For example, another popular adjustment scheme is to add a matrix representing model error covariances to the sampling covariance matrix of forecast ensemble (e.g., Bai and Li, 2011). Even in inflation adjustment schemes, there are also other alternative schemes for the estimation of inflation factor, e.g., Bayesian scheme proposed by Anderson (2007, 2009) and Miyoshi (2011). To our knowledge, problems exist in all the schemes (e.g., all the existing Bayesian schemes suppose that observation errors are spatially independent in the statistical sense). The work devoted to the comparison among the schemes is also inadequate. It is reasonable to believe that the development of other schemes is still necessary despite the availability of certain schemes.

This chapter is structured in this way: after the brief introduction in Sec. 8.1, Sec. 8.2 describes the algorithm of EnKF and several methods available for estimating inflation factors of error covariance matrices, including Wang and Bishop's method, Li *et al.*'s method, and our method. In Sec. 8.3, we present two simplified ideal models, i.e., Lorenz-96 model and two-dimensional Shallow Water Equation (SWE) model and the corresponding observation system, which are used to verify various estimation methods for inflation factors. In Secs. 8.4 and 8.5, we further discuss the verification results in cases of linear and nonlinear observation operators, respectively. Section 8.6 states our major conclusions.

8.2 Inflation Adjustment on Error Covariance Matrices in EnKF

8.2.1 *Ensemble Kalman filter*

Compared with linear Kalman filter, Ensemble Kalman filter (EnKF) is designed specifically for the nonlinear forecast and observation system. Following the notations proposed by Ide *et al.* (1997), a nonlinear forecast and observation system can be represented as

$$\boldsymbol{x}_t^t = M_{t-1}(\boldsymbol{x}_{t-1}^t) + \omega_i \tag{1}$$

$$\boldsymbol{y}_t = H_t(\boldsymbol{x}_t^t) + \varepsilon_t \tag{2}$$

where $\boldsymbol{t}$ is time step index; $\boldsymbol{x}_t^t$ is n-dimensional true state vector at time step t; M_t is nonlinear forecast operator (e.g., weather forecast model); $\boldsymbol{y}_t$ is m_t-dimensional observation vector; H_t is nonlinear observation operator; $\boldsymbol{\omega}_i$ and $\boldsymbol{\varepsilon}_t$ are model error and observation error vectors, with zero mean and covariance matrices $\mathbf{Q}_t$ and $\mathbf{R}_t$, respectively, and are assumed to be statistically independent of each other and time-uncorrelated. The objective of EnKF assimilation is to find a series of analysis states $\boldsymbol{x}_t^a$ that are sufficiently close to the true state $\boldsymbol{x}_t^t$, applying the information provided by M_t and $\boldsymbol{y}_t$.

The procedure of EnKF is as follows. Suppose perturbed analysis states at the last time step $\boldsymbol{x}_{t-1,i}^a$ are known (where i stands for a perturbation, counting from 1 to the number of ensemble members N), then the perturbed analysis states at the next time step $\boldsymbol{x}_{t,i}^a$ are derived by the following steps:

Step (1). Run the full model forward in time to get the perturbed forecast states:

$$\boldsymbol{x}_{t,i}^f = M_{t-1}(\boldsymbol{x}_{t-1,i}^a) \tag{3}$$

Forecast state $\boldsymbol{x}_t^f$ is defined as ensemble mean of $\boldsymbol{x}_{t,i}^f$, i.e., $\boldsymbol{x}_t^f = \frac{1}{N}\sum_{i=1}^N \boldsymbol{x}_{t,i}^f$.

Step (2). Estimate the perturbed innovation as

$$\mathbf{d}_{t,i} = \boldsymbol{y}_t + \varepsilon'_{t,i} - H_t(\boldsymbol{x}^f_{t,i}) \tag{4}$$

where $\varepsilon'_{t,i}$ is normally distributed with zero mean and covariance matrix $\mathbf{R}_t$.

Step (3). Compute perturbed analysis states at time step t:

$$\boldsymbol{x}^a_{t,i} = \boldsymbol{x}^f_{t,i} + \mathbf{P}^f_t H^T_t (H_t \mathbf{P}^f_t H^T_t + \mathbf{R}_t)^{-1} \mathbf{d}_{t,i} \tag{5}$$

where

$$\mathbf{P}^f_t H^T_t \equiv \frac{1}{N-1} \sum_{i=1}^{N} (\boldsymbol{x}^f_{t,i} - \boldsymbol{x}^f_t) \cdot (H_t(\boldsymbol{x}^f_{t,i}) - H_t(\boldsymbol{x}^f_t))^T \tag{6}$$

and

$$H_t \mathbf{P}^f_t H^T_t \equiv \frac{1}{N-1} \sum_{i=1}^{N} (H_t(\boldsymbol{x}^f_{t,i}) - H_t(\boldsymbol{x}^f_t)) \cdot (H_t(\boldsymbol{x}^f_{t,i}) - H_t(\boldsymbol{x}^f_t))^T \tag{7}$$

Now repeat the assimilation cycle from Step (1) to compute perturbed forecast states at the next time step. Then ensemble mean of perturbed analysis states $\boldsymbol{x}^a_{t,i}$ is estimated as analysis state $\boldsymbol{x}^a_t$, i.e., $\boldsymbol{x}^a_t = \frac{1}{N}\sum_{i=1}^{N} \boldsymbol{x}^a_{t,i}$.

If the dimension m_t of observation state is high (e.g., high-resolution satellite images), the calculation of inverse matrix $(H_t \mathbf{P}^f_t H^T_t + \mathbf{R}_t)^{-1}$ can be very expensive. In Eq. (5), if the inverse of observation error covariance matrix $\mathbf{R}^{-1}_t$ is easy to calculate, then $(H_t \mathbf{P}^f_t H^T_t + \mathbf{R}_t)^{-1}$ can be calculated by Sherman–Morrison–Woodbury formula (Sherman and Morrison, 1950; Golub and Van Loan, 1996; Tippett *et al.*, 2003).

$$(H_t \mathbf{P}^f_t H^T_t + \mathbf{R}_t)^{-1} = \mathbf{R}^{-1}_t - \mathbf{R}^{-1}_t \mathbf{Z}_t (\mathbf{I} + \mathbf{Z}^T_t \mathbf{R}^{-1}_t \mathbf{Z}_t)^{-1} \mathbf{Z}^T_t \mathbf{R}^{-1}_t \tag{8}$$

where

$$\mathbf{Z}_t \equiv \frac{1}{\sqrt{N-1}} (H_t(\boldsymbol{x}^f_{t,1}) - H_t(\boldsymbol{x}^f_t), H_t(\boldsymbol{x}^f_{t,2}) - H_t(\boldsymbol{x}^f_t), \ldots, H_t(\boldsymbol{x}^f_{t,N}) - H_t(\boldsymbol{x}^f_t)),$$

i.e., $H_t \mathbf{P}^f_t H^T_t \equiv \mathbf{Z}_t \mathbf{Z}^T_t$. The matrix $\mathbf{I} + \mathbf{Z}^T_t \mathbf{R}^{-1}_t \mathbf{Z}_t$ in Eq. (8) is N-by-N dimension. In practical problems, the number of perturbed ensemble members N may not be very large, and therefore it is not difficult to calculate the inverse of $\mathbf{I} + \mathbf{Z}^T_t \mathbf{R}^{-1}_t \mathbf{Z}_t$. This means that it is feasible to compute $(H_t \mathbf{P}^f_t H^T_t + \mathbf{R}_t)^{-1}$ with Eq. (8).

8.2.2 *Inflation adjustment on error covariance matrices in the case of linear observation operator*

In EnKF assimilation algorithm, the sampling error covariance matrix of forecast state $\boldsymbol{x}_t^f$ is defined as

$$\mathbf{P}_t^f = \frac{1}{N-1}\sum_{i=1}^{N}(\boldsymbol{x}_{t,i}^f - \boldsymbol{x}_t^f)\cdot(\boldsymbol{x}_{t,i}^f - \boldsymbol{x}_t^f)^T \tag{9}$$

It is well known that if $\mathbf{P}_t^f$ is far from the true error covariance matrix of $\boldsymbol{x}_t^f$, the analysis states induced by Eq. (5) may be also far from the true states. A common method to mitigate this problem is inflation adjustment on forecast error covariance matrix, i.e., to multiply $\mathbf{P}_t^f$ by an inflation factor λ_t to adjust it to $\lambda_t\mathbf{P}_t^f$ (Zheng, 2009; Li *et al.*, 2009; Liang *et al.*, 2011).

Initially, inflation factor λ_t is given empirically by trial. Under the condition that observation operator H_t is linear (i.e., H_t can be represented as a matrix $\mathbf{H}_t$), Li *et al.* (2009) proposed an estimation of inflation factor λ_t on forecast error covariance matrix, i.e.,

$$\lambda_t \approx \frac{\mathbf{d}_t^T\mathbf{d}_t - \mathrm{tr}(\mathbf{R}_t)}{\mathrm{tr}(\mathbf{H}_t\mathbf{P}_t^f\mathbf{H}_t^T)} \tag{10}$$

where the so-called innovation $\mathbf{d}_t$ is defined as:

$$\mathbf{d}_t \equiv \boldsymbol{y}_t - \mathbf{H}_t\boldsymbol{x}_t^f \tag{11}$$

In fact, this estimation is based on the first-order moment estimation of the statistics $\mathbf{d}_t$ (refer to Appendix A for more details).

Based on the work of Dee *et al.* (Dee and Da Silva, 1999; Dee *et al.*, 1999; Zheng, 2009 and Liang *et al.*, 2011) suggested that λ_t can be obtained by minimizing the following objective function

$$-2L_t(\lambda_t) = \ln[\det(\mathbf{H}_t\lambda_t\mathbf{P}_t^f\mathbf{H}_t^T + \mathbf{R}_t)] + \mathbf{d}_t^T(\mathbf{H}_t\lambda_t\mathbf{P}_t^f\mathbf{H}_t^T + \mathbf{R}_t)^{-1}\mathbf{d}_t \tag{12}$$

where "det" represents determinant of matrix. Equation (12) is the -2 log-likelihood of innovation, where the determinant is not difficult to calculate (see the discussions in Appendix B).

In the early inflation adjustment research, the inflation factor was assumed to be time-uncorrelated, indicating that λ is always a constant for all time steps. In such cases, λ can be estimated by minimizing the following objective function:

$$-2\sum_{t} L_t(\lambda) \tag{13}$$

where $L_t(\lambda)$ is defined in the same way as in Eq. (12).

If observation error covariance matrix $\mathbf{R}_t$ also needs to be inflated, its inflation factors $\boldsymbol{\mu}_t$ and $\boldsymbol{\lambda}_t$ can be simultaneously estimated by minimizing the following objective function:

$$\begin{aligned} -2L_t(\boldsymbol{\lambda}_t, \boldsymbol{\mu}_t) &= \ln[\det(\mathbf{H}_t\boldsymbol{\lambda}_t\mathbf{P}_t^f\mathbf{H}_t^T + \boldsymbol{\mu}_t\mathbf{R}_t)] \\ &\quad + \mathbf{d}_t^T(\mathbf{H}_t\boldsymbol{\lambda}_t\mathbf{P}_t^f\mathbf{H}_t^T + \boldsymbol{\mu}_t\mathbf{R}_t)^{-1}\mathbf{d}_t \end{aligned} \tag{14}$$

In this respect, our estimation method has some advantages over that of Li *et al.*

Initially, the inflation factor of observation error covariance was also assumed to be time-uncorrelated. That is to say, $\boldsymbol{\mu}$ is a constant at all time steps, particularly when the dimension of observation state m_t is small. In this case, $\boldsymbol{\mu}$ can be estimated by minimizing the following objective function:

$$-2\sum_{t} \min_{\lambda_t} L_t(\boldsymbol{\lambda}_t, \boldsymbol{\mu}) \tag{15}$$

where $L_t(\lambda, \boldsymbol{\mu})$ is defined in the same way as in Eq. (14).

8.2.3 *Inflation adjustment on error covariance matrices in the case of nonlinear observation operator*

The estimation method of inflation factors introduced in Sec. 8.2.2 can be easily extended to the case of tangent linear observation operator. Observation operator H_t is tangent linear with respect to state vector $\boldsymbol{x}_0$, if for any state vector $\boldsymbol{x}$,

$$H_t(\boldsymbol{x}) \approx H_t(\boldsymbol{x}_0) + \dot{\mathbf{H}}_t(\boldsymbol{x}_0)(\boldsymbol{x} - \boldsymbol{x}_0) \tag{16}$$

where

$$\dot{\mathbf{H}}_t(\boldsymbol{x}) = \frac{\partial H_t(\boldsymbol{x})}{\partial \boldsymbol{x}} \tag{17}$$

is the tangent direction of operator H_t at point $\boldsymbol{x}$.

For diagonal observation error covariance matrix, Li *et al.* (2009) extended the estimation method of Wang and Bishop (2003). When the observation operator is tangent linear, two estimations of λ_t were proposed:

$$\boldsymbol{\lambda}_t \approx \frac{\mathbf{d}_t^T \mathbf{d}_t - \mathrm{tr}(\mathbf{R}_t)}{\mathrm{tr}(\dot{\mathbf{H}}_t(\boldsymbol{x}_t^f)\mathbf{P}_t^f \dot{\mathbf{H}}_t^T(\boldsymbol{x}_t^f))} \tag{18}$$

and

$$\boldsymbol{\lambda}_t \approx \frac{(H_t(\boldsymbol{x}_t^a) - H_t(\boldsymbol{x}_t^f))^T \mathbf{d}_t}{\mathrm{tr}(\dot{\mathbf{H}}_t(\boldsymbol{x}_t^f)\mathbf{P}_t^f \dot{\mathbf{H}}_t^T(\boldsymbol{x}_t^f))} \tag{19}$$

where $\boldsymbol{x}_t^a$ is the analysis state of EnKF.

Wu *et al.* (2011) extended the maximum likelihood estimation method proposed by Liang *et al.* (2011). We have $\boldsymbol{y}_t - H_t(\boldsymbol{x}_t^t) = \varepsilon_t$ from Eq. (2) and $H_t(\boldsymbol{x}_t^t) - H_t(\boldsymbol{x}_t^f) \approx \dot{\mathbf{H}}_t(\boldsymbol{x}_t^f)(\boldsymbol{x}_t^t - \boldsymbol{x}_t^f)$ according to Eq. (16). Then the innovation statistics can be approximately represented as follows:

$$\begin{aligned} \mathbf{d}_t &= \boldsymbol{y}_t - H_t(\boldsymbol{x}_t^f) \\ &= \boldsymbol{y}_t - H_t(\boldsymbol{x}_t^t) + H_t(\boldsymbol{x}_t^t) - H_t(\boldsymbol{x}_t^f) \\ &\approx \varepsilon_t + \dot{\mathbf{H}}_t(\boldsymbol{x}_t^f)(\boldsymbol{x}_t^t - \boldsymbol{x}_t^f) \end{aligned} \tag{20}$$

In EnKF, observation error and forecast error are assumed to be Gaussian distribution and are statistically independent of each other. Thus with the given forecast state $\boldsymbol{x}_t^f$, the conditional distribution of innovation $\mathbf{d}_t$ is close to a Gaussian distribution with zero mean and covariance matrix $\dot{\mathbf{H}}_t(\boldsymbol{x}_t^f)\mathbf{P}_t^f \dot{\mathbf{H}}_t^T(\boldsymbol{x}_t^f) + \mathbf{R}_t$. Therefore, after forecast error covariance matrix is inflated, -2log-likelihood of innovation can be written as:

$$\begin{aligned} -2L_t(\lambda_t) = {} & \ln[\det(\dot{\mathbf{H}}_t(\boldsymbol{x}_t^f)\lambda_t \mathbf{P}_t^f \dot{\mathbf{H}}_t^T(\boldsymbol{x}_t^f) + \mathbf{R}_t)] \\ & + \mathbf{d}_t^T(\dot{\mathbf{H}}_t(\boldsymbol{x}_t^f)\lambda_t \mathbf{P}_t^f \dot{\mathbf{H}}_t^T(\boldsymbol{x}_t^f) + \mathbf{R}_t)^{-1}\mathbf{d}_t \end{aligned} \tag{21}$$

Then λ_t can be estimated by minimizing the objective function (21).

Similar to the case of linear observation, if observation error covariance matrix $\mathbf{R}_t$ also needs to be inflated, the inflation factors $\boldsymbol{\mu}_t$ on observation error covariance matrix and $\boldsymbol{\lambda}_t$ can be simultaneously obtained by minimizing -2log-likelihood of innovation

$$\begin{aligned} -2L_t(\boldsymbol{\lambda}_t, \boldsymbol{\mu}_t) = {} & \ln[\det(\dot{\mathbf{H}}_t(x_t^f)\boldsymbol{\lambda}_t \mathbf{P}_t^f \dot{\mathbf{H}}_t^T(x_t^f) + \boldsymbol{\mu}_t \mathbf{R}_t) \\ & + \mathbf{d}_t^T(\dot{\mathbf{H}}_t(x_t^f)\boldsymbol{\lambda}_t \mathbf{P}_t^f \dot{\mathbf{H}}_t^T(x_t^f) + \boldsymbol{\mu}_t \mathbf{R}_t)^{-1}\mathbf{d}_t] \end{aligned} \tag{22}$$

In practical problems, tangent linear operator is usually difficult to compute. To avoid using tangent linear operator in Eq. (21), we note the following fact: adjusting sampling forecast error covariance matrix to $\boldsymbol{\lambda}_t \mathbf{P}_t^f$

is equal to adjusting the perturbed forecast ensemble to $\sqrt{\boldsymbol{\lambda}_t}(\boldsymbol{x}_{t,i}^f - \boldsymbol{x}_t^f) + \boldsymbol{x}_t^f, (i = 1, 2, \ldots, N)$ (Anderson and Anderson, 1999). Similar to Eqs. (6) and (7), approximating the first-order derivative in Eq. (21) using difference, we get

$$-2L_t(\boldsymbol{\lambda}_t) = \ln[\det(H_t\boldsymbol{\lambda}_t\mathbf{P}_t^f H_t^T + \mathbf{R}_t)] + \mathbf{d}_t^T(H_t\boldsymbol{\lambda}_t\mathbf{P}_t^f H_t^T + \mathbf{R}_t)^{-1}\mathbf{d}_t \tag{23}$$

where

$$\begin{aligned}\boldsymbol{\lambda}_t\mathbf{P}_t^f H_t^T \equiv \frac{1}{N-1}\sum_{i=1}^{N}&[\sqrt{\boldsymbol{\lambda}_t}(\boldsymbol{x}_{t,i}^f - \boldsymbol{x}_t^f)]\\ &\cdot [H_t(\sqrt{\boldsymbol{\lambda}_t}(\boldsymbol{x}_{t,i}^f - \boldsymbol{x}_t^f) + \boldsymbol{x}_t^f) - H_t(\boldsymbol{x}_t^f)]^T\end{aligned} \tag{24}$$

and

$$\begin{aligned}H_t\boldsymbol{\lambda}_t\mathbf{P}_t^f H_t^T \equiv \frac{1}{N-1}\sum_{i=1}^{N}&[H_t(\sqrt{\lambda_t}(\boldsymbol{x}_{t,i}^f - \boldsymbol{x}_t^f)+\boldsymbol{x}_t^f) - H_t(\boldsymbol{x}_t^f)]\\ &\cdot[H_t(\sqrt{\boldsymbol{\lambda}_t}(\boldsymbol{x}_{t,i}^f - \boldsymbol{x}_t^f) + \boldsymbol{x}_t^f) - H_t(\boldsymbol{x}_t^f)]^T\end{aligned} \tag{25}$$

Then inflation factor λ_t can be estimated by minimizing Eq. (23).

8.2.4 *Statistics to verify assimilation methods*

Root-mean-square error of the analysis state (A-RMSE) is a statistics to verify whether the analysis state is close to the true state, defined as

$$\text{A-RMSE} = \sqrt{\frac{1}{n}\sum_{k=1}^{n}\left(\frac{1}{N}\sum_{i=1}^{N}\boldsymbol{x}_{t,i}^a(k) - \boldsymbol{x}_t^t(k)\right)^2} \tag{26}$$

Smaller A-RMSE indicates better effect of assimilation. In practical application, however, the true value is not available, so how to construct the statistics to verify the assimilation methods is a challenge.

To verify whether the perturbed forecast state $\boldsymbol{x}_{t,i}^f$ has the same probability distribution as the true state $\boldsymbol{x}_t^t$, we introduce root mean square error of the forecast state (F-RMSE):

$$\text{F-RMSE} = \sqrt{\frac{1}{n}\sum_{k=1}^{n}\left(\boldsymbol{x}_t^t(k) - \frac{1}{N}\sum_{i=1}^{N}\boldsymbol{x}_{t,i}^f(k)\right)^2} \tag{27}$$

and sampling mean variance of forecast state (F-Spread):

$$\text{F-Spread} = \sqrt{\frac{1}{n(N-1)} \sum_{k=1}^{n} \sum_{i=1}^{N} \left(\boldsymbol{x}_{t,i}^{f}(k) - \frac{1}{N} \sum_{l=1}^{N} \boldsymbol{x}_{t,l}^{f}(k) \right)^2} \tag{28}$$

If the true state $\boldsymbol{x}_t^t$ and perturbed forecast state $\boldsymbol{x}_{t,i}^f$ have the same mean value of $\frac{1}{N}\sum_{i=1}^{N} \boldsymbol{x}_{t,i}^f$ and identical covariance, then F-RMSE and F-Spread should be close (Anderson, 2007).

8.3 Introduction of Simplified Ideal Models for Verification

8.3.1 *Lorenz-96 model*

Lorenz-96 model (Lorenz, 1996) is a second-order strongly nonlinear dynamic system and its governing equation is

$$\frac{\mathbf{dX}_k}{\mathbf{d}t} = (\mathbf{X}_{k+1} - \mathbf{X}_{k-2})\mathbf{X}_{k-1} - \mathbf{X}_k + F \tag{29}$$

where $k = 1, 2, \ldots, \mathbf{K}$ ($\mathbf{K} = 40$ is the total number of state variables); the boundary conditions are defined as $\mathbf{X}_{-1} = \mathbf{X}_{K-1}$, $\mathbf{X}_0 = \mathbf{X}_K$, $\mathbf{X}_{K+1} = \mathbf{X}_1$. Lorenz-96 model is derived by simplifying the governing equation of meteorology. The three items on the right side of Eq. (29) qualitatively represent nonlinear advection, damping and external forcing respectively.

In our assimilation system, the true value of factor F is 8, and then Lorenz-96 model is chaotic (Lorenz and Emanuel, 1998). By calculating the relevant Lyapunov exponent, the error-doubling time is about 8 times of the time step, but the fractal dimension of the attractor is about 27.1 (Lorenz and Emanuel, 1998). We use fourth-order Runge–Kutta integration scheme (Butcher, 2003) to solve Eq. (29) numerically, with a time step of 0.05 non-dimensional units, which is about 6 h in real-world time. We integrate the model for 2000 steps.

Synthetic observations are available for every model grid point and generated by adding random noises which are normally distributed with variance $\sigma_\mathrm{o}^2 = 1$. The diagonal elements of $\mathbf{R}_t$ are $\sigma_\mathrm{o}^2 = 1$ and the off-diagonal elements $f(d, \rho)$ are defined as

$$f(d, \boldsymbol{\rho}) = \sigma_o^2 \times \boldsymbol{\rho}^d \tag{30}$$

where $d = \min(|z_i - z_j|, 40 - |z_i - z_j|)$ is the shortest distance between ith and jth observation site. $0 < \rho < 1$ is a scalar to ensure the correlation coefficient between the two observations decreasing with the increasing distance. In this chapter, ρ is set to 0.5. Observations were assimilated every four steps for 2000 steps. However, we only reported the results for the last 1000 steps, while the first 1000 steps were treated as spin-up.

In EnKF assimilation, the ensemble size of forecast state is set to be 30. F is set to 8 when simulating true state, but in assimilation process we used $F' = 5$ unless otherwise specified, with the purpose to introduce model error in the assimilation.

8.3.2 *Two-dimensional SWE model*

In this section we introduce a system with model and observation dimensions larger than those in Sec. 8.3.1, i.e., barotropic nonlinear SWE (Lei and Stauffer, 2009) in the form of

$$\begin{gathered}\frac{\partial u}{\partial t} + u\frac{\partial u}{\partial x} + v\frac{\partial u}{\partial y} - fv = -g\frac{\partial h}{\partial x} + k\nabla^2 u \\ \frac{\partial v}{\partial t} + u\frac{\partial v}{\partial x} + v\frac{\partial v}{\partial y} + fu = -g\frac{\partial h}{\partial y} + k\nabla^2 v \\ \frac{\partial h}{\partial t} + u\frac{\partial h}{\partial x} + v\frac{\partial h}{\partial y} = -h\left(\frac{\partial u}{\partial x} + \frac{\partial v}{\partial y}\right) + k\nabla^2 h \\ 0 \le x \le L, \quad 0 \le y \le D \end{gathered} \tag{31}$$

where the model variables are u, v, and h, respectively, representing the fluid velocities in the directions of x and y and the fluid height; $g = 9.8\,\mathrm{ms^{-2}}$ is acceleration of gravity; $f = 10^{-4}\,\mathrm{s^{-1}}$ is Coriolis parameter using f-plane approximation; $k = 10^4\,\mathrm{m^2\,s^{-1}}$ is diffusion coefficient. $L = 500\,\mathrm{km}$ and $D = 300\,\mathrm{km}$ are the domain dimensions of model integration. The model is discretized with a uniform grid spacing of 10 km in x- and y- directions and integrated using the Lax–Wendroff method with a time step of 30 s. Periodic boundary conditions are used at the x boundaries, and a free-slip rigid wall boundary condition, where u and h are defined from the values one point inside the boundary, is used at the y boundaries. Thus the model has a dimension of 50×31 for each state variable. The initial height (depth) field is given by:

$$\begin{aligned} h(x, y) = H_0 + H_1 \times \tanh\left\{\frac{9 \times (D/2 - y)}{2D}\right\} \\ + H_2 \times \sec h^2\left\{\frac{9 \times (D/2 - y)}{D}\right\} \times \sin\left(\frac{2\pi}{L}x\right) \end{aligned} \tag{32}$$

where H_0, H_1, and H_2 are set to 50.0 m, 5.5 m, and 3.325 m, respectively. The initial velocity field is derived from the initial height field with the geostrophic relation.

In our experimental design, the SWE is integrated for 72 h and the results are taken as the "true states". We then set up synthetic observations of u, v, and h at 310 randomly located grid points for every 3 h in an 48-h assimilation period. A 24-h forecast is run after the assimilation. The observation error variances are specified as $\boldsymbol{\sigma}_{ou}^2 = 0.5\,\mathrm{m}^2\mathrm{s}^{-2}$, $\boldsymbol{\sigma}_{ov}^2 = 0.5\,\mathrm{m}^2\mathrm{s}^{-2}$, and $\boldsymbol{\sigma}_{oh}^2 = 1.0\,\mathrm{m}^2$. The observation errors of each variable are spatially correlated. The correlation coefficient between grid points z_k and $z_{k'}$ is $\boldsymbol{\rho}^{d(z_k, z_{k'})}$, where $d(z_k, z_{k'})$ is the distance between grid points z_k and $z_{k'}$; ρ is set to 0.5, but observation errors of different variables are assumed to be independent of each other. To simulate "model error", we used $k = 5 \times 10^4\,\mathrm{m}^2\,\mathrm{s}^{-1}$ in the assimilation. Ensemble size is 100.

8.4 Verification Results Using Linear Observation

Most inflation schemes on error covariance matrices assume that observation operator is linear. Based on this assumption, this section verifies our inflation scheme using the two simplified ideal models proposed in Sec. 8.3. We will discuss the two different adjustment schemes, i.e., time-dependent and time-independent inflation factors. The influence of wrong observation error covariance matrix on the assimilation effect and its correction methods will also be discussed. The observation operator in this section can be represented by identity matrix.

8.4.1 *The case of time-dependent inflation*

In the case of time-dependent inflation factor, we first apply our inflation scheme in the assimilation using Lorenz-96 model and then compare it with EnKF without inflation. Figure 8.1 shows the A-RMSE and −2log-likelihood of d_t for the EnKF with and without inflation. Compared with traditional EnKF, A-RMSE of EnKF with inflation is significantly reduced, with smaller −2log-likelihood. Figure 8.2 shows the ratio of F-RMSE to F-Spread in these two schemes. It can be seen that the ratio in EnKF with inflation is close to 1, while that in EnKF without inflation is much larger than 1.

Although both our approach and the Li *et al.*'s method adopt the statistics of d_t to estimate time-dependent inflation factors, these two methods are different. The former is based on the maximum likelihood

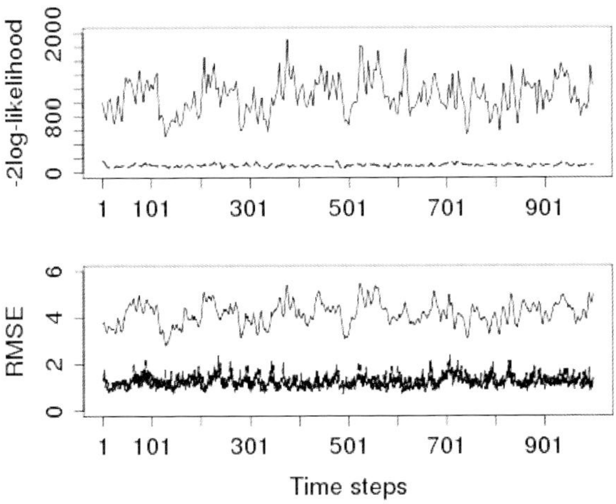

Fig. 8.1. −2log-likelihood and A-RMSE of the analysis state of the traditional EnKF (solid line) and the EnKF after our maximum likelihood inflation adjustment (dashed line). Most of the inflation factors are between 1 and 4 and the model parameter is $F = 5$.

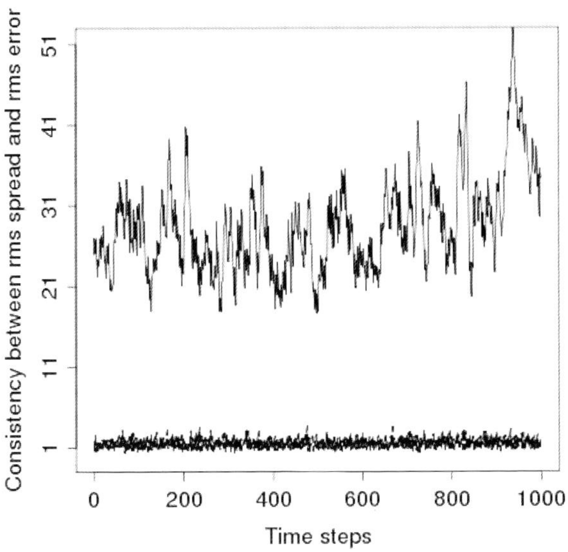

Fig. 8.2. The ratio of F-RMSE to F-Spread of the traditional EnKF (solid line) and EnKF after our maximum likelihood inflation adjustment (dashed line). The model parameter is $F = 5$.

Table 8.2. Time-mean values of inflation factors, A-RMSE, F-RMSE, and F-Spread, when inflation factor is dependent on time.

	Maximum likelihood adjustment method	
	$F = 5$	$F = 12$
Average λ_i	3.0	4.9
A-RMSE	1.25	1.69
F-RMSE	1.43	1.95
F-Spread	1.05	1.87

Note: The model parameters are $F = 5$ and $F = 12$, respectively.

Table 8.3. Time-mean values of inflation factors, −2log-likelihood, and A-RMSE of the three cases:

	λ_t	−2log-likelihood	A-RMSE	μ
Case 1	3.01	97.67	1.25	—
Case 2	1.90	109.69	1.79	—
Case 3	2.85	96.83	1.31	0.3

Note: Case 1: Correct observation error covariance matrix ($\sigma_o^2 = 1$) is used, with inflation adjustment only on forecast error covariance matrix. Case 2: Four times of the correct observation error covariance matrix ($\sigma_o^2 = 4$) is used, with inflation adjustment only on forecast error covariance matrix. Case 3: Four times of the correct observation error covariance matrix ($\sigma_o^2 = 4$) is used, with inflation adjustment on both forecast error covariance matrix and observation error covariance matrix. The model parameter is $F = 5$.

observation error variance is accurately specified ($\boldsymbol{\sigma}_o^2 = 1$); while in the second test, the observation error variance is erroneously specified as $\boldsymbol{\sigma}_{ow}^2 = 4$. The results are shown in Table 8.3, which indicates that A-RMSE in the first test ($\boldsymbol{\sigma}_o^2 = 1$) is smaller than that in the second test ($\boldsymbol{\sigma}_{ow}^2 = 4$). This suggests that the accuracy of observation error covariance matrix has impacts on the assimilation result. The closer the observation error covariance matrix to the true value, the better the effect of assimilation is, with smaller −2log-likelihood value of innovation.

To reduce the influence of inaccurate observation error variance on the effect of assimilation, we introduce another inflation factor μ which is time-uncorrelated, i.e., parameterize the observation error variance as $\boldsymbol{\mu}\boldsymbol{\sigma}_{ow}^2$. The inflation factors $\boldsymbol{\lambda}_t$ and $\boldsymbol{\mu}$ on forecast and observation error covariance matrices are simultaneously estimated by minimizing the objective function in Eq. (15). The results are also listed in Table 8.3. By introducing inflation factor $\boldsymbol{\mu}$, the observation error variance is corrected as $1.2(0.3 \times 4)$, closer to the true value $\boldsymbol{\sigma}_o^2 = 1$. Moreover, the assimilation error based on the corrected variance is close to that obtained in the first test.

We also study the sensitivity of assimilation effect to observation error variance by analyzing the influence of the forecast error covariance matrix adjustment on the filter result under conditions that the observation error variance has some errors. We suppose: (a) the observation error variance adopted in the assimilation is accurate; (b) the observation error variance adopted in the assimilation is only half of the accurate value; and (c) the observation error variance adopted in the assimilation is 1.5 times of the accurate value. Then the assimilation results of EnKF with inflation on forecast error covariance are compared with those of EnKF without inflation, as shown in Table 8.4. It is noted that under the three conditions, A-RMSE in EnKF with inflation is smaller than that in EnKF without inflation. Regardless of the accuracy of observation error variance, the inflation adjustment on forecast error covariance is useful. In addition, when inflating on the forecast error covariance, A-RMSE in (a) is smaller than that in (b) and (c), indicating that the more accurate the observation error variance adopted in the assimilation, the smaller the A-RMSE is.

Table 8.4. Time-mean values of A-RMSE of the three cases in case that inflation factor is time-dependent and inflation adjustment is only conducted on forecast error covariance matrix:

	Traditional EnKF	EnKF after maximum likelihood adjustment
Case (a)	2.24	0.90
Case (b)	3.81	1.23
Case (c)	1.38	0.97

Note: Case (a): Correct observation error covariance matrix ($\boldsymbol{\sigma}_o^2 = 1$) is used; Case (b): 0.5 times of correct observation error covariance matrix ($\boldsymbol{\sigma}_o^2 = 0.5$) is used. Case (c): 1.5 times of correct observation error covariance matrix ($\boldsymbol{\sigma}_o^2 = 1.5$) is used. The model parameter is $F = 7.6$.

8.4.4 *The case of time-dependent inflation factor of observation error covariance matrix*

In Sec. 8.4.3, we only assume that the inflation factor μ on observation error covariance matrix is independent with respect to time. We speculate that, to obtain the time-dependent inflation factor of observation error covariance matrix, we may need model and observation systems of higher dimensions. The dimension of Lorenz-96 model in Sec. 8.4.3 is low, making it unsuitable for the research in this respect. Therefore, we choose the two-dimensional SWE model with a higher dimension instead.

We first consider the case of accurately specified observation error. Figure 8.4 shows the assimilation results of u and h for EnKF with and without inflation. The assimilation result of v is similar to that of u. We can see that in 48-h assimilation period and the later 24-h forecast period, estimating inflation factor based on the maximum likelihood method is successful in obtaining better results than traditional EnKF assimilation.

When the error variances of u, v, and h are erroneously specified as $\sigma^2_{ouw} = 5\,\mathrm{m^2\,s^{-2}}, \sigma^2_{ovw} = 5\,\mathrm{m^2\,s^{-2}}$, and $\sigma^2_{ohw} = 10\,\mathrm{m^2}$, respectively, we introduce a common inflation factor μ to be estimated using MLE that is similar to the test using Lorenz-96 model in Sec. 8.4.3. The difference is that μ is not constant in this test but changes with time, so we minimize the -2log-likelihood in Eq. (14) to simultaneously estimate λ_t and μ_t at

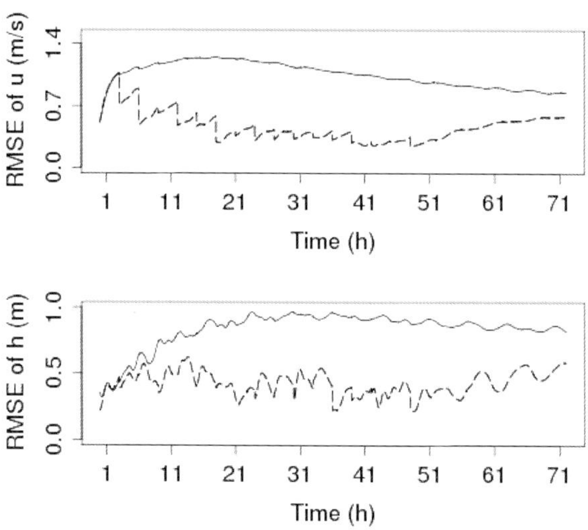

Fig. 8.4. A-RMSE of u and h of the traditional EnKF (solid line) and EnKF after our maximum likelihood inflation adjustment (dashed line).

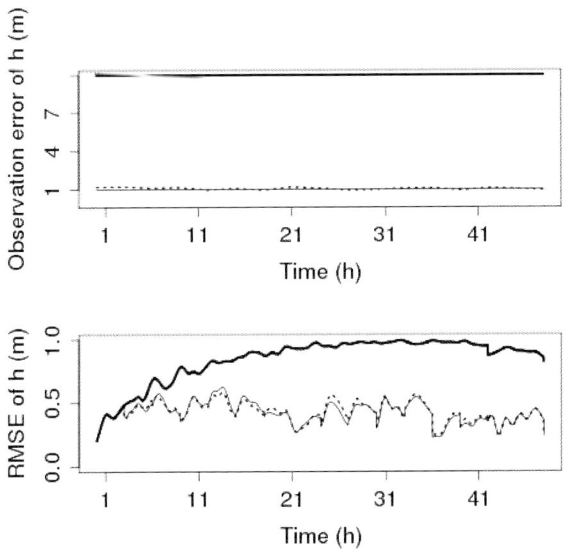

Fig. 8.5. Observation error variance and A-RMSE of h in the following cases: correct observation error variance is used, with inflation adjustment only on forecast error covariance matrix (thin solid line); incorrect observation error variance is used, with inflation adjustment only on forecast error covariance matrix (thick solid line); incorrect observation error variance is used, with inflation adjustment on both forecast error covariance matrix and observation error covariance matrix (dotted line).

every time step. Because the assimilation results for u, v, and h are very similar, we only show the results of h in Fig. 8.5. When the observation error is erroneously specified, MLE successfully estimates the inflation factor of observation error matrix. Although the adjusted observation error variance is not completely equal to the true observation error variance, the assimilation results (A-RMSE) are very close.

We also apply time-dependent $\boldsymbol{\mu}$ to the observation error covariance matrix in Lorenz-96 model, but the estimated inflation factor $\boldsymbol{\mu}_t$ shows larger oscillation, which may be due to the low observation dimension (40), resulting in less stable estimation of the two parameters. Two-dimensional shallow-water equation model has a higher dimension (310) and can ensure the estimation precision of the two parameters.

8.5 Verification Results Using Nonlinear Observation

In Sec. 8.4, we discuss the effect of inflation on EnKF assimilation with linear observation operator. In this section, we further investigate the case of nonlinear observation operator.

8.5.1 Sensitivity analysis on the degree of tangent linearity of observation operator

Section 8.2.3 gives several inflation schemes on forecast error covariance matrix in the case of tangent linear observation operator. Here, we begin with the discussion of the sensitivity of these inflation schemes to tangent linear requirement.

In this section, the nonlinear observation equation is defined as

$$\boldsymbol{y}_k = 10\exp\{\alpha \boldsymbol{x}_k^t\} + \varepsilon_k \tag{33}$$

where k represents the kth grid point of model, and α is a parameter.

Taylor expansion is performed on nonlinear observation operator $10\exp\{\alpha \boldsymbol{x}_k^t\}$ with respect to forecast state $\boldsymbol{x}_k^t$

$$10\exp\{\alpha \boldsymbol{x}_k^t\} - 10\exp\{\alpha \boldsymbol{x}_k^f\} = 10\exp\{\alpha \boldsymbol{x}_k^f\}\cdot\alpha(\boldsymbol{x}_k^t - \boldsymbol{x}_k^f) + 10\exp\{\alpha \boldsymbol{x}_k^f\} \cdot \frac{1}{2}(\alpha(\boldsymbol{x}_k^t - \boldsymbol{x}_k^f))^2 + \cdots \tag{34}$$

Apparently, smaller α means more feasibility of omitting the high-order items on the right side of Eq. (34) and more rationality of the assumption of tangent linearity for this function.

To further study the influence of nonlinear observation operator on inflation, we use Lorenz-96 model as the platform to conduct the following test. In this test, the observation error variance adopted in the assimilation is accurate and we use $F = 6$ to simulate model error. The ensemble size is 30, and inflation factor λ_t is estimated with maximum likelihood. The parameters in observation operators are successively specified as $\alpha = 0.10, 0.13, 0.17$, and 0.20.

Figure 8.6 shows the histograms of forecast error $\alpha(\boldsymbol{x}_k^t - \boldsymbol{x}_k^f)$ generated in 2000 steps when $\alpha = 0.10, 0.13, 0.17, 0.20$, respectively. We can see that the forecast error $\boldsymbol{x}_k^t - \boldsymbol{x}_k^f$ is normally distributed. Figure 8.6 also draws the interval of $[-1, 1]$ for each histogram. When $\alpha = 0.10, 0.13, 0.17, 0.20$, the probabilities that $\alpha(\boldsymbol{x}_k^t - \boldsymbol{x}_k^f)$ falls outside the interval of $[-1, 1]$ are respectively $0.0002, 0.0238, 0.0892$, and 0.1862. With the increase of α, the probability that the absolute value of $\alpha(\boldsymbol{x}_k^t - \boldsymbol{x}_k^f)$ is larger than 1 also increases. It becomes increasingly difficult to omit the high-order items in Eq. (34), and hence the assumption of tangent linearity for the function $10\exp\{\alpha \boldsymbol{x}_k^t\}$ becomes more unreasonable.

Inflation adjustment is conducted on forecast error covariance by minimizing Eq. (23), with $\alpha = 0.10, 0.13, 0.17, 0.20$, respectively. A-RMSE

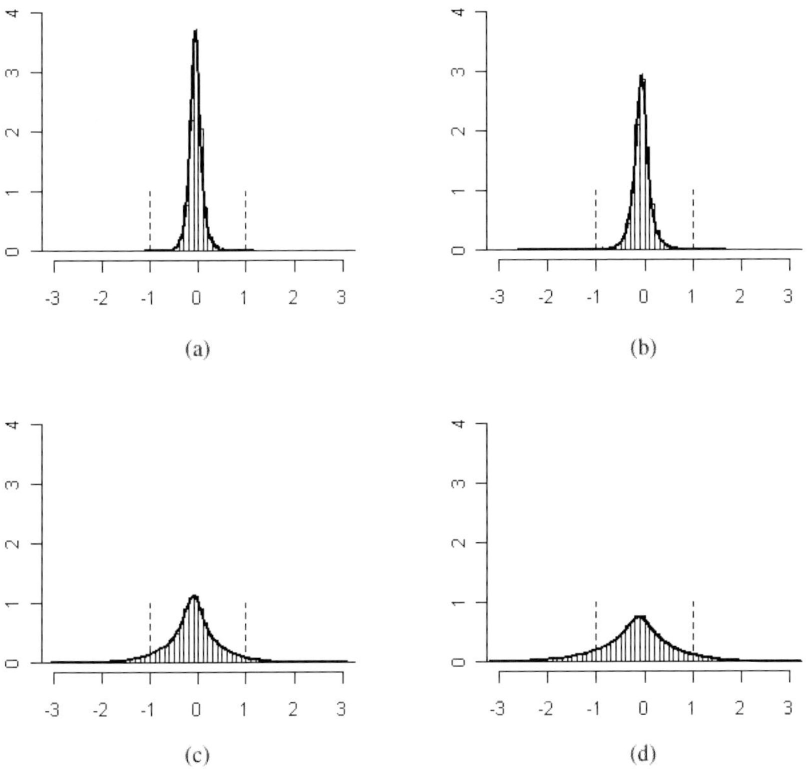

Fig. 8.6. Histogram of $\alpha(x_k^t - x_k^f)$ for 2000 time steps, where (a) $\alpha = 0.10$; (b) $\alpha = 0.13$; (c) $\alpha = 0.17$ and (d) $\alpha = 0.20$. The thick solid line is kernel density estimation curve.

obtained is reduced respectively by 75%, 73%, 38%, and 21% compared with that in the scheme without inflation. Apparently, if observation operator does not meet the tangent linear requirement, it will result in the poor effect of assimilation. Based on the above analysis, we choose $\alpha = 0.10$ in Sec. 8.5.2 in order to ensure the tangent linearity of observation operator $10\exp\{\alpha \boldsymbol{x}_k^t\}$.

8.5.2 *Comparison of several inflation adjustment schemes in the case of tangent linear observation operator*

Section 8.2.3 provides four inflation schemes in the case of tangent linear observation operator, of which two are proposed by Li *et al.* (2009), i.e., Eqs. (18) and (19) (respectively denoted as OMB and AMB), while the other two are proposed by Wu *et al.* (2011), i.e., estimating inflation factors by minimizing objective functions (21) and (23) (respectively denoted as EnKF1 and EnKF2). This section will compare these methods using Lorenz-96 model as in Sec. 8.5.1.

Equation (33) is selected as the nonlinear observation equation. We use $\alpha = 0.10$ to ensure that the tangent linear condition holds based on the discussion in the last section. The model error is unavoidable in practical dynamic system and we need to study the influence of various inflation schemes on the assimilation results in the presence of model error. We choose $F = 4, 5, \ldots, 12$ in the assimilation test. We do not use $F \leq 3$ and $F \geq 13$ because the model error would be too large, resulting in highly unstable numerical simulation. The ensemble size is 30.

Figure 8.7 shows the time-mean A-RMSE of the four methods. We can see when the model error increases (when F becomes increasingly distant

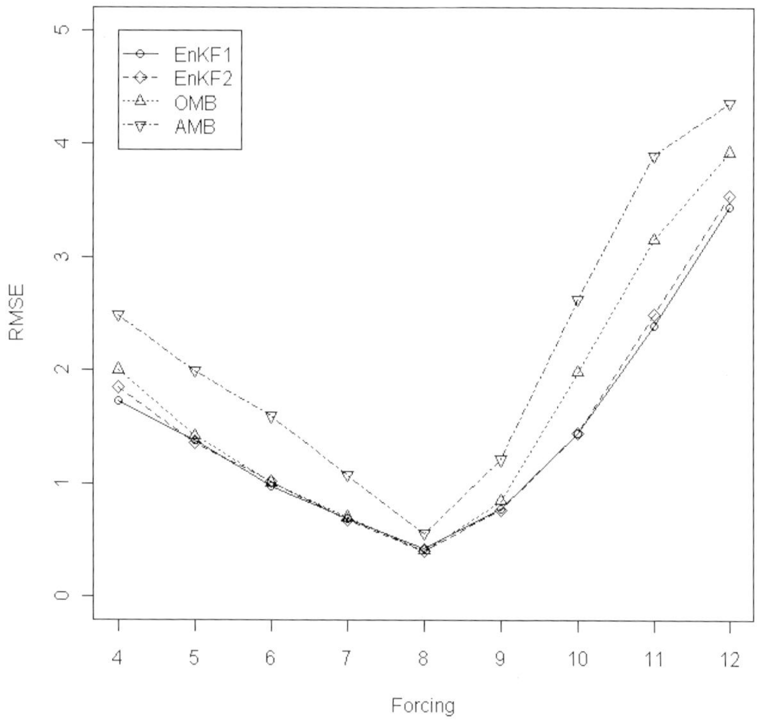

Fig. 8.7. Time-mean values of A-RMSE of our maximum likelihood adjustment methods (EnKF1 and EnKF2 indicate estimations of inflation factor by minimizing Eqs. (21) and (23), respectively) and Li *et al.*'s OMB and OMA adjustment methods, as a function of forcing F.

from 8), A-RMSE also increases with it, but the two methods proposed by Wu *et al.* (2011) increases more slowly than the two methods proposed by Li *et al.* (2009), which may indicate the two estimation methods proposed by Wu *et al.* (2011) are more stable than the two estimation methods (specially for AMB method) proposed by Li *et al.* (2009).

Under the condition that the observation operator is tangent linear, we carry out discussions as in Secs. 8.4.2–8.4.4, with its results similar to that under the condition that the observation operator is linear. Therefore, we do not need to expand the discussion anymore.

8.6 Discussion and Main Conclusion

In traditional EnKF assimilation, forecast error covariance matrix is estimated as sampling covariance matrix of ensemble forecast states. But many people found this estimation to be too low and very easy, resulting in degeneration of the filter. The research in this text further proves this point.

A common method to overcome degeneration of filter is to make inflation adjustment on forecast error covariance matrix. Incited by Dee's early work, this text proposes a specific algorithm to estimate the inflation factor of forecast error covariance matrix by optimizing the likelihood function of innovation. Whether the inflation factor is dependent or independent on time, it can be estimated with this method, which can also be used in estimating the inflation factor of observation error covariance matrix, whether it is dependent or independent on time.

If the true state has been known, Anderson and Anderson (1999) proposed a best-tuned method of estimating inflation factor, which is to multiply a time-independent inflation factor λ to forecast error covariance matrix, and the resulted analysis state has minimum root-mean-square error (A-RMSE). However, the true state cannot be known in advance in reality, so the best-tuned method can only be discussed in simplified ideal model and cannot be operated in practical application. In this text, we estimate λ by minimizing objective function (13), namely the total -2log-likelihood of innovation over the full time period. In order to study the effectiveness of our approach, we compare the relevant assimilation results with those of the best-tuned method.

Wang and Bishop (2003) and Li *et al.* (2009) adopted the method of first-order moment estimation of innovation to estimate the time-dependent inflation factor of forecast error covariance matrix. From the experiment result of simplified ideal model, the assimilation result of the estimation method proposed by us is better, especially when the model error is large. In addition, the main advantage of our estimation method lies in its applicability in estimating the inflation factor of observation error covariance matrix, which cannot be realized in other methods.

An early method of inflation adjustment is to multiply the sampling covariance matrix of ensemble forecast states by a constant greater than 1 and then determine this constant empirically according to the assimilation result of EnKF (e.g., Anderson and Anderson, 1999). Theoretically, such estimation method of inflation factor is not superior to the best-tuned method (Anderson and Anderson, 1999). According to the experiment, the analysis error created by our inflation adjustment method differs little from that created by the best-tuned method, but the probability distribution of ensemble forecast created from our approach is closer to the probability distribution of the true value.

It can also be speculated from the experiment that our approach can effectively estimate the time-independent inflation factor of observation covariance matrix. If the dimension of observation vector is large, the inflation factor can also be time-dependent. The important premise is that the inflation factor of observation error covariance matrix must be estimated simultaneously with that of forecast error covariance matrix. If only the inflation factor of observation error covariance matrix is adjusted, the assimilation effect will be very poor, which also indicates the following phenomenon: although the forecast error covariance matrix after inflation adjustment is not necessarily the true forecast error covariance matrix, their closeness makes the estimation of inflation factor of observation error covariance matrix not that sensitive.

At last, we emphasize that all inflation adjustment methods discussed in this text require linear or tangent linear observation operator. Otherwise, the inflation adjustment method may be not helpful to EnKF assimilation. How to solve this problem is a topic to be explored in our future study.

Appendix A: Li *et al.*'s Estimation of the Inflation Factor of Forecast Error Covariance Matrix $\boldsymbol{\lambda}_t$

$$\langle \tilde{\mathbf{d}}_t^T \tilde{\mathbf{d}}_t \rangle = \mathrm{tr}(\langle \tilde{\mathbf{d}}_t \tilde{\mathbf{d}}_t^T \rangle) = \mathrm{tr}(\mathbf{H}_t \lambda_t \mathbf{P}_t^f \mathbf{H}_t^T + \mathbf{R}_t) = \lambda_t \mathrm{tr}(\mathbf{H}_t \mathbf{P}_t^f \mathbf{H}_t^T) + \mathrm{tr}(\mathbf{R}_t) \tag{A.1}$$

where $\langle\rangle$ represents the mathematic expectation operator, and $\tilde{\mathbf{d}}_t$ represents the random vector of relevant state. Based on the assumption that $\boldsymbol{\lambda}_t\mathbf{P}_t^f$ is error covariance matrix of $\boldsymbol{x}_t^f$, the covariance matrix of $\tilde{\mathbf{d}}_t$ is $\mathbf{H}_t\boldsymbol{\lambda}_t\mathbf{P}_t^f\mathbf{H}_t^T + \mathbf{R}_t$, so the second equality sign of Eq. (A.1) is set up. $\tilde{\mathbf{d}}_t^T\tilde{\mathbf{d}}_t$ is an unbiased estimation of $\langle\tilde{\mathbf{d}}_t^T\tilde{\mathbf{d}}_t\rangle$. Substituting the above equator with $\tilde{\mathbf{d}}_t^T\tilde{\mathbf{d}}_t$ and solve $\boldsymbol{\lambda}_t$ and then the estimation of Eq. (10) is obtained.

Appendix B: A Calculation Method of Determinant $\mathbf{dot}(\mathbf{H}_t\boldsymbol{\lambda}_t\mathbf{P}_t^f\mathbf{H}_t^T + \mathbf{R}_t)$

By decomposing the observation error covariance matrix $\mathbf{R}_t$ into the following form,

$$\mathbf{R}_t = \mathbf{W}_t\mathbf{W}_t^T \tag{B.1}$$

we obtain the following equation according to Eq. (B.1) and the property of determinant

$$\begin{aligned}
&\det(\mathbf{H}_t\boldsymbol{\lambda}_t\mathbf{P}_t^f\mathbf{H}_t^T + \mathbf{R}_t)\\
&\quad= \det\{\mathbf{W}_t^{-1}(\mathbf{H}_t\boldsymbol{\lambda}_t\mathbf{P}_t^f\mathbf{H}_t^T + \mathbf{W}_t\mathbf{W}_t^{-1})\}\det(\mathbf{W}_t)\det(\mathbf{W}_t^T)\\
&\quad= \det\{\boldsymbol{\lambda}_t\mathbf{W}_t^{-1}\mathbf{H}_t\mathbf{P}_t^f(\mathbf{W}_t^{-1}\mathbf{H}_t)^T + \mathbf{I}\}\det(\mathbf{W}_t\mathbf{W}_t^T)\\
&\quad= \det\{\boldsymbol{\lambda}_t(\mathbf{W}_t^{-1}\mathbf{H}_t\mathbf{L}_t)(\mathbf{W}_t^{-1}\mathbf{H}_t\mathbf{L}_t)^T + \mathbf{I}\}\det(\mathbf{R}_t)
\end{aligned} \tag{B.2}$$

where

$$\begin{aligned}
L_t &= \frac{1}{\sqrt{N-1}}\\
&\quad\times\left(\boldsymbol{x}_{t,1}^f - \frac{1}{N}\sum_{l=1}^{N}\boldsymbol{x}_{t,1}^f, \boldsymbol{x}_{t,2}^f - \frac{1}{N}\sum_{l=1}^{N}\boldsymbol{x}_{t,l}^f, \ldots, \boldsymbol{x}_{t,N}^f - \frac{1}{N}\sum_{l=1}^{N}\boldsymbol{x}_{t,l}^f\right)
\end{aligned} \tag{B.3}$$

is the square root of matrix $\mathbf{P}_t^f$, i.e., $\mathbf{L}_t\mathbf{L}_t^T = \mathbf{P}_t^f$, and $\mathbf{I}$ is the identity matrix of dimension $m_t \times m_t$.

By conducting singular value decomposition for $m_t \times N$ dimensional matrix $\mathbf{W}_t^{-1}\mathbf{H}_t\mathbf{L}_t$, we obtain

$$\mathbf{W}_t^{-1}\mathbf{H}_t\mathbf{L}_t = \mathbf{U}_t\mathbf{B}_t\mathbf{V}_t \tag{B.4}$$

where $\mathbf{U}_t$ and $\mathbf{V}_t$ are orthogonal matrices and $\mathbf{B}_t$ is $m_t \times N$ dimensional diagonal matrix, wherein the diagonal elements are the singular values of matrix $\mathbf{W}_t^{-1}\mathbf{H}_t\mathbf{L}_t$.

According to the property of determinant,

$$\begin{aligned}
&\det(\mathbf{H}_t\boldsymbol{\lambda}_t\mathbf{P}_t^f\mathbf{H}_t^T + R_t)\\
&\quad= \det\{\boldsymbol{\lambda}_t(\mathbf{U}_t\mathbf{B}_t\mathbf{V}_t)(\mathbf{U}_t\mathbf{B}_t\mathbf{V}_t)^T + \mathbf{I}\}\det(\mathbf{R}_t)\\
&\quad= \det\{\mathbf{U}_t(\boldsymbol{\lambda}_t\mathbf{B}_t\mathbf{B}_t^T + \mathbf{I})\mathbf{U}_t^T\}\det(\mathbf{R}_t)\\
&\quad= \det\{(\boldsymbol{\lambda}_t\mathbf{B}_t\mathbf{B}_t^T + \mathbf{I})\det(\mathbf{U}_t)\det(\mathbf{U}_t^T)\det(\mathbf{R}_t)\\
&\quad= \det(\boldsymbol{\lambda}_t\mathbf{B}_t\mathbf{B}_t^T + \mathbf{I})\det(\mathbf{R}_t)\\
&\quad= \det(\mathbf{R}_t) \times \prod_{i=1}^{N}(\boldsymbol{\lambda}_t b_{t,i}^2 + 1)
\end{aligned} \tag{B.5}$$

where $b_{t,i}$ is the ith diagonal element of $\mathbf{B}_t$, so only the calculation of determinant of $\mathbf{R}_t$ is necessary. Generally in practical, it can be supposed that $\mathbf{R}_t$ is a diagonal matrix or independent of time, by which $\det(\mathbf{R}_t)$ may possibly be calculated economically.

References

Anderson, J. L. and Anderson, S. L. (1999). A Monte Carlo implementation of the non-linear filtering problem to produce ensemble assimilations and forecasts. *Monthly Weather Review*, **127**, 2741–2758.

Anderson J. L. (2007). An adaptive covariance inflation error correction algorithm for ensemble filters. *Tellus*, **59A**, 210–224.

Anderson J. L. (2009). Spatially and temporally varying adaptive covariance inflation for ensemble filters. *Tellus*, **61A**, 72–83.

Bai, Y. L. and Li, X. (2011). Evolutionary algorithm-based error parameterization methods for data assimilation. *Monthly Weather Review*, **139**(8), 2668–2685.

Butcher, J. C. (2003). Numerical methods for ordinary differential equations. Hoboken, New Jersey: John Wiley & Sons.

Constantinescu, M., Sandu, A., Chai, T. and Carmichael, G. R. (2007). Ensemble-based chemical data assimilation, Part I: General approach. *Quarterly Journal of the Royal Meteorological Society*, **133**, 1229–1243.

Dee, D. P. (1995). On-line estimation of error covariance parameters for atmospheric data assimilation. *Monthly Weather Review*, **123**, 1128–1145.

Dee D. P. and da Silva, A. M. (1999). Maximum-likelihood estimation of forecast and observation error covariance parameters, Part I: Methodology. *Monthly Weather Review*, **127**, 1822–1834.

Dee D. P, Gaspari, G., Redder, C., Rukhovets, L. and da Silva, A. M. (1999). Maximum-likelihood estimation of forecast and observation error covariance parameters, Part II: Applications. *Monthly Weather Review*, **127**, 1835–1849.

Evensen, G. (1994a). Sequential data assimilation with a nonlinear quasi-geostrophic model using Monte Carlo methods to forecast error statistics. *Journal of Geophysical Research*, **99**(C5), 10143–10162.

Evensen, G. (1994b). Inverse methods and data assimilation in nonlinear ocean models. *Physica D*, **77**, 108–129.

Golub, G. H. and Van Loan, C. F. (1996). *Matrix Computations*, 694 pp. Maryland, USA: The Johns Hopkins University Press.

Ide, K., Courtier, P., Michael, G. and Lorenc, A. C. (1997). Unified notation for data assimilation: Operational, sequential and variational. *Journal of the Meteorological Society of Japan*, **75**, 181–189.

Lei, L. L. and Stauffer, D. R. (2009). A hybrid ensemble Kalman filter approach to data assimilation in a two-dimensional shallow water model. In *Proceedings of 23rd Conference on Weather Analysis and Forecasting/19th Conference on Numerical Weather Prediction*, Omaha, NE 1 June 5, 2009. Boston: American Meteorological Society.

Li H., Kalnay, E. and Miyoshi, T. (2009). Simultaneous estimation of covariance inflation and observation errors within an ensemble Kalman filter. *Quarterly Journal of the Royal Meteorological Society*, **135**, 523–533.

Liang X., Zheng, X. G., Zhang, S. P., Wu, G. C., Dai, Y. J. and Li, Y. (2011). Maximum likelihood estimation of inflation factors on forecast error covariance matrix for ensemble Kalman filter assimilation. *Quarterly Journal of the Royal Meteorological Society*, **137**, 263–273.

Lorenz, E. N. (1996). Predictability — A problem partly solved. In *Proceedings of the Seminar on Predictability, ECMWF*, Shinfield Park, Reading, Berkshire, UK.

Lorenz, E. N. and Emanuel, K. A. (1998). Optimal sites for supplementary weather observations: Simulation with a small model. *Journal of the Atmospheric Sciences*, **55**, 399–414.

Miyoshi, T. (2011). The Gaussian approach to adaptive covariance inflation and its implementation with the local ensemble transform Kalman filter. *Monthly Weather Review*, **139**, 1519–1535.

Sénégas, J. H. H., Wackernagel, W., Rosenthal, W. and Wolf, T. (2001). Error covariance modeling in sequential data assimilation. *Stochastic Environmental Research and Risk Assessment*, **15**, 65–86.

Sherman, J. and Morrison, W. J. (1950). Adjustment of an inverse matrix corresponding to a change in one element of a given matrix. *Annals of Mathematical Statistics*, **21**, 124–127.

Tippett, M. K., Anderson, J. L. Bishop, C. H., Hamill, T. M. and Whitaker, J. S. (2003). Notes and correspondence: Ensemble square root filters. *Monthly Weather Review*, **131**, 1485–1490.

Wang, X. and Bishop, C. H. (2003). A comparison of breeding and ensemble transform Kalman filter ensemble forecast schemes. *Journal of the Atmospheric Sciences*, **60**, 1140–1158.

Wu, G., Zheng, X. and Li, Y. (2011). Inflation adjustment on error covariance matrix of ensemble Kalman filter. *The 2nd International Conference on Multimedia Technology*, **3**, 2160–2163.

Zheng, X. G. (2009). An adaptive estimation of forecast error covariance parameters for Kalman filtering data assimilation. *Advances in Atmospheric Sciences*, **26**, 154–160.

Chapter 9

A Review of Error Estimation in Land Data Assimilation Systems

Yulong Bai[*,†], Xin Li[*] and Qianlong Chai[†]

[*]*Cold and Arid Regions Environmental and Engineering Research Institute, Chinese Academy of Sciences, Lanzhou 730000, China*

[†]*College of Physics and Electrical Engineering, Northwest Normal University, Lanzhou 730070, China*

Error estimation issues regarding observing and/or modeling are thought to be a bottleneck for improving the performance of data assimilation (DA) systems. The main purpose of this chapter is to review the error-related problems for land DA systems and to expose the basic ingredients of practical estimation algorithms developed for different types of DA systems, focusing mainly on error descriptions and parameterizations. Several novel methods for estimating model errors are reviewed in three parts: model input error estimation, model parameter error estimation, and model structural error estimation. For observation errors, the error sources can be divided according to observation algorithm errors, representative errors, and instrument errors. To reduce the errors for ensemble DA systems, common error parameterization methods, such as multiplicative inflation, additive inflation, and the "relaxation-to-prior" methods, are introduced. Moreover, a new evolutionary algorithm-based error parameterization method is proposed and compared with the traditional methods mentioned above. Finally, the discussion will conclude with a number of key questions regarding the DA challenges of the next decade.

9.1 Introduction

Land models and other similar models (e.g., atmospheric, hydrological, and oceanic) have large but unknown deficiencies that are major obstacles to further progress in data assimilation (DA) research, as they can seriously undermine research results (Reichle, 2008; Li *et al.*, 2007). As an important methodology for optimally merging Earth observation information

S. Liang, X. Li & X. Xie (eds.), *Land Surface Observation, Modeling and Data Assimilation*, 233–272.

and geophysical model dynamics, DA has played an important role in Earth science. Accurate estimates of forecast error statistics are considered to be crucial for the success of the assimilation process, so they have received much attention over the last decade or so. Given the complexity of research objectives, significant developments have been achieved, while some topics require further attention (e.g., Reichle, 2008; Li *et al.*, 2007; Evensen, 2007; Anderson, 2007). Error parameterization for DA systems is one of these topics. Reichle (2008) summarized modern DA methods and their applications and pertinent research questions in the Earth sciences and emphasized that current critical DA research topics include the construction of the covariance matrix model, the quality control of remotely sensed data, and the bias estimation of DA systems. McLaughlin (2002) noted that the essence of the DA problem is the emphasis placed on the uncertainty from multiple sources within the framework of probabilities. In the future, the main points regarding DA research in hydrology will be the high dimension problem and the accurate estimation of the observation error. Concentrating on land DA systems, Crow and Van Loon (2006), Crow and Bolten (2007), and Crow and Reichle (2008) have all discussed error problems in recent years. Because DA approaches require some type of state forecast error covariance information, the Ensemble Kalman filter (EnKF) dynamically derives such information through a Monte Carlo approach and the introduction of random noise in model states, fluxes, and/or forcing data. However, in land DA, relatively little guidance exists concerning strategies for selecting the appropriate magnitude and/or type of introduced model noise. In addition, little is known about the sensitivity of the filter prediction accuracy to (potentially) inappropriate assumptions concerning the source and magnitude of the modeling error. Crow and Van Loon (2006) conducted a series of synthetic identical twin experiments to explore the consequences of making incorrect assumptions concerning the source and magnitude of model error on the efficiency of assimilating surface soil moisture observations to constrain deeper root-zone soil moisture predictions made by a land surface model (LSM). The results suggested that inappropriate model error assumptions can lead to circumstances in which the assimilation of surface soil moisture observations actually degrades the performance of a LSM (relative to open-loop assimilations that lack a DA component). Prospects for diagnosing such circumstances and adaptively correcting the culpable model error assumptions using filter innovations are discussed (Crow and Van Loon, 2006).

Poor error estimation or parameterization will lead to filter divergence (Jazwinski, 1970), which means the filter forecasting will divert from the true value with the assimilation progress due to model structure errors, uncertain model inputs, and observation errors. To resolve this issue, some explorative works have been performed in the atmospheric and oceanic fields, such as multiplicative inflation (Anderson and Anderson, 1999), additive inflation (Hamill and Whitaker, 2005), and the "relaxation-to-prior" method (Zhang *et al.*, 2004). These methods focus on ameliorating the bias error in the ensemble second moment. As an extended application of the maximum likelihood theory developed in Dee (1995) and Dee and da Silva (1999), Zheng (2009) proposed a "multivariate covariance inflation" to extend the inflation factor to a time-dependent diagonal matrix. However, only a simple model and independent observation errors were tested for that work. Liang *et al.* (2011) conducted a study to further develop the work of Zheng (2009), where the inflation method was tested on more realistic models with much higher dimensions and using spatially correlated observation errors. They also demonstrated the capability of their approach to simultaneously inflate both forecast error and observation error covariance matrices when the observation error variance is wrongly specified. Motivated by the "relaxation-to-prior" method (Zhang *et al.*, 2004), Bai and Li (2011) proposed a new method to generate the ensemble perturbation based on the crossover principles originally developed in intelligent computing research. The essence of this method is to generate proper ensemble perturbation to inflate the covariance matrix and to obtain appropriate weights for innovation in the EnKF by which filter divergence could be effectively mitigated. Combined with the factor search frameworks, the best error adjustment factors were found to guarantee the best DA performance in the corresponding circumstance.

The objective of this chapter is twofold: first, to review the fundamental error problems for land surface DA systems; and second, to expose the basic ingredients of practical assimilation algorithms, mainly focusing on error problems. This chapter is organized as follows. Section 9.2 is dedicated to the fundamentals of error problems in modern DA methods. Error estimation issues will then be discussed in Sec. 9.3. Section 9.4 will provide several error-handling methods in ensemble DA. The traditional inflation methods and the new evolutionary-based methods are illustrated based on the classical nonlinear Lorenz model. In Sec. 9.5, we discuss various approaches to reducing the error of land surface DA systems. Finally, the discussion will conclude with a number of key questions related to the DA challenges of the next decade.

9.2 Error Problems in Modern DA Methods

9.2.1 *Error definitions and their sources*

In this section, we introduce the basic assimilation problem in the state space using the conventional notations proposed by Ide *et al.* (1997). The goal here is not to present a rigorous and comprehensive derivation of the DA algorithms, which can be found elsewhere in dedicated textbooks (e.g., Gelb, 1974; Evensen, 2007), but rather to introduce a simplified framework that still contains the essential characteristics needed to illustrate the more advanced concepts and implementation issues discussed in the following sections.

According to the state-space methods, state prediction and observation models can be derived directly for illustrating dynamics systems, which are also named model operators and observation operators in DA systems.

The nonlinear forecast model of state space, the model operator, can be denoted as:

$$\mathbf{X}^t(t_k) = M_k(\mathbf{X}^t(t_{k-1}), u_k, \theta) + \omega(t_k) \quad \mathbf{X} \in \Re^{n\times 1},\ \omega \in \Re^{n\times 1} \tag{1}$$

where $\mathbf{X}$ is the state vector of the system (e.g., soil moisture profile); n is the dimension of the systems; superscript t denotes the true value; $\omega(t_k)$ is the model errors; t_k, which is behind the state vector, or subscript k of M, denotes the time; u is the external force data (e.g., atmospheric drive data); θ is the model parameter (e.g., soil hydraulic and thermodynamic parameters); and M is the nonlinear model, $M(\cdot) : \Re^{n\times 1} \times \Re^{n\times 1} \to \Re^{n\times 1}$. M may be dependent on time, so the time subscript k is used. However, in DA systems, M generally does not change with time.

Because DA methods are used to assimilate the observation data into the dynamic models, the observation model is denoted as:

$$\mathbf{Y}^o(t_k) = H_k(\mathbf{X}^t(t_k), \varphi) + \varepsilon(t_k) \quad \mathbf{Y} \in \Re^{m\times 1},\ \varepsilon \in \Re^{m\times 1} \tag{2}$$

where $\mathbf{Y}^o(t_k)$ is the observation at time t_k; $\varepsilon(t_k)$ is the observation error; m is the dimension of the observation vector; H is the observation operator, which satisfies: $H(\cdot) : \Re^{n\times 1} \times \Re^{m\times 1} \to \Re^{m\times 1}$; and φ is the parameters of the observation operator.

According to the above system equations, the error sources of land DA systems include model errors, observation errors, and the assimilation algorithm errors. Among these types of errors, model errors are composed of the model structural errors, parameter errors, model calculation errors, and the errors caused by forcing data. The model calculation error refers to spatial and temporal differentials, truncation errors, and the other errors shown in the process of calculation. The observation error is composed of the observation operator error, instrument errors, and representative errors. Moreover, different DA methods will bring about different kinds of assimilation algorithm error. For example, sampling errors are caused by finite sample numbers in various Monte Carlo methods. Detailed discussions will be presented Sec. 9.3.

In conclusion, taking ensemble DA systems as an example, Fig. 9.1 shows the basic composition of DA systems and the algorithm procedures from the perspective of the model evolution and the error characteristics. The arrows and the serial numbers in Fig. 9.1 denote the assimilation process of the model's forward evolution. The block diagram shows the error characteristics in each step.

The ultimate purpose of DA is to balance information from different sources: modeled outputs and observations. Effective land surface DA in operational settings need to be based on realistic descriptions of uncertainties; thus, research may need to consider the way in which uncertainties are actually represented. As far as land surface DA systems are concerned,

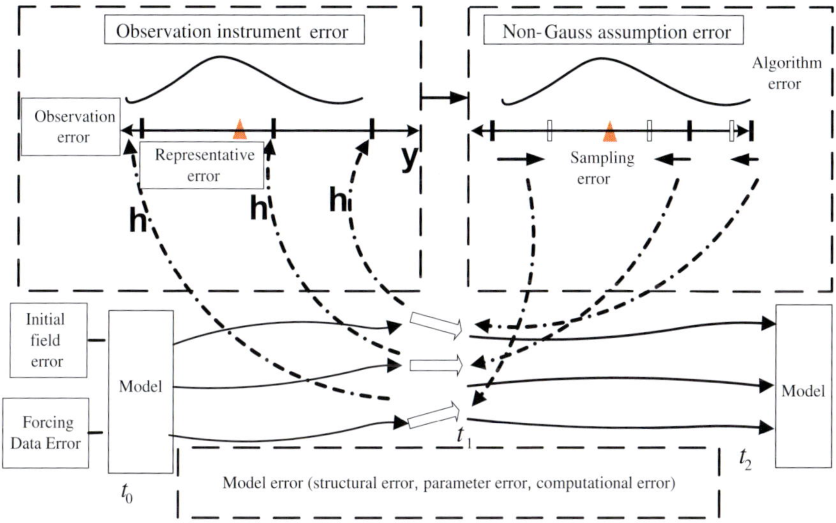

Fig. 9.1. Error sources and error characteristics of DA systems.

the ultimate goal is to "balance" multiple sources of information such as model outputs and observations. In the dynamic framework of the LSM, DA is a method to merge all available observations with a dynamic model to produce spatiotemporally consistent system states that are as accurate as possible by considering the observation and model errors (Talagrand, 1997). Therefore, the research of error problems is a key issue in all studies of DA methods. One of the main research themes of the progress of DA methods is "how to present all kinds of error information of the assimilation system correctly and reasonably". The theory of DA error research in land surfaces rests on the mathematical framework of estimation theory, control theory, and optimization theory.

At present, DA methods are classified into two categories: variational DA and sequential DA. While variational DA is performed by fitting a dynamic model to all of the available observations during a period of interest, sequential DA is an online approach that updates the estimation of a state at each observation time. The problem of error definitions and error processing will be discussed as follows from the perspective of the DA method (Li and Bai, 2010).

9.2.2 *Error definitions in sequential DA methods*

Sequential DA consists of a process whereby the system state is recursively estimated/corrected at each time when an observation becomes available. These methods include the classical Kalman filter (KF), the extended Kalman filter (EKF), and the EnKF. The application areas of those methods range from ideal linear systems to nonlinear systems, and the error assumptions for applying these methods change from the simple Gauss error assumption to the complex non-Gaussian error with the consideration of high-efficiency problems for processing error information.

9.2.2.1 *Classical Kalman filter*

Determining the forecast covariance matrix correctly is the key factor in the classical KF. The forecast and analysis error covariance matrices are evolved with time by:

$$\mathbf{P}_k^f = \mathbf{M}_k \mathbf{P}_{k-1}^a \mathbf{M}_k^T + \mathbf{Q}_k \tag{3}$$

$$\mathbf{P}_k^a = (\mathbf{I} - \mathbf{K}_k \mathbf{H}_k)\mathbf{P}_k^f \tag{4}$$

where $\mathbf{M}_k$ is the tangent linear model of the nonlinear dynamics, and $\mathbf{P}_{k-1}^a$ is the analysis error covariance matrix at time $k-1$. The model structure covariance matrix $\mathbf{Q}_k$ can be given as the following equations:

$$\omega_k = \mathbf{X}_k^t - M(\mathbf{X}_{k-1}^t) \tag{5}$$

$$\mathbf{Q}_k = < \omega_k \omega_k^T > \tag{6}$$

where ω_k denotes the model errors, and $\mathbf{X}_k^t$ represents the real state value at the specific analysis time k. Similar to Li (2007), the forecast errors are composed of two parts: one is called the "internal error", which is the error in the initial state and its dynamical growth during the forecast period, i.e., $\mathbf{M}_k\mathbf{P}_{k-1}^a\mathbf{M}_k^T$; the other is the "external error" due to the model deficiencies, i.e., $\mathbf{Q}_k$. Based on the analysis of Brasseur (2010), $\mathbf{Q}_k$ includes all the errors associated with the various physical parameterizations necessary in the model (e.g., mixing, diffusion, hydrostatic approximation); the errors in the atmospheric forcings and, more generally, in the boundary conditions; and the errors due to the numerical discretization on the horizontal and vertical dimensions. Note that Q represents the model error statistics accumulated during an assimilation interval and should not be confused with errors generated at every time step. Those errors are clearly distributed over a wide spectrum of spatial scales, and it would be extremely difficult to prescribe a full Q matrix. By considering the prior misfit between the model simulations and observations, it is possible to derive some general properties of the model errors as manifested in the observation space. These difficulties promote the adoption of simplified parameterizations for Q following, for instance, the approach proposed by Dee (1995), and to adjust those parameterizations by sensitivity experiments (Brasseur, 2010).

In the classical KF, the model error is assumed to be a Gaussian white noise with a zero-mean. In a linear system, if both model and observation errors are zero-mean (i.e., unbiased) and uncorrelated in time, the KF is the best (i.e., minimum variance) linear unbiased estimator (BLUE) among any linear combinations of model forecasts and measurements. Additionally, if the errors are Gaussian, the KF is the minimum variance unbiased estimator. From the Bayesian view, the KF is both the minimum mean squared error (MMSE) estimation (i.e., conditional mean) and the maximum *a posteriori* (MAP) probability estimation (i.e., conditional mode), provided that all the assumptions on errors as stated above are valid (Wang and Cai, 2008). However, for land surface systems, accurate simulation models are not readily available, and the statistical nature of the uncertainties impinged in those systems is rarely understood completely. Thus, the limitations of applying KF and its extensions are exposed around the nature of the errors. Another limitation of KF is that the statistics of errors must be known in advance, but the difficulty in obtaining such measurements (Vrugt *et al.*, 2005) and quantifying the model error statistics is always a challenge, as argued by Reichle *et al.* (2002): "Any DA approach that provides for model error faces the serious challenge of determining the true model error covariance matrix in operational applications (Wang and Cai, 2008)."

9.2.2.2 *Extended Kalman filter*

For nonlinear models, the extended Kalman filter (EKF) approximation uses the nonlinear model in the forecast step to advance the background state, but the covariance is advanced using the model liberalized around the trajectory and its adjoint. The cost of advancing the background error covariance with the linear tangent and adjoint models makes the EKF computationally unfeasible for any land models of realistic size without major simplifications. Therefore, these methods have not been widely used in high-dimensional land DA systems. Because the calculation of the error covariance in the EKF for large-dimensional systems is computationally demanding, computationally efficient forms of the EKF for complex models, such as the Reduced Rank Square Root (RRSQRT) Kalman filter and the Singular Evolutive Extended Kalman (SEEK) filter, have been made available. All the above filters are approximations of the optimal Bayesian solution, including the particle filters, which will be introduced below.

9.2.2.3 *Ensemble Kalman filter*

The general formulation of the Ensemble Kalman filter (EnKF) was initially proposed by Evensen (1994) and Burgers *et al.* (1998), and it has been widely used as a filter algorithm in the field of sequential DA. The key principle of the EnKF is to use Monte Carlo methods to realize the ensemble forecast, ensemble filtering, and calculations of the error matrix. On one hand, the EnKF corrects the shortcomings of the EKF, which requires a higher computational cost to calculate the error matrix; on the other hand, it avoids the linearization of the model and retains all dynamic characteristics of the model. A detailed introduction can be found in Evensen (2003, 2007).

In EnKF, the "internal error" can be represented by an ensemble integral. The ensemble covariance matrix is defined as:

$$\mathbf{P}_e^f(t_k) = \frac{1}{N-1}\sum_{i=1}^{N}[\mathbf{X}_i^f(t_k) - \overline{\mathbf{X}^f(t_k)}][\mathbf{X}_i^f(t_k) - \overline{\mathbf{X}^f(t_k)}]^T \quad \mathbf{P} \in \Re^{n\times n} \tag{7}$$

where $\mathbf{X}^f(t_k) = \frac{1}{N}\sum_{i=1}^{N}\mathbf{X}_i^f(t_k)$

Theoretically speaking, if the designed ensemble could meet the all possible realizations of all real geographical states, $\mathbf{P}_e^f$ is the best estimation of the "true" but unknown forecast error covariance. However, determining the error covariances matrix for the model fields could be an even greater challenge. According to Reichle (2008), $\mathbf{P}_e^f$ is typically based on the following: (i) comparisons of model simulations with observations outside the DA system; (ii) comparisons of model forecasts with different lead times at the same verifying time; (iii) analysis of the assimilation increments or innovations; or (iv) ensemble integrations.

Because the model error covariance Q has to be specified, the whole forecast error covariance matrix should be denoted as:

$$\mathbf{P}^f = \mathbf{P}_e^f + \mathbf{Q} \tag{8}$$

The model structure error covariance in Eq. (8) should be explicitly expressed during the DA process. Because little is actually known about model error statistics in complex systems, a crude way to account for model error is to add additional noise with a zero-mean and a prescribed covariance Q to each forecast ensemble member. However, in complex Earth surface systems, it is difficult to determine Q. For this reason, some model error parameterization methods have been developed in the atmospheric field, such as multiplicative inflation methods, additive inflation methods, and the "relaxation-to-prior" methods. The related problem will be deeply explored in Sec. 9.4.

9.2.2.4 *Nonlinear and non-Gaussian methods*

The Kalman filter and its variants have dominated this field for decades, but they are limited by the linear Gaussian assumption. To better handle the shortcomings of the KF methods, in recent years, several Bayesian filtering algorithms have been proposed and successfully applied to the nonlinear and non-Gaussian processes. Particle filtering (PF) is one of the commonly used DA algorithms for the recursive estimation of model states. The algorithm is also known as bootstrap filtering, the condensation algorithm, sequential Monte Carlo (SMC) sampling, interacting particle approximations, and survival of the fittest (Arulampalam *et al.*, 2002). The particle filter is a typical Bayes filter method based on Monte Carlo simulation. The key of the algorithm is Sequential Importance Sampling (SIS), where the main idea is to represent the required posterior possibility distributions by a set of random samples with associated weights and to compute the estimation based on these samples and weights. Compared with other filter algorithms, the advantages of the particle filter are that it can be used without the constraint of the model's linearity and Gauss assumption and it is suitable for any nonlinear/non-Gaussian dynamic systems (Han and Li, 2008a). There is no need to calculate the invert matrix; as a result, the error propagation is easy to realize. Moreover, PF methods are more sensitive for the tail part of error distribution than EnKF methods because the full probability distribution is used to represent model errors. This method has proved to be more effective than EnKF for special cases of land surface process research (e.g., flood forecasting). In Qin *et al.* (2009), a variant of the particle filtering method is developed for the joint estimation of soil moisture and soil parameters such as texture and porosity. However, the applications of PF methods require more ensemble members than EnKF methods to obtain more reliable model error estimations. The result is a large computational burden and difficulty for real application in real time.

9.2.3 *Error definitions in variational DA methods*

From a signal processing point of view, DA methods can be classified into two major categories: continuous DA and sequential DA (Daley, 1991; Li and Bai, 2010). Continuous DA merges the model outputs with all observations in an assimilation window by iteratively and constantly adjusting the model initiation conditions using optimization algorithms. Continuous DA can be further classified into variational methods and Monte Carlo global optimization methods. The former, represented by 3DVar and 4DVar, use variational methods to solve objective functions, and the so-called adjoint model of the numerical model is needed to develop during the realization (Cou 1994; Courtier, 1997).

In the standard formulation of 4DVar (e.g., Courtier, 1997), the analysis is obtained by minimizing a cost function:

$$J(x_0) = \frac{1}{2}(x_0 - x_0^b)^T \mathbf{B}^{-1}(x_0 - x_0^b) + \frac{1}{2}\sum_{k=0}^{N}[H_k(x_k) - y_k]^T \mathbf{R}_k^{-1}[H_k(x_k) - y_k] \qquad (9)$$

Computed over an assimilation window of length $t_N - t_0$, where x_0^b is the background or first guess at y_k; $\mathbf{R}_k$ is the vector of observations made at time t_k and its corresponding observation error covariance; $\mathbf{B}$ is the background error covariance; and $x_k = M_k(x_0)$ is the model state at the observation time obtained by integrating the nonlinear model.

With the application of the VAR methods, there is only one time prior error estimation for model covariance error in the whole 3DVar analysis process. Usually, the homogenous and orthogonal assumptions are applied for using 3DVar, which are far from the truth of the real atmospheric conditions or the land surface situation. Thus, the assimilation results are far from better. Several centers (ECMWF, France, UK, Japan, and Canada) have switched to 4DVar, which requires the development and maintenance of an adjoint model. Although the computation requirement is higher than 3DVar, it has been proved to be significantly more accurate than 3DVar in preoperational tests leading to their implementation. In addition to its demonstrated higher accuracy, 4DVar was developed and implemented because it allows the assimilation of synoptic data, such as satellite radiances at their correct observation time, and because further improvements, such as weak constraint formulations, can be incorporated later. One of the potentially promising extensions of 4DVar is the use of reduced-rank Kalman filters to estimate the analysis error covariance, but tests in a high-resolution NWP system have shown no significant benefit (Bannister, 2008). Bannister (2008) reviewed the characteristics of forecast error statistics in meteorological DA from the substantial literature on this subject. It was shown how forecast error statistics appear in the DA problem through the background error covariance matrix, **B**. This review presents the status of the latest research of forecast error covariance statistics in atmospheric variational DA.

9.3 Error Estimation Issues

Error estimation research is a key part of modern DA research. As mentioned before, if the input error covariances for the model and the observations are known in some cases, minimizing the objective function is a fairly technical aspect that can be accomplished with existing engineering solutions. Therefore, determining how to reduce uncertainties in the forecasting part is the main objective in land surface process research. Based on the error definitions and characteristics discussed in the last section, several error estimation methods will be reviewed as follows.

9.3.1 *Model error*

Many complex land surface parameterization schemes have been developed for existing global and regional climate and land process models, but the accuracy of the model simulations is still low. For instance, in the global soil wetness project (GSWP), 10 representative LSMs, including the Simple Biosphere Model 2 (SiB2), the Simplified Simple Biosphere Model (SSiB), and the Biosphere-Atmosphere Transfer Scheme (BATS), were applied to simulate the global distribution of soil moisture in 1987 and 1988. However, it has been concluded that none of these models can simulate soil moisture accurately in any place for two reasons. On one hand, the existing models are far from perfect due to the errors in the physical process formulations and parameterization schemes. On the other hand, it is difficult to determine the initial conditions of model state and the parameters of soil and vegetation in certain regions. Therefore, the model errors of land DA systems could be clarified as the following parts.

9.3.1.1 *Input model error estimation*

The errors in the input part of DA systems include forcing data errors and initial field errors. To produce reliable forecasts, land surfaces models must be initialized with conditions that represent the actual state of the Earth surface as accurately as possible. Due to the properties of land dynamics, the forecast range cannot be extended beyond the limit of predictability of the system, and the model has to be reinitialized intermittently by correcting the forecast with the most recent observations. Forcing data errors have a great impact on the assimilation results. For example, the spatial and temporal variants of precipitation rainfall data can contradict the stable fixed observation site (Goodrich *et al.*, 1995). Rainfall data is the main source of the input model error. Meanwhile, the error sources of forcing data are not at all clear. The spatial variance situation, the watershed characteristics, and the formal facing data conditions are all uncertain for the model input.

A reasonable method for estimating the input model error is to use random variables as the input information. In the early stage of land DA research, all error items were handled using disturbed ensembles, in which a random number is added to each forecast or analysis ensemble to enlarge the diversity of the ensemble. This method not only introduces non-additive noise but also maintains the diverse distributions of the ensembles. Additionally, knowledge of geostatistical theory has played an important role in the enhanced performance of the assimilation system (Han and Li, 2008b).

9.3.1.2 *Model parameter error estimation*

To some extent, the accuracy of model parameters determines the accuracy of the model forecast. For this reason, it is important to estimate the model parameter error correctly. In LSMs, there are two kinds of model parameters: some can be derived from dynamic equations with the explicated physical meaning; and the others are parameters with experimental concepts, which can only obtained by experience. Therefore, the model parameter errors have different characteristics from case to case. The sensitivity study of the model parameters is necessary. For example, the variance-based sensitivity analysis method can be applied to determine the global sensitivity of the parameters, which is briefly introduced in Helton *et al.* (2006).

Moreover, Beven and Binley (1992) introduced the generalized likelihood uncertainty estimation (GLUE) methodology, where multiple competing model structures and parameter sets are allowed to account for the possibility of the equifinality of models, producing a likelihood-weighted probability distribution of output predictions. GLUE methods represent recent advances in the field of parameter error estimation. However, the high computation requirements and the hard selection needed for the parameter set dimension are still problems for future research.

9.3.1.3 *Model structural error*

Structural, parameter, and input errors collectively lead to uncertainties in land predictions of model outputs and states. Among these three types of errors, structural errors are generally the most poorly understood and the most difficult to cope with; nevertheless, their impacts on land predictions can be far more detrimental than those of parameter errors and data errors (Abramowitz *et al.*, 2006) for the following reasons.

(1) Our knowledge of the real physical processes is always poor. The uncertainty of the model is the more general and important issue in land DA systems. Normally, all mathematical models of land DA systems are abstractions and approximations of a complex physical environment. The designs of these models are not completely derived from rigorous natural laws, and hence, the mathematical descriptions of the biological processes are not universal. Thus, we are uncertain not only about the values of the numerous model parameters but also the model parameterizations and errors.
(2) The more recent interest in Earth sciences (numerical weather prediction or oceanography) has raised new issues related to the enormous number of degrees of freedom taken into account by the models, with consequences for the size of the discretized operators and the quantity of information to be manipulated.

(3) Model errors were initially small in scale. As the time span of the forecast over which model errors are computed increased, the model errors grew in size and scale.

LSMs are generally one-dimensional, so error correlation matrices are usually set in an empirical way. Even if the spatial correlation is taken into account, the traditional spatial correlation functions in atmosphere data analysis are used as the primary means. With the development of land DA systems (LDAS), especially the application of distributed hydrological models in LDAS, the spatial correlation of errors in land surface state variables and remote sensing observations will consequentially be given more attention.

Ultimately, the specification of model structural error remains largely subjective.

9.3.2 *Observation error*

A measurement error usually consists of three components: observation operator error, representativeness errors, and instrument errors. Instrument errors are caused by imperfect measurement devices that do not accurately record the variables they are designed to measure. Representativeness errors come from scale incompatibility or differences (in time or space) between the variable measured by a device and the corresponding model variable. Representativeness errors can be discussed in terms of spacing (distance or interval between samples), extent (overall coverage of measurements in space or time), and support (averaging volume or area of samples) (Liu and Gupta, 2007). The two error components tend to have very different characteristics, which may vary from one variable to another. To effectively quantify or reduce uncertainty in the predictions, the statistics of both errors should be considered and adequately specified.

As a typical source of land surface observation, remote sensing itself cannot provide a continuous spatiotemporal evolution of land surface states for four reasons.

(1) The remote sensing observation is instantaneous, but the land surface process is continuous in both space and time.
(2) Remote sensing is an indirect measurement because the relationship between remote sensing observations and land surface variables is implicit. In general, we can establish a forward model of the relationship between land surface variables and satellite observations, such as

brightness temperatures; however, the number of observations is usually less than that of land surface variables, and the forward models are often complex and nonlinear. These factors make the inversion difficult, and even an ill-posed problem. Therefore, *a priori* knowledge should be used to improve the possibility and accuracy of inversion. The LSM, as a physical constraint, can provide *a priori* knowledge for remote sensing inversions.

(3) Most remote sensors cannot detect subsurface information. Microwave-based sensors can only sense the top few centimeters of wet soil. However, for hydrological predictions, the soil moisture in the root zone and deeper soil layer is just as important.

(4) There are errors in remote sensing observations, including instrumental errors and representative errors caused by instrument inaccuracy and surface heterogeneity, respectively.

9.3.3 *Algorithm errors in ensemble DA*

Normally, algorithm errors include the sampling errors using Monte Carlo methods and the computation errors in all kinds of numerical methods.

Sampling error is the main source of errors in ensemble DA, and the reason for this kind of error is the ensemble numbers chosen in the process. As many investigations have shown, the increase of the ensemble number will reduce the sampling error but with an increase of the computation cost. The error-handling methods introduced in Sec. 9.4 are meant to improve the quality of assimilation without increasing the ensemble numbers.

9.3.4 *Summary for error estimation*

All errors mentioned above will be present simultaneously during the DA process, so it is extremely difficult to quantify those errors and their related values individually. In ensemble DA, the "net" effect of all errors will cause a small bias in the error covariance matrix. Meanwhile, the new character space is created in the forecast error. To better handle the error problems, several kinds of methods are imported.

9.4 *Error Handling Methods* in Ensemble DA

As discussed earlier, it is difficult to exactly estimate model error or observation error in real DA systems, but a series of *error-handling methods* have been developed in modern DA methods. First, most of these methods do not estimate the prior model error or observation error directly. Instead, they are used to reduce the overall influence of the errors on the DA systems. For example, (1) in ensemble DA, "inflation" methods could be used to reduce all sorts of errors, including multiplicative inflation to eliminate the error in the ensemble second moment, additive inflation, and "relaxation-to-prior". By tuning the ensemble covariance, these methods can achieve better assimilation results with a small ensemble size. This method has been used in land DA systems. (2) To deal with model bias, Baek *et al.* (2006) proposed a scheme for the EnKF to obtain the optimal estimation of the state and model bias simultaneously based on the state space augmentation method (Jazwinski 1970; Dee and da Silva 1998). In this method, both the states and the model biases have ensemble members. For each member, the state vector is augmented with the uncertain model bias vector. The bias is updated by state observation through the cross-correlation between the forecast state and the bias. Li *et al.* (2009) tested all the common methods in an atmospheric model named Speedy, in which the assimilation results were compared in different ways. In Sec. 9.4.1, the traditional error-handling methods and evolutionary algorithm-based error parameterization methods proposed by Bai and Li (2011) will be introduced briefly. The corresponding experiments are designed to test and compare our results with the former research works (Evensen 2007; Kalnay *et al.*, 2007). For detailed information and comparisons regarding these methods, please refer to Bai and Li (2011).

9.4.1 *Multiplicative inflation methods*

Multiplicative inflation was proposed by Anderson and Anderson (1999). This method can ameliorate the sample error caused by a small ensemble number with the inflation of the forecast covariance without changing the ensemble mean. Before the first observation is assimilated, the ensemble member's covariance will be enlarged by a constant r that is slightly larger than zero, as follows.

$$\tilde{P}_e^f \leftarrow (1+r) \times P_e^f = P_e^f + rP_e^f = P_e^f + Q_{\mathrm{multi}} \tag{10}$$

The symbol "$\leftarrow$" means the replacement of the former value. From the above formula, we can see that Q_{multi} denotes the model error in the multiplicative inflation methods. Because Q_{multi} and P_e^f have the same error subspace, the DA system will be more stable, and the performance of the filter will be improved (Ott *et al.*, 2004).

9.4.2 *Additive inflation methods*

Compared with multiplicative inflation, the additive inflation method is a more accurate model error parameterization (Hamill and Whitaker, 2005). The application of this approach to EnKF involves the addition of random perturbations to each ensemble member, which can be sampled from a distribution with known covariance statistics. Before the update cycle starts, the noise r^* with the same dimension as the model states is added to each ensemble member background forecast:

$$\tilde{x}_n^f = x_n^f + r^* \tag{11}$$

Moreover, the model forecast error covariance matrix will turn into

$$\tilde{P}_e^f = \frac{1}{N-1}\sum_{n=1}^{N}(\bar{\tilde{x}}_{e(n)}^f - \bar{\tilde{x}}_e^{f-})(\bar{\tilde{x}}_{e(n)}^f - \bar{\tilde{x}}_e^f)^T = P_e^f + Q_{\text{add}}, \tag{12}$$

where N is the number of ensemble members and Q_{add} is the model error simulated by this method.

9.4.3 *The relaxation-to-prior method*

The relaxation-to-prior method was proposed by Zhang *et al.* (2004) as an alternative to the multiplicative inflation methods. The name refers to the fact that it relaxes the analysis perturbations back toward the prior perturbations independently at each analysis point via

$$x_{\text{new}}^{a'} = (1-\alpha)x^{a'} + \alpha x^{f'}, \tag{13}$$

where $x_{\text{new}}^{a'}$ is the new analysis perturbation, $x^{a'}$ represents the original perturbations, and $x^{f'}$ is the background perturbations; α is the error adjustment factor and has a value between zero and one. The notation used here is similar to the original paper from Zhang *et al.* (2004), who further explored the potential and behavior of the EnKF at convective scales by considering more realistic initial analyzes and variations in the availability and quality of radar observations. The EnKF provides an algorithm for updating the forecast deviation $x^{f'}$, for each member to produce an analysis deviation $x^{a'}$. (Typically, $x^{a'}$ is smaller than $x^{f'}$, reflecting the reduction of uncertainty after assimilating the observations.)

Moreover, as suggested by Whitaker *et al.* (2008), more tunable parameters can be added to the parameterization to force the structures to match, or new parameterizations can be developed that more accurately reflect the structure of the underlying system error covariance. Therefore, the theories developed in the control engineering field will be implemented in the DA system. With further study of the relaxation-to-prior method, Bai and Li (2011) applied the crossover principles of genetic algorithms (GAs) to DA systems to create new error parameterization methods. Via a comparison with the traditional methods, the relaxation-to-prior method will serve as a bridge between the new methods and the former methods.

9.4.4 *Evolutionary algorithm-based error parameterization methods*

Motivated by the relaxation-to-prior method (Zhang *et al.*, 2004), Bai and Li (2011) proposed a new method to generate the ensemble perturbation based on the crossover principles originally developed in intelligent computing research. The essence of this method is to inflate the covariance matrix by generating proper ensemble perturbation and to obtain the appropriate weights for the innovation in EnKF, thereby effectively handling filter divergence. Combined with the factor search frameworks, the best error adjustment factors could be found to guarantee the best DA performance in the corresponding circumstance.

Based on the crossover principle of a GA, a new method is developed to deal with the model error within the framework of sequential DA. The parent populations will be chosen as each column of the forecast ensemble perturbation matrix and each column of the analysis ensemble perturbations matrix after the one-step assimilation. More specifically, one parent comes from the forecast ensemble perturbation matrix and another comes from the analysis ensemble perturbations matrix. The number of parents is the same as the number of ensembles. Figure 9.2 indicates the principle behind this new method. The traditional sequential DA only propagates forward based on the analysis ensemble after each assimilation step, while the new method will use the crossover principle to find the best crossover position, where the background parent's population and the analysis parent's population will have the evolutionary action to generate new offspring. The offspring will take the place of the original analysis ensemble to enter the next cycle of assimilation as the background information. In a sense, this method could be considered an alternative to the relaxation-to-prior method, with a more clearly physical background. Moreover, different crossover principles that have been developed in the field of intelligent computing will be used in the DA system.

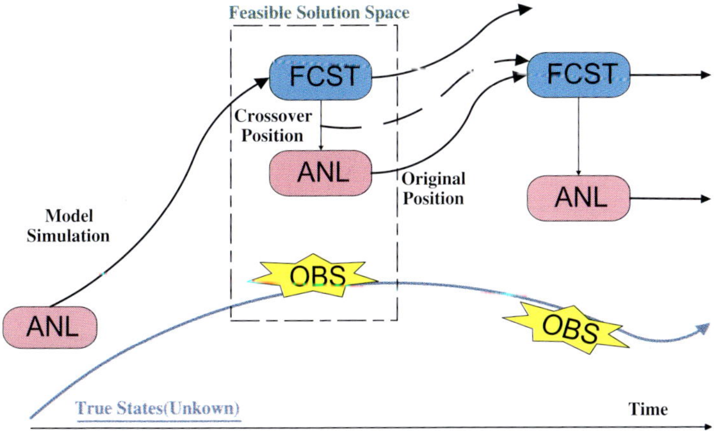

Fig. 9.2. Applications of crossover principles in the DA systems are shown. "ANL" indicates the elements of the analysis ensemble perturbations, "FCST" represents the elements of the forecast ensemble perturbations, and "OBS" represents the observation. "Crossover position" indicates the offspring solutions obtained after the evolutionary algorithms. "Original position" indicates the normal DA results without the evolutionary operation.

In a real application, the method is applied to the posterior ensemble after the computation of the analysis increment and before running the forecasts to be used as the first guess for the next state update. This step is performed to more easily accommodate the time interpolation in the forward operator because time interpolation requires the use of the first-guess ensemble at several forecast times. If the system error parameterization were applied to the prior ensemble instead of the posterior ensemble, the parameterization would have to be applied at each forecast time used in the time interpolation. The application of this parameterization to the posterior ensemble is justified by the interpretation of the model error as a system error reflecting an accumulation of errors arising from several components of the DA system and not the forecast model alone (Houtekamer and Mitchell 2005; Anderson, 2007).

From the EnKF point of view, the DA system with the crossover principle will relax the analysis perturbations back toward the prior perturbations independently at each analysis point. Figure 9.3 shows the schematic of this method with the concepts of the ensemble. At the time of $t1$, under the premise that the ensemble mean will remain the same after each assimilation step, the crossover operation will be taken between the original background perturbations and the analysis perturbations. New

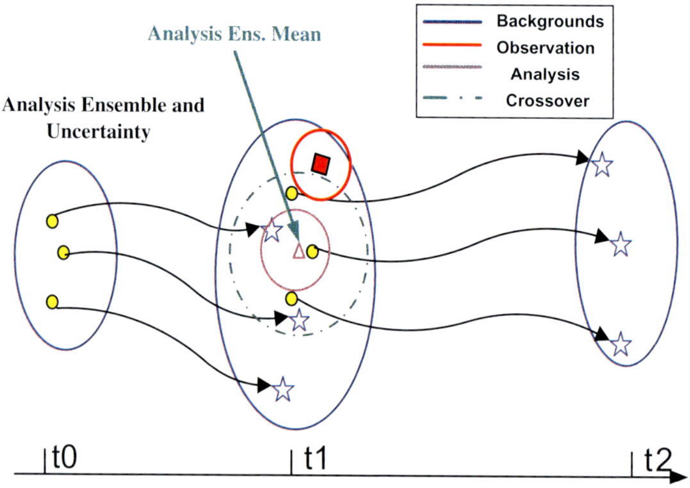

Fig. 9.3. Schematic diagram of the crossover methods shown according to ensembles. "Analysis Ensemble mean" is the mean value after the assimilation operation, which will not change during the evolution. "Crossover" indicates the ensemble perturbations after the evolution algorithm.

offspring after the crossover will be added with the original ensemble mean to be used as the new ensemble to enter the next assimilation cycle. This method not only imports the "survival of the fittest" principle of the evolutionary algorithm but also includes the feedback mechanism of control theory through the crossover factor. During the offline search process, the best combination of the error factors could be found, and the optimal assimilation results could consequently be obtained.

9.4.5 *Experiments designed with crossover error parameterization methods*

Error parameterization methods with the crossover principle take each ensemble of the EnKF as the individual in the evolutionary algorithms, while the integrations of the ensemble are thought of as the evolution of the individual. During parameterization, each background ensemble perturbation and analysis perturbation are considered the parent generations, and the mean of the analysis is kept the same. With the different crossover principles, the offspring after the crossover become the final analysis ensemble perturbation prior to entering the next assimilation step. Because there are different crossover factors to control the position and the performance of the crossover in each crossover method, the import of the crossover principle not only optimizes the ensemble population but also introduces the feedback mechanism to the ensemble DA. In the process of choosing the crossover error factor, a new DA system coupled with a fast GA is designed. To set up the unified fitness function, the best error adjustment can be obtained and the optimal accuracy of the DA output can be achieved. Further details about the theory of crossover principles and the algorithm of the Ensemble Transform Kalman Filter (ETKF) can be found in Bai and Li (2011).

9.4.5.1 *Error factor search methods coupled with the GA*

Genetic algorithm (GA) is an evolutionary computation technique used for searching and optimization that is inspired by biological evolution (Holland, 1992). GA can be used to find near-optimal solutions to highly nonlinear optimization problems, and it has received considerable attention (Lee *et al.*, 2006; Ma *et al.*, 2009). Compared with traditional optimization methods, GA is more appropriate when the function includes some complexities and/or discontinuities (Barth, 1992). GA consists of three basic operations: selection, crossover, and mutation. In this algorithm, each estimated parameter is encoded as a string of binary or real numbers called a chromosome. Each possible combination of estimated parameters is thus represented by a chromosome. These chromosomes are evaluated on their performance (or "fitness") with respect to the objective function, which minimizes the discrepancies between the modeled values and the observed data. Using the fitness value, the chromosomes compete in a selection tournament, where chromosomes having high fitness values enter the mating population and chromosomes with low fitness values are killed off. The surviving chromosomes are randomly assigned a mating partner from within the mating population, and a random crossover location on the chromosomes is selected (Whitley, 2001; Ma *et al.*, 2009).

Similar to the DA system proposed by Wang and Cai (2007), Fig. 9.4 shows the schematic diagram coupled with the GA. Described by the language of control theory, the system is a complex system with feed-forward systems and feedback systems. The feed-forward system involves DA systems, while GAs are taken as the offline searching method. The initial state of the system operation is the same as that of the normal DA systems. In accordance with the set-up procedure for the GA, error adjustment factors are set up randomly within the searching range as the initial population. In each time step of sequential data assimilation, individuals of the population will be evaluated by fitness functions.

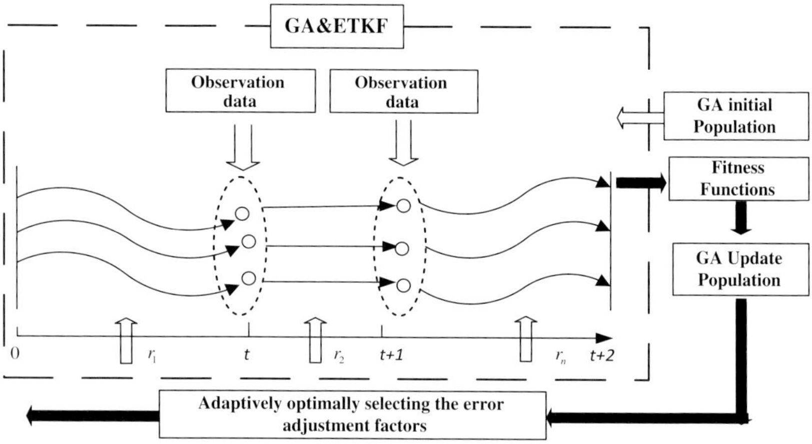

Fig. 9.4. Schematic diagram of the DA system coupled with the GA. The feed-forward system shows the data assimilation system, while the GAs will be taken as out-loop offline searching methods.

With the help of the evaluation, the new population will be generated by crossover and mutation operations. The purpose of the crossover and mutation in this part is to find the diversity of the individuals in the colony, avoid local solutions, and obtain the optimal error adjustment factors, which will update the parameters in the error parameterization methods. When the new observation is available, the forecast model state and the error covariance will be updated synchronously. After several iterations, the best individual will be kept as the best value for the DA systems.

9.4.5.2 *Experiment model*

The Lorenz model consists of a system of three coupled and nonlinear ordinary differential equations (Lorenz, 1963) as follows:

$$\frac{dx}{dt} = \sigma(y - x) + q^x$$
$$\frac{dy}{dt} = rx - y - xz + q^y$$
$$\frac{dz}{dt} = xy - bz + q^z \tag{14}$$

where $x(t)$, $y(t)$, and $z(t)$ are the dependent variables. We have chosen the following commonly used values for the parameters in the equation: $\sigma = 10$, $r = 28$, $b = 3/8$. The terms q^x, q^y, q^z are assumed to represent the unknown model errors. In Sec. 9.4.4.4, these three terms are set to zero to compare the results with those of Kalnay *et al.* (2007).

9.4.5.3 *The fitness function*

A fitness function is a method to evaluate the performance of each individual in a population in the GA; it is the only way to realize the selection step in the algorithm. For the Lorenz model in this paper, the fitness function is the mean square error. The target of the search is the minimum of the fitness function. The background mean square error fitness function ($\mathrm{RMSE_b}$) and the analysis ($\mathrm{RMSE_a}$) function are defined as the follows, respectively:

$$\mathrm{RMSE_a} = \sqrt{\frac{1}{N}\left(\sum_{i=1}^{N}(x_{\mathrm{ana},i} - x_{\mathrm{sim},i})^2\right)} \tag{15}$$

$$\mathrm{RMSE_b} = \sqrt{\frac{1}{N}\left(\sum_{i=1}^{N}(x_{\mathrm{for},i} - x_{\mathrm{sim},i})^2\right)} \tag{16}$$

where N is the number of the ensemble, t is the step subscriber at each time, $x_{\mathrm{for},i}$ is the forecast at each time, $x_{\mathrm{sim},i}$ is the real value and $x_{\mathrm{ana},i}$ is the analysis value.

9.4.5.4 *Numerical experiments*

At present, the error estimation methods in DA systems are mainly concerned with theoretical research, while the classical nonlinear Lorenz model is often taken as the test bed. We will choose the ETKF as the basic DA algorithm to realize the above methods. With regard to the problems for the best error factor search methods, the fast GA is selected as the search algorithm with the mean square errors as the fitness function. For the different error parameterization methods discussed above, the resulting best combinations of the error adjustment factors will be found with the GA. Meanwhile, the DA algorithm will propagate forward with the observational updating. With the adaptive updating of the error adjustment factors, the procedure will stop when the GAs stoppage conditions are met.

(a) Traditional method experiments

In the first part of the experiment, the three error parameterization methods introduced in Sec. 9.1 will be tested with the Lorenz model as the

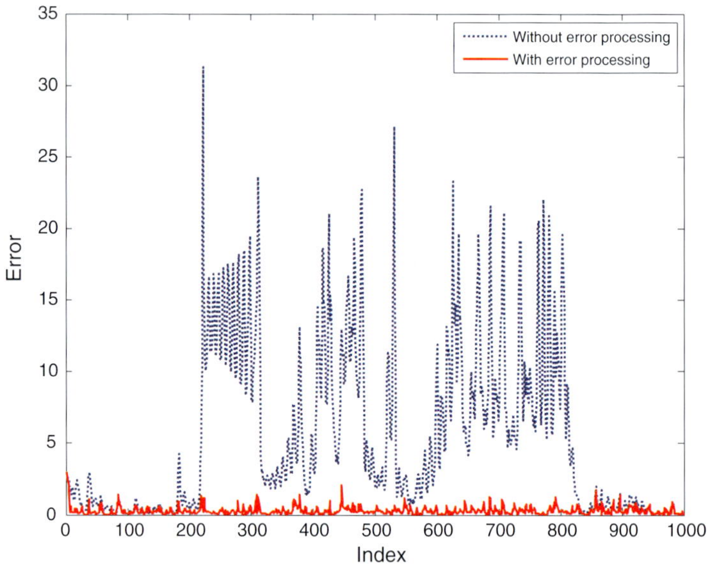

Fig. 9.5. The results of the comparisons between the methods using error parameterization and not using error parameterization.

traditional method. Comparisons between the methods with and without error parameterization are investigated. Then, the error adjustment factors are studied, and GAs are used to search for the best value based on the methods discussed above. Lastly, a new blending method is proposed to combine the advantages of the different methods.

In light of the Lorenz model data assimilation system, the initial condition is set as $x_0 = [8.0\ 0.0\ 30.0]$; the time increment is $dt = 0.01$; the observation window is $bst = 8$; and the observation error is $R_0 = [2\ 2\ 2]$. To show the performance of the error parameterization methods clearly, the number of ensembles is set as $K = 3$ for all the experiments, except in some special cases. Figure 9.5 indicates the results of the comparisons. It can be clearly seen that the errors of each step during the assimilation are larger without than with the error parameterization. The largest amplitude is approximately 30, and it shows heterogeneity along the time steps. Meanwhile, the model output and the assimilation results seriously depart from one another. With multiplicative inflation methods to address the model error, the error amplitude after the assimilation is greatly improved. The largest amplitude is smaller than three, and the mean square error is reduced to 0.381.

Table 9.1. The assimilation results with different multiplicative inflation factors $e1$.

Inflation factor $e1$	0.557	1.0	1.018	1.02	1.03	1.04	1.08	1.18	1.28
$RMSE_b$	10.600	2.176	0.301	4.766	0.458	0.379	0.407	0.509	0.661
$RMSE_a$	10.597	2.053	0.288	4.451	0.381	0.315	0.335	0.410	0.492

The experiments show that the multiplicative inflation factor can significantly affect the accuracy of the data assimilation systems. Table 9.1 lists the different assimilation results with the different multiplicative inflation factors.

Figure 9.6 shows the variation of the RMSE within different ranges. The results indicate that the multiplicative inflation factor is slightly larger than one if we take the RMSE as the optimization index.

The following four characteristics are found. (1) When the multiplicative inflation factor is smaller than one, the RMSE of the DA systems significantly increases, which means that the performance of the EnKF will rapidly deteriorate when the ensemble's discrete covariance values are reduced. (2) When the multiplicative inflation factor is larger than one, the RMSE significantly decreases. A better filter performance can be obtained within the best range of 1.01~1.2. Beyond this range, the optimization is no longer obvious. These results are similar to those of Kalnay *et al.* (2007). (3) Within the best numerical range, the target destination functions change rapidly, with obvious multi-hill and multi-valley phenomena, which can be taken as evidence that the target functions are convex functions. (4) At each assimilation step, the analyzed RMSE value is always smaller than the background RMSE, which means that the data assimilation process progresses with step-by-step improvement. All the results obtained here could be compared with those of Kalnay *et al.* (2007) to confirm their accuracies.

Because of the multi-hill characteristic shown by the error adjustment factor, a rapid GA is employed here as the adaptive searching method (Ma *et al.*, 2009). In the experiments, the RMSE functions are selected as the fitness function, and the initial population of the GA is composed of multiplicative inflation factors within the feasible space. DA systems take this initial value combined with models and assimilation algorithms for filtering. After the one-step sequential assimilation, the fitness functions will feedback the information to help the GA generate the next generations until the end conditions are met. The parameters of the GA are set as the following: the population is pop = 100; the largest iteration is ite = 50; the

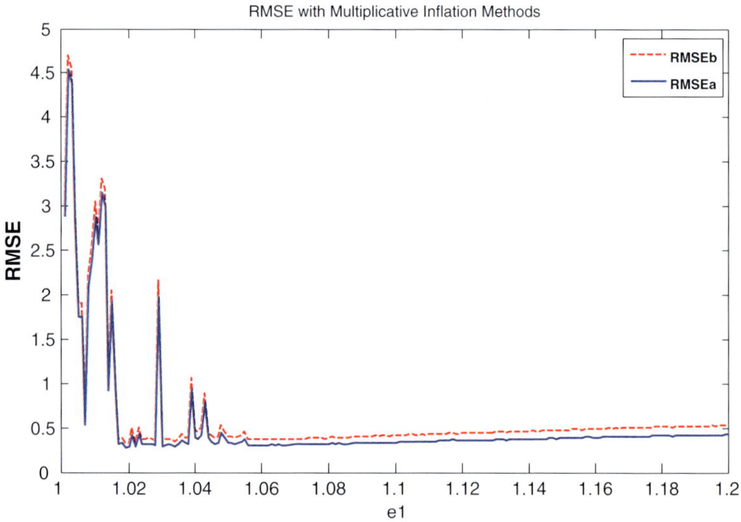

(a) Inflation factor in [1, 1.2]

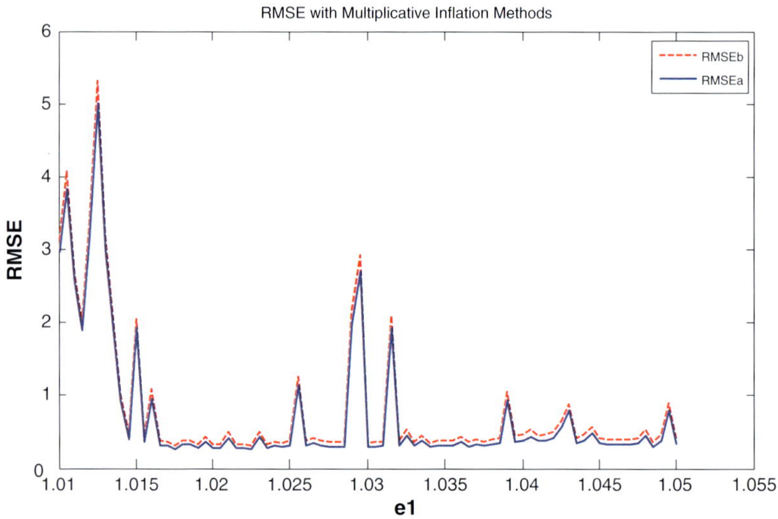

(b) Inflation factor in [1.01, 1.055]

Fig. 9.6. (a) The variation of RMSE within the range of the inflation factor in [1, 1.2]. (b) The variation of RMSE within the range of the inflation factor in [1.01, 1.055].

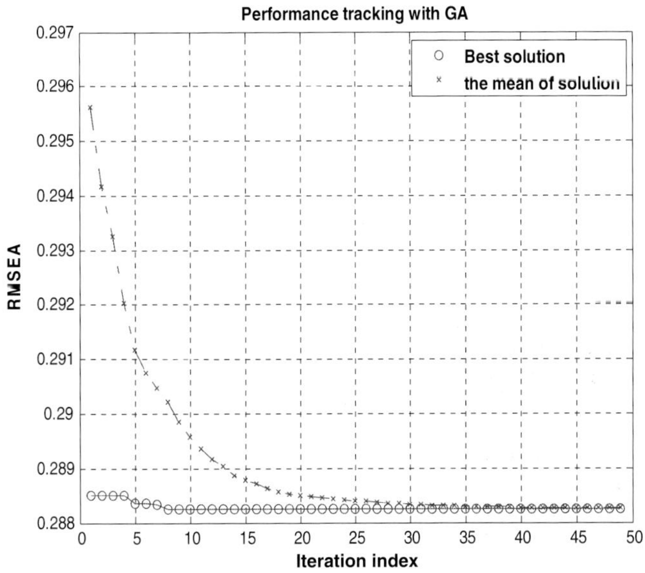

Fig. 9.7. The results of the GA search. The best solution can be obtained after 10 iterations.

variable dimension is $d = 1$; the generation gap is $\mathrm{GA} = 0.9$; the crossover probability is $pc = 0.8$; and the mutation probability is $pb = 0.3$.

Figure 9.7 shows the results of the GA search. After 10 iterations, the best inflation factor for this case was found with $e1 = 1.018$, which coincides with the results of Table 9.1.

The other two traditional error parameterization methods, such as the additive inflation method and the relaxation-to-prior method, were tested according to the same procedure as the above methods, and the optimal error factors were found by GAs with the constraint of the fitness function.

(b) Experiments with the crossover principle

To use DA algorithms coupled with crossover principals as the basic algorithm, the convex crossover, the affine crossover, the blending crossover, and the direction-based crossover deal with the errors in the second moment of the ensembles. From the EnKF point of view, the application progress of the crossover is investigated first. In light of the normal DA system, the feedback mechanism for the control theory is introduced with a GA as the feedback operator.

As discussed above, if we choose the background ensemble perturbations and the analysis ensemble perturbations after one-step assimilation as the crossover population, the different crossover algorithms will determine the position of the final solutions within the scope of the feasible space. With the constraint of the fitness function, the best error factors for different methods can be obtained offline. To choose the ensemble number $N = 10$, Fig. 9.8 shows the schematic drawings of the ensemble perturbations for the convex crossover and the affine crossover. Figure 9.8(a) shows that the perturbation after the crossover is located between the background and the original analysis, which means that the convex crossover action takes a smooth action with respect to the original analysis so that the perturbation after the crossover is near the original one, but with better solutions. Figure 9.8(b) shows that the analysis perturbation after the affine crossover is located beyond the line of the background and the original analysis, which means that the affine crossover operation takes the widely searched solution as the better solution in the feasible space.

(c) Comparisons of different error parameterization methods

To show the effectiveness of all the methods with respect to data assimilation, we compared them under the same conditions. By applying the fast searching ability of the GA, the optimal error adjustment factors could be obtained for different crossover principals. The standard experimental parameters are set as the following: initial value $x_0 = [8.0\ 0.0\ 30.0]$, time increment $dt = 0.01$, observational windows $bst = 8$, observational error $R_0 = [2\ 2\ 2]$, and the ensemble number $K = 3$. Table 9.2 indicates the search results for different error parameterization methods.

From Table 9.2, we can observe the following: (1) the best RMSE could be obtained with the affine crossover method, but the running time is longer than that of the others; (2) the optimal error factor of the direction-based crossover is located in $e1 = 0$, which means that the crossover principle based on the directional methods is not suitable for the error process in the DA system; (3) the error parameterization with the crossover principals model is able to obtain a better assimilation result than the traditional multiplicative inflation methods, while the internal physical mechanism can be partially explained; and (4) from the location of the optimal error adjustment factor in each type of crossover experiment, the best crossover solution (crossover offspring) is located on the line between each parent's background perturbations and the analysis perturbation after the assimilation. Combined with the GAs' searching ability, the optimal

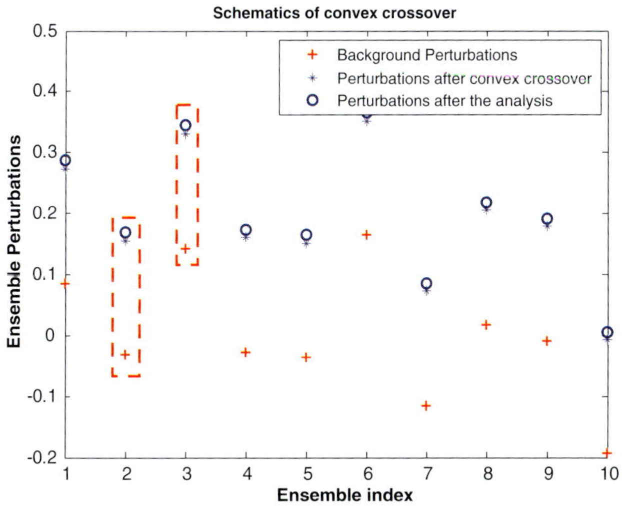

(a) Schematics diagrams of convex crossover

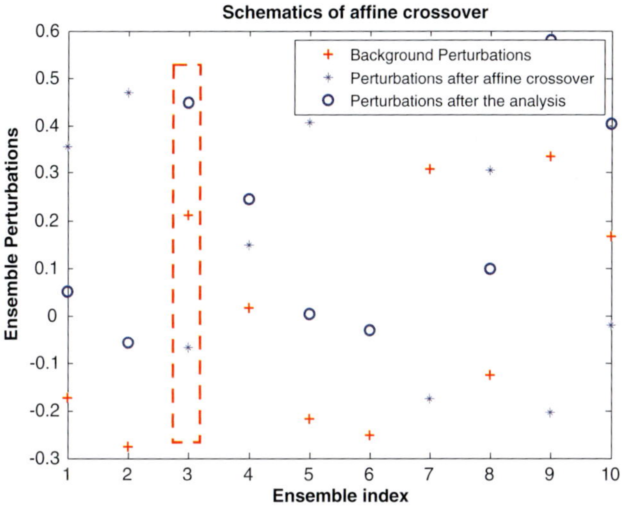

(b) Schematics diagrams of affine crossover

Fig. 9.8. To choose the ensemble number $N = 10$, the schematic diagrams of the ensemble perturbations for the convex crossover and the affine crossover operations are shown. The resulting position of the offspring coincides with the theoretical results.

Table 9.2. Search results with different error parameterization methods.

	Optimal error adjustment factor	Optimal RMSE	Running time (s)
Convex crossover experiments	$e1 = 0.67$	RMSEa = 0.6425	725.84
Affine crossover experiments	$e1 = -1$ $e2 = -0.1$	RMSEa = 0.5192	1558.64
Blending crossover experiments	$e1 = 0.2$ $e2 = 1$	RMSEa = 0.6192	1672.63
Direction-based crossover experiments	$e1 = 0$	RMSEa = 0.7556	763.24
Multiplicative inflation experiments	$e1 = 1.075$	RMSEa = 0.5862	660.77

error adjustment factor can be obtained offline. Therefore, the feedback DA systems are also available offline.

9.5 Summary and Discussions

Data assimilation combines, in an objective way, information from observations and dynamic system models (e.g., atmosphere, land, and ocean) while taking account of the errors in the observations and the models (Nichols, 2009). Beyond numerical weather prediction, DA also plays a key role in the organization and evaluation of observational and model information of various elements of the Earth system and is increasingly used to assess and prepare for future Earth observation missions (see various chapters in Lahoz *et al.*, 2009). The only practical way to observe the land surface on continental to global scales is by satellite remote sensing. However, this approach cannot provide information on the entire system, and measurements only represent a snapshot in time. LSMs can predict spatial or temporal land system variations, but these predictions are often poor due to model initialization, parameter, forcing errors, inadequate model physics, and/or resolution.

With regard to fast-developed ensemble data assimilation research, error estimation and error-handling problems are becoming a bottleneck constraining the development of data assimilation systems. Therefore, the representation, estimation, and parameterization of errors as well as controlling error propagation are fundamental to DA. One needs to consider errors in observations, background information, and models. In general, the error sources of land data assimilation systems include model errors, observation errors, and the algorithm's errors. The observation error covariance matrix is typically assumed to be diagonal, although this is not always justified. The matrix includes errors of the measurements themselves and errors of representativeness. Estimating the background error covariance matrix is a key part of the DA method, and its off-diagonal elements determine how information is spread spatially from observation locations. In most data assimilation techniques, errors in model structures are usually accounted for by adding an (unbiased) error term to the model transition equation. However, because of the equifinality of models, a full consideration of the model structure error requires involving at least some "independent" alternative model structures that encompass a range of different assumptions (Beven and Young, 2003). In this sense, a multi-model approach based on a suite of conceptual models is better suited for handling uncertainty associated with model structure errors than single-model approaches (National Research Council, 2001; Liu and Gupta, 2007). In typical data assimilation methods research, in which the Kalman filter method is used, a main focus is estimating model error Q. In DA, errors are generally assumed to be Gaussian. The most fundamental justification for assuming Gaussian errors, which is entirely pragmatic, is the relative simplicity and ease of implementation of statistical linear estimation under these conditions. Because Gaussian probability distribution functions (PDFs) are fully determined by their mean and their variance, the solution of the DA problem becomes computationally practical. Typically, there are biases between different observations and between the observations and the model. These biases vary spatially and temporally, and it is a major challenge to estimate and correct these biases. Nevertheless, and mainly for pragmatic reasons, in DA it is often assumed that the errors are unbiased. For NWP, many assimilation schemes now incorporate a bias correction, and various techniques have been developed to correct observations to remove biases; these methods are now applied to land DA (DeLannoy *et al.*, 2007a, b). Possibility distributions (more general than PDFs) have been used in the retrieval of information from satellite imagery to account for incomplete information (Verhoest *et al.*, 2007) and for non-Gaussian errors, which can occur with land surface variables.

Recently, Meng and Zhang (2007) commented on the approaches for model error quantification in ensemble forecasts. In their opinion, one popular (yet *ad hoc*) approach involves the use of different forecast models (e.g., Evans *et al.*, 2000; Krishnamurti *et al.*, 2000) or different physical parameterization schemes (e.g., Stensrud *et al.*, 2000). Other ways

Goodrich, D. C., Faures, J., Woolhiser, D. A., *et al.* (1995). Measurement and analysis of small-scale convective storm rainfall variability. *Journal of Hydrology*, **173**(4), 283–308.

Grell, G. A. and Devenyi, D. (2002). A generalized approach to parameterizing convection combining ensemble and data assimilation techniques. *Geophysical Research Letters*, **29**, 1693.

Hamill, T. M. and Whitaker, J. S. (2005). Accounting for the error due to unresolved scales in ensemble data assimilation: A comparison of different approaches. *Monthly Weather Review*, **133**, 3132–3147.

Han, X. J. and Li, X. (2008a). An evaluation of the nonlinear/non-Gaussian filters for the sequential data assimilation. *Remote Sensing of Environment*, **112**(4), 1434–1449.

Han, X. J. and Li, X. (2008b). Nonlinear filter methods and land data assimilation. *Advances in Earth Science*, **23**(8), 813–820 (in Chinese).

Hansen, J. A. (2002). Accounting for model error in ensemble based state estimation and forecasting. *Monthly Weather Review*, **130**, 2373–2391.

Helton, J. C. *et al.* (2006). Survey of sampling-based methods for uncertainty and sensitivity analysis. *Reliability Engineering and System Safety*, **91**, 1175–1209.

Holland, J. H. (1992). *Adaptation in Natural and Artificial Systems*, 2nd edn., Cambridge, MA: MIT Press.

Houtekamer, P. L., Herschel, L., Mitchell, *et al.* (2009). Model error representation in an operational ensemble Kalman filter. *Monthly Weather Review*, **137**(7), 2126–2143.

Houtekamer, P. L. and Mitchell, H. L. (2005) Ensemble Kalman filering. *Quarterly Journal of the Royal Meteorological Society*, **131**, 3269–3289.

Ide, K., Courtier, P., Ghil, M. and Lorenc, A. C. (1997). Unified notation for data assimilation: Operational, sequential and variational. *Journal of the Meteorological Society of Japan*, **75**(1B), 181–189.

Jazwinski, A. H. (1970). *Stochastic Processes and Filtering Theory*, 376 pp. New York: Academic Press.

Keppenne, C. L. and Rienecker, M. M. (2003) Assimilation of temperature into an isopycnal ocean general circulation model using a parallel ensemble Kalman filter. *Journal of Marine Systems*, **40–41**, 363–380.

Kalnay, E., Li, H., Miyoshi, T., Yang, S. C. and Ballabrera P. J. (2007). 4D-Var or ensemble Kalman filter? *Tellus A*, **59**, 758–773.

Krishnamurti, T. N., Krishtawal, C. M., Zhang, Z., LaRow, T., Bachiochi, D., Williford, E., Gadgil, S. and Surendran, S. (2000). Multimodel ensemble forecasts for weather and seasonal climate. *Journal of Climate*, **13**, 4196–4216.

Lahoz, W. A. *et al.* (2009). *Data Assimilation: Making Sense of Observations*, Lahoz, W. A., Park, S. K. and Ménard, R. (eds.), Berlin: Springer, 2010, **732**.

Lee, Y. H., Park, S. K. and Chang, D. E. (2006). Parameter estimation using the genetic algorithm and its impact on quantitative precipitation forecast. *Annales Geophosicae*, **24**, 3185–3189.

Li, H. (2007). *Local Ensemble Transform Kalman Filter with Realistic Observations*. Maryland: University of Maryland.

Li, X., Huang, C. L., Che, T., Jin, R., Wang, S. G., Wang, J. M., Gao, F., Zhang, S. W., Qiu, C. J. and Wang, C. H. (2007). Development of a Chinese land data assimilation system: Its progress and prospects. *Progress in Natural Science*, **17**(8), 881–892.

Li, H., Kalnay, E., Miyoshi, T. and Danforth, C. M. (2009). Accounting for model errors in ensemble data assimilation. *Monthly Weather Review*, **137**, 3407–3419.

Li, X. and Bai, Y. L. (2010). A Bayesian filter framework for sequential data assimilation. *Advances in Earth Science*, **25**(5), 515–523 (in Chinese)

Liang, X., Zheng, X., Zhang, S., Wu, G., Dai, Y., Li, Y. (2011). Maximum likelihood estimation of inflation factors on error covariance matrices for ensemble Kalman filter assimilation. Quarterly Journal of the Royal Meteorological Society. **138**, 263–273. doi: 10.1002/qj.912.

Liu, Y. Q. and Gupta, H. V. (2007). Uncertainty in hydrologic modeling: Toward an integrated data assimilation framework. *Water Resources Research*, **43**, W07401.

Lorenz, E. N. (1963). Deterministic nonperiodic flow. *J Atmos. Sci.*, **20**, 130–141.

Ma, Y. J., Bai, Y. L. and Jiang, Z. Y. (2009). Fast multi-objective constrained evolutionary algorithm and its convergence. *Systems Engineering–Theory & Practice*, **29**, 149-157.

McLaughlin, D. (2002). An integrated approach to hydrologic data assimilation: Interpolation, smoothing, and filtering. *Advance in Water Resources*, **25**, 1275–1286.

Meng, Z. and Zhang, F. (2007). Tests of an ensemble Kalman filter for mesoscale and regional-scale data assimilation, Part II: Imperfect model experiments. *Monthly Weather Review*, **135**, 1403–1423.

Mitchell, H. L., Houtekamer, P. L. and Pellerin, G. (2002). Ensemble size, balance, and model-error representation in an ensemble Kalman filter. *Monthly Weather Review*, **130**, 2791–2808.

National Research Council. (2001). *Conceptual Models of Flow and Transport in the Vadose Zone*, 392 pp., Washington, DC: National Academic Press.

Nichols, N. K. (2009). Mathematical concepts of data assimilation. In *Data Assimilation: Making Sense of Observations*, Eds. W. A. Lahoz, B. Khattatov and R. Ménard, Springer.

Ott, E., Hunt, B. R., Szunyogh, I., Zimin, A. V., Kostelich, E. J., Corazza, M. and Kalnay, E. (2004). A local ensemble Kalman filter for atmospheric data assimilation, *Tellus A*, **56**, 415–428.

Palmer, T. N. (2001). A nonlinear dynamical perspective on model error: A proposal for non-local stochastic-dynamic parameterization in weather and climate prediction models. *Quarterly Journal of the Royal Meteorological Society*, **127**, 279–304.

Qin, J., Liang, S., Yang, K. *et al.* (2009). Simultaneous estimation of both soil moisture and model parameters using particle filtering method through the assimilation of microwave signal. *Journal of Geophysical Research*, **114**, D15103, doi: 10.1029/2008JD011358.

Reichle, R. H., McLaughlin, D. and Entekhabi, D. (2002). Hydrologic data assimilation with the ensemble Kalman filter. *Monthly Weather Review*, **130**, 103–114.

Reichle, R. H. (2008). Data assimilation methods in the Earth Science, *Advance in Water Resources*, **31**, 1411–1418.

Stensrud, D. J., Bao, J. W. and Warner, T. T. (2000). Using initial condition and model physics perturbations in short-range ensemble simulations of mesoscale convective systems. *Monthly Weather Review*, **128**, 2077–2107.

Talagrand, O. (1997). Assimilation of observations, an introduction. *Journal of the Meteorological Society of Japan*, **75**(1B), 191–209.

Verhoest, N. E. C. *et al.* (2007). A possibilistic approach to soil moisture retrieval from ERS synthetic aperture radar backscattering under soil roughness uncertainty. *Water Resources Research*, **43**, WR005295.

Vrugt, I. A., Diks, C. G. H., Gupta, H. V., Bouten, W. and Verstraten, J. M. (2005). Improved treatment of uncertainty in hydrologic modeling: Combining the strengths of global optimization and data assimilation. *Water Resources Research*, **41**, W01017.

Walters, G. A. and Smith, D. K. (1995). Evolutionary design algorithm for optimal layout of tree networks. *Engineering Optimization*, **24**, 261–281.

Wang, D. B. and Cai, X. M. (2007). Optimal estimation of irrigation schedule — An example of quantifying human interferences to hydrologic process. *Advance in Water Resources*, **30**, 1844–1857.

Wang, D. B. and Cai, X. M. (2008). Robust data assimilation in hydrological modeling — A comparison of Kalman and H-infinity filters. *Advance in Water Resources*, **31**, 455–472.

Whitaker, J. S., Hamill, T. M., Wei, X., Song, Y. and Toth, Z. (2002). Ensemble data assimilation without perturbed observation. *Monthly Weather Review*, **130**, 1913–1924.

Whitaker, J. S., Compo, G. P., Wei, X. and Hamill, T. M. (2004). Reanalysis without radiosondes using ensemble data assimilation. *Monthly Weather Review*, **132**, 1190–1200.

Whitaker, J. S., Hamill, T. M., Wei, X., Song, Y. and Toth, Z. (2008). Ensemble data assimilation with NCEP global forecast system. *Monthly Weather Review*, **136**, 463–482.

Whitley, D. (2001). An overview of evolutionary algorithms: Practical issues and common pitfalls. *Information and Software Technology*, **43**, 817–831.

Yannick, T. (2007). Model-error estimation in 4-Var. Quarterly Journal of the Royal Meteorological Society, **132**, 2483–2504.

Zhang, F., Snyder, C. and Sun, J. (2004). Impacts of initial estimate and observation availability on convective-scale data assimilation with an ensemble Kalman filter. *Monthly Weather Review*, **132**, 1238–1253.

Zheng, X. G. (2009). An Adaptive Estimation of Forecast Error Covariance Parameters for Kalman Filtering Data Assimilation. Advances in Atmospheric Sciences, **26**(1), 154–160.

Zupanski, D. and Zupanski, M. (2006). Model Error estimation employing an ensemble data assimilation approach. *Monthly Weather Review*, **134**(5), 1337–1354.

Chapter 10

An Introduction to Multi-scale Kalman Smoother-Based Framework and Its Application to Data Assimilation

Daniel E. Salas* and Xu Liang*,†

*Department of Civil and Environmental Engineering, University of Pittsburgh, Pittsburgh, PA, USA

†State Key Laboratory of Hydraulics and Mountain River Engineering, Sichuan University, Chengdu, Sichuan, China

In this chapter, an extended Multi-scale Kalman Smoother-based (MKS) framework (Parada and Liang, 2004) will be described. It is a modification of the traditional Kalman filter and is used to estimate the probability distribution of hydrological variables, given predictions from a model, observations, and MKS parameters. One of the applications of the MKS-based framework is data assimilation, in which observations at different spatial scales are used to improve model predictions. If no observations are available, the MKS-based framework can be used to effectively estimate the model simulations at different spatial scales or fuse model predictions of multiple models (e.g., for data fusion). The Expectation-Maximization (EM) algorithm for obtaining optimal parameter estimations are coupled to the MKS-based framework and will also be described and discussed in this chapter. The EM algorithm is employed to improve the estimations of the MKS parameters. The extended MKS-based framework will be presented in this chapter in a way which is used for data assimilation. Also, an example to illustrate this framework for data assimilation will be provided.

10.1 Introduction

The technique of data assimilation (DA) has been widely used in numerous disciplines, such as atmospheric sciences, hydrology, ecology, engineering, etc. In hydrology, it has been applied to improve simulations of hydrological cycle and the exchange of energy fluxes of the atmosphere–land–soil–vegetation continuum. There are a number of data assimilation algorithms

S. Liang, X. Li & X. Xie (eds.), *Land Surface Observation, Modeling and Data Assimilation*, 273–334.

used to improve land surface model (LSM) simulations, such as to improve model simulations on soil moisture, snow water equivalent, etc. The commonly used data assimilation algorithms can be classified into three categories: (1) traditional Kalman filtering (Kalman, 1960; Galantowicz *et al.*, 1999), extended Kalman filtering (EKF; e.g., Walker and Houser, 2001; Boulet *et al.*, 2002; Kumar and Kaleita, 2003; Montaldo and Albertson, 2003; Sun *et al.*, 2004), ensemble Kalman filtering (EnKF; e.g., Hoeben and Troch, 2000; Reichle *et al.*, 2002a; Crow and Wood, 2003; Margulis *et al.*, 2003), and their variations; (2) variational data assimilation via the adjoint method (e.g., Reichle *et al.*, 2002a, 2002b); and (3) heuristic approaches such as statistical correction and Newtonian nudging (e.g., Houser *et al.*, 1998; Pauwels *et al.*, 2001, 2002). Despite significant progress in evaluating these techniques for assimilation of observations (such as near-surface soil moisture or radio brightness temperature, snow water equivalent, etc.) into LSMs, and in assessing the potential gains associated with them, there still remain several fundamental challenges about these widely used assimilation techniques. Some of the important challenges are associated with the difficulty in representing explicitly the spatial data distribution (e.g., near-surface soil moisture) of individual data images of a study region and the scaling properties of the fields of the state variables in question (e.g., near-surface soil moisture), in dealing with remote sensing retrievals (e.g., near-surface soil moisture data) and LSM predictions at dissimilar spatial scales (both gridded and non-gridded), and in quantifying both satellite data errors/uncertainties and model errors/uncertainties required for data assimilation by commonly used assimilation frameworks (e.g., Ni-Meister *et al.*, 2005).

Parada and Liang (2004) developed a multi-scale Kalman Smoother-based (MKS-based) framework which addresses some of these challenges. This MKS-based framework is an extension of the MKS algorithm (Chou *et al.*, 1994; Fieguth *et al.*, 1995; Luettgen and Willsky, 1995; Kumar, 1999). It combines an extended MKS algorithm with an Expectation Maximization (EM) algorithm for parameter estimation and can be applied to improve model simulations of soil moisture at different soil layers through data assimilation. Although the MKS algorithm has gained more attention in recent years (e.g., Pan *et al.*, 2009; De Lannoy *et al.*, 2010), the main difference between Parada and Liang (2004) and others (e.g., Pan *et al.*, 2009) is that the former allows for the updating of both surface and subsurface soil moisture simultaneously.

In this chapter, we focus on describing the basic concepts and procedures included in the MKS-based framework developed by Parada and Liang (2004) under a context of applying it for soil moisture data assimilation. An example with detailed step-by-step procedures is constructed to illustrate an application of the MKS-based framework using synthetic soil moisture data. Readers are referred to Parada and Liang (2004) for more complete and theoretical descriptions of the framework.

The MKS-based framework (Parada and Liang, 2004) has several important characteristics and are briefly summarized below:

(1) It has the EM algorithm (Kannan *et al.*, 2000) tightly coupled with an extended MKS algorithm. The statistical parameters inherent to the MKS algorithm (e.g., observed data and model error noise variances) are allowed to vary over time.
(2) It allows explicit and efficient modeling of the spatial data distribution (e.g., near-surface soil moisture) of individual data images of a study region and scaling properties of the state variable such as soil moisture (Parada and Liang, 2003a, b).
(3) It has the ability to cope with model predictions and observations made at different spatial scales and to assimilate both gridded and non-gridded observations into a model simultaneously.
(4) For applications to soil moisture data assimilations, it allows for simultaneously updating both the near-surface soil moisture, which has multiple spatial scales (e.g., observations from different sources with different spatial scales), and the sub-surface soil moisture, which typically has only a single spatial scale.

Using the soil moisture data assimilation process as an example, the MKS algorithm can be briefly summarized as follows:

(1) The framework is applied, through its Kalman filter (KF), to the bottom soil layer and is worked up to the top soil layer. For the soil layer with multiple spatial scales, the framework is applied from the finest scale to the coarsest scale.
(2) After (1), the Kalman smoother is used in the reverse order. These processes, called upward and downward sweeps, together make up one MKS iteration. Model simulated soil moisture is treated as an observation in the MKS-based framework.
(3) After each MKS iteration (i.e., (1) and (2) above) is completed, the EM algorithm is used to improve the parameters inherent to the MKS

algorithm, which include:

(a) Unconditional variance of the hidden state variable, for the root node in the multi-scale layer.
(b) Variances of predictions, per layer and scale.
(c) Variances of observations, per layer and scale where observations exist.

These steps are repeated until the improvement produced by the EM algorithm is insignificant. At that point, the best estimation for the hidden variable (e.g., soil moisture) is achieved in every soil layer and at every spatial scale for a given time step over which the MKS is executed and the available observations and model predictions are used.

10.2 Traditional Kalman Filter

The Kalman filter is a statistical method that uses observations and predictions (both are associated with noise and inaccuracies) to obtain an improved estimation toward the true value of a hidden variable.

Traditionally, it is used in time series, where the first prediction is made *a priori* (unconditional). Then, if there are observations available, they are used together with the prediction to obtain an updated estimation of the hidden variable. The updated estimation is the input for the dynamic equation, which encapsulates the physical rules to predict the next time step. With the new prediction, the cycle can be repeated through the time series.

The dynamic equation encloses physics rules that represent a transformation of the hidden state variable (x) from one time step to the next. It has the form (i.e., Eq. (1), dynamic equation)

$$x_{t+1} = F_t x_t + G_t + w_t \tag{1}$$

where F, G are the parameters of the linear transformation from x_t to x_{t+1}; and w is the white noise with variance Q'.

The observation equation represents the measurements obtained. It depends on the real value of the hidden state variable and includes some noise. The observation equation has the form

$$y_t = C_t x_t + D_t + v_t \tag{2}$$

where C, D are the parameters of a linear transformation from the hidden state variable (x) to the observation (y) and v is the white noise with variance R'.

Inputs of the KF include:

(a) Unconditional distribution of the hidden variable, its mean and variance.
(b) Observations of the hidden variable.
(c) Q' — the variance of the noise in the dynamic equation.
(d) R' — the variance of the noise in the observation equation.

The outputs include:

(a) The updated estimations of the hidden states, with mean and variance (P) for each time step.
(b) The predictions of the next time step with mean and variance (P) obtained at every time step.

10.3 MKS and Its Extension

The MKS is a modification of the traditional KF, in which the algorithm is not applied through time but through different soil layers and/or different spatial scales. Also, while the filter is used from coarser scales and deeper layers to finer scales and upper layers, a smother is also performed in the reverse direction. Figure 10.1 shows the hierarchical data configuration in which an extended MKS-based framework performs the filtering and smoothing in space. It is made up of soil layers and spatial scales. For the purposes of this description, it is assumed that three soil layers are being modeled (i.e., $L = 3$).

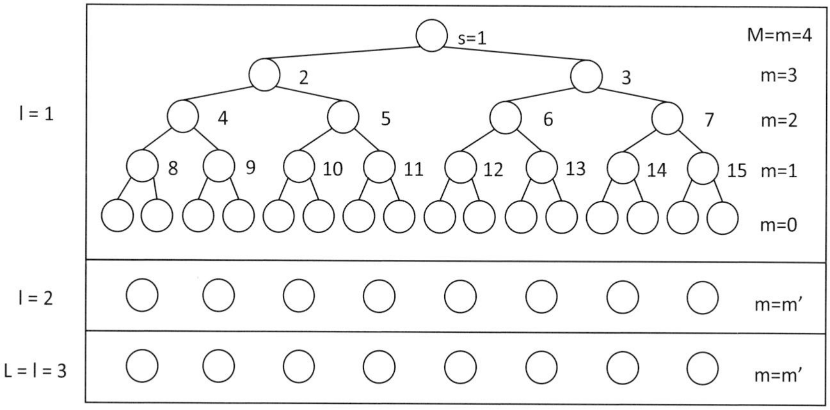

Fig. 10.1. MKS filtering and smoothing in space.

The top soil layer is layer 1 (i.e., $l = 1$), which is also the layer having multiple scales. This means that for layer $l = 1$, it is possible to have observations, predictions, and estimations at different spatial scales. Layer $l = 1$ has 5 spatial scales in the configuration as shown in Fig. 10.1, enumerated from 0 to 4 (i.e., $M = 4$). Scale $m = 4$ is the coarsest one, while scale $m = 0$ is the finest one. The other two soil layers are single-scale. The scale of the single-scale layers is labeled m', and $m' = 1$ in this case.

Each node in Fig. 10.1 represents a hidden state variable (e.g., soil moisture) for a given area. Its mean and variance are represented by x and P, respectively. In the coarsest spatial scale of soil layer 1, the root node represents the hidden state variable for the entire study area. Each child node represents the hidden state variable over a portion of the same study area. The area represented by a parent is equal to the area represented by all of its children together. Thus, in the MKS space associated with the soil layer, every spatial scale represents the same study area, where each node is one cell in its corresponding spatial scale.

To represent variables defined in the nodes of a multi-scale tree, the following notation is used:

any-var(s, l). Any variable defined in cell s located in the soil layer l. "s" can be located at any level of the multi-scale tree.

Figure 10.2 shows a multi-scale tree (Parada and Liang, 2004). The operator γ is used to represent the parent of a node, and the operators α_1 to α_4 are used to represent each of the four children of a node. The

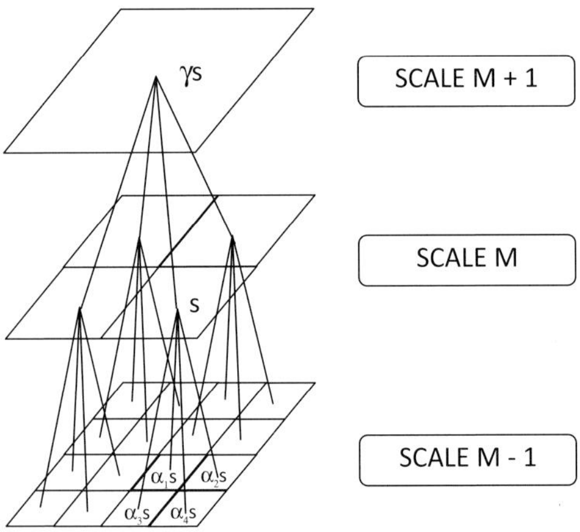

Fig. 10.2. Multi-scale tree. Copyright 2004, John Wiley & Sons, Inc.

dynamic equation describing the evolution of the multi-scale process (e.g., near-surface soil moisture) from coarser to finer scales is expressed as,

$$x(s, l = 1) = A(s, l = 1) \cdot x(\gamma s, l = 1) + w(s, l = 1)$$

where $x(s, l = 1)$ is the hidden variable (soil moisture), $w(s, l = 1)$ is the white noise distribution $N(0, Q(s, l = 1))$ with variance $Q(s, l = 1)$, and A is a parameter that scales the variable x when the area represented is split in 4 cells. For soil moisture which is a rate of water per volume unit, $A = 1$. But for bulk variables like 'amount of water in the cell' $A = 1/4$.

The dynamic equation describing the evolution from shallower to deeper layers is expressed as,

$$\begin{aligned} x(s, l+1) = F(s, l+1) \cdot x(s, l) + G(s, l+1) \\ + w(s, l+1), \quad l = 1, \ldots, L-1 \end{aligned}$$

where $x(s, l+1)$ is the hidden variable of layer $l+1$, $w(s, l+1)$ is the white noise distribution $N(0, Q(s, l+1))$ with variance $Q(s, l+1)$, and F and G are parameters that can be specified based on the underlying physics describing soil moisture dynamics in an LSM.

The observation equation related to either the multi-scale or single-scale hidden states can be expressed as,

$$y(s, l) = C(s, l) \cdot x(\gamma s, l) + D(s, l) + v(s, l), \quad l = 1, \ldots, L$$

$y(s, l)$ is the observation of layer l at node s, $v(s, l)$ is the observation white noise distribution $N(0, R(s, l))$ with variance $R(s, l)$, and $D(s, l = 1)$ is a parameter used to capture the biases in the land surface model predictions and the real observations in the multi-scale.

Figure 10.3 shows the MKS algorithm as a modification of the traditional KF. The time steps have been replaced by the soil layers and spatial scales. The first process is the upward sweep (in which filtering occurs) that starts in the lowest soil layer (single-scale in this illustration case). It goes upward to the upper soil layers. The top-soil layer, which is a multi-scale soil layer, requires steps for each spatial scale. Once the root node of the scale tree is reached, the downward (smoothing) process starts. It finishes when the lowest soil layer is reached again.

The dynamic equation of the KF is replaced by two equations here. The first is used to predict a hidden state variable from one layer to the other (with parameters F and G). The second is used to predict a hidden state variable from one spatial scale to the other (with parameter A).

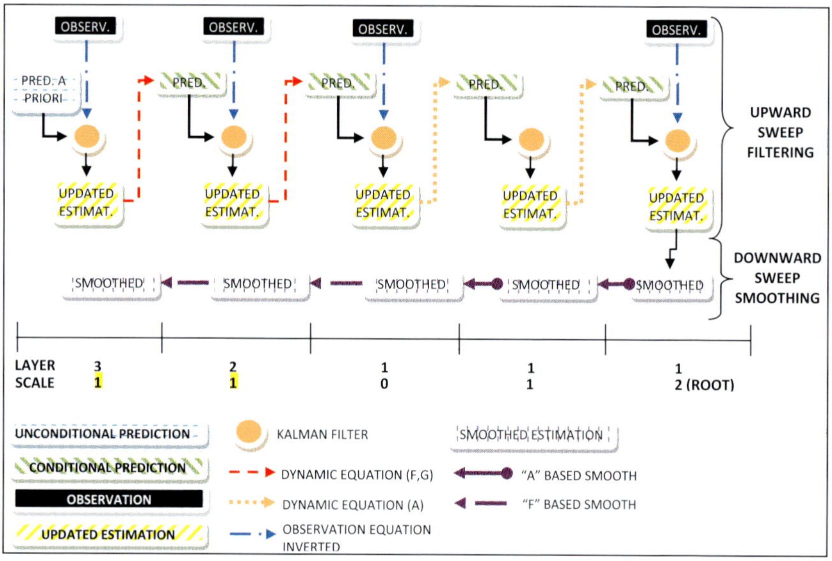

Fig. 10.3. MKS framework showing one iteration.

The dynamic equation of the traditional KF linearly predicts the value of the hidden state variable (soil moisture in this case) of the next time step with parameters F and G (i.e., Eq. (1)). In the MKS framework, the parameters are used to linearly predict the value of the hidden state variable over different soil layers. If the hidden state variable is the soil moisture, the equation that relates its value from one layer to its upper layer is expressed as Eq. (3) if the VIC-3L model, a three-layer hydrologically based Variable Infiltration Capacity LSM (e.g., Liang *et al.*, 1994, 1996a, b, 1999, 2003; Cherkauer and Lettenmaier, 1999, 2003; Liang and Xie, 2001), is used. VIC-3L has been extensively tested and successfully applied to various basins of different scales with good performance (e.g., Nijssen *et al.*, 1997; Wood *et al.*, 1997). The VIC-3L model has also performed well in the various phases of the project for intercomparison of land surface parameterization schemes (PILPS) (Chen *et al.*, 1997; Liang *et al.*, 1998; Lohmann *et al.*, 1998; Wood *et al.*, 1998; Bowling *et al.*, 2003). Furthermore, the VIC-3L model has been applied to a wide range of studies, including soil moisture estimation (Nijssen *et al.*, 2001a), streamflow forecasting (Nijssen *et al.*, 2001b), climate change impact analyzes (Leung *et al.*, 1999), and LUCC impact analyzes (Mao and Cherkauer, 2009).

Equation (3) is the mass balance equation which is consistent with those used in VIC-3L (Parada and Liang, 2004):

$$\theta_{i+1}^{t+1} = -\frac{z_i - z_{i-1}}{z_{i+1} - z_i}\theta_i^{t+1} + \frac{h_{i,i+1}^t \Delta t}{z_{i+1} - z_i} \tag{3}$$

where θ_{i+1}^{t+1} is the soil moisture at time $t + 1$ in the layer of $i + 1$, where $i = 1$ or 2; z_i is the deepest edge of the layer i, and $z_0 = 0$; and $h_{i,i+1}^t$ is the pressure head between layers i and $i + 1$ at time t.

Equation (3) is a piece-wise approximation from the mass balance equations in VIC-3L which is obtained from an explicit finite difference approach. In case of having a nonlinear form of Eq. (3), the recommendations are either (1) to obtain a piece-wise approximation or (2) to use EnKF, approach to propagate between single-scale layers. For option (1), one can use more soil layers to obtain a better approximation.

From Eq. (3), we can obtain Eq. (4) with parameters F and G in the adapted dynamic equation expressed as,

$$F = -\frac{z_i - z_{i-1}}{z_{i+1} - z_i}, \quad G = \frac{h_{i,i+1}^t \Delta t}{z_{i+1} - z_i}, \quad i = 1 \text{ or } 2 \tag{4}$$

It is worth noting that if another LSM rather than the VIC-3L model is used, an equation similar to Eq. (3) would be obtained to relate the soil moisture between the soil layers. The dynamic equation from one scale to the other is based on the multiplier "A" that changes the value of the hidden state variable when the scale changes.

It is important to remark that, in the case of the traditional KF, the physical time dependence of the variables is included in the dynamic equation. In the case of the MKS, however, the physical time dependence of the hydrological variables is not included. Results of the mathematical model for each time-step must be included as observations in the MKS, and a full MKS algorithm must be run for each time-step. Therefore, it should be noted that in the MKS-based framework, both the LSM predictions and the real observations are treated as observation sources. By using F and G as defined in Eq. (4), the physics involved in the VIC-3L model is partially included in the framework. Thus, it is adequate to treat the VIC-3L model predictions of the top soil layer as observations, and not to treat the model soil moisture predictions of layers 2 and 3 as observations.

The MKS-based framework can be summarized in the following steps.

(1) Initialization of input parameters

(a) $\hat{x}(root, 1)$: Unconditional mean of the estimated hidden state variable in the coarsest scale (root) of the top-soil layer (i.e., layer 1). It can be estimated as the average of the averages of all observations in the multi-scale layer or it can be treated as an input.

(b) $P(root, 1)$: Unconditional predicted variance of the hidden state variable in the top soil layer (i.e., layer 1) for the coarsest scale (the root of the scale tree).

(c) $Q(m, l)$: Variance of the noise in each of the multi-scale or single-scale dynamic equations.

(d) $R(m, l)$: Variance of the noise in each of the multi-scale or single-scale observation equations.

(2) Upward sweep

It passes the effect of the observations from the finer scales and deeper layers to coarser scales and shallower layers.

(a) Tree initialization: Propagate the unconditional distributions from the root to the leaves in the multi-scale layer and then to the single-scale layers.

(b) For the first iteration only, use the values from the unconditional prediction as conditional predictions. They are considered the first prediction. This step is executed only in the finest scale of the deepest layer.

(c) Use the latest prediction and the observations at this scale or this soil layer to perform a KF to obtain updated conditional distributions. If no observations are available, the updated distributions are the incoming predictions.

(d) Use the updated conditional distributions to compute predictions of the parents in the next scale (i.e., a coarser scale) or in the next shallower soil layer based on the inverse dynamic equations. If each parent has n children, each cell in the next coarser scale will have n predictions. These are fine-to-coarse or deep-to-shallow predictions.

(e) Fuse the predictions from all children to obtain an integrated prediction of each cell in the next coarser scale (this step is unnecessary for the single-scale soil layers). Move up to the next scale or shallower soil layer and go to step 3 again until the root node (i.e., the coarsest scale) or the top-soil layer (i.e., near-surface soil layer) is reached.

(3) Downward sweep

It is used to propagate the effect of the observations from the coarser scales or shallower soil layers down to finer scales or deeper layers. In this way every estimation will be smoothed given the full set of observations.

(a) The latest update in the root from the upward sweep is an estimation of the hidden variable in the coarsest scale given all the observations and model predictions. Consider this as the smoothed estimation and start the downward sweep at this layer.

(b) Pass down to the layer below (i.e., deeper soil layer) or scale below (i.e., finer scale). Use the smoothed estimation from the layer or scale above to compute the smoothed estimation in the current layer or scale.

(c) Repeat step 2 iteratively for each underlying layer until the deepest layer (i.e., the bottom of the soil layer) and the leaves (i.e., the finest spatial scale) are reached.

At the end, we have estimations for the hidden variables in all the soil layers and spatial scales given all the available observations. **These estimations are valid for the given parameter inputs and the current time-step!**

10.3.1 *Upward sweep*

In this section, the upward sweep is described in detail. First, the unconditional predictions are initialized. After that, the filtering process (see Fig. 10.3) is executed. The filtering process executes two steps in each layer and/or scale. The first is the KF step, which uses the prediction available from the previous layer/scale (or the unconditional prediction) and the observations to generate an updated estimation. The second is the prediction step that uses the updated estimation to generate a prediction for the next layer/scale.

10.3.1.1 *Initialize unconditional distributions*

This process assigns the mean and variance for the unconditional Gaussian probability distribution that represents the first prediction of the hidden state variable. The first node to be assigned is the root of the scale tree in the multi-scale layer. The mean and variance for the root are inputs. Their values are propagated in the scales and later in the layers (see Fig. 10.1).

10.3.1.1.1 Multi-scale layers

Propagate the unconditional distribution from the root to the leaves using Eq. (5) (i.e., "dynamic" equation between scales) and Eq. (6) (i.e., to compute variance from one scale to the other).

$$\hat{x}(s,1) = A(s,1)\hat{x}(\gamma s,1) \tag{5}$$
$$P(s,1) = A(s,1)P(\gamma s,1)A^T(s,1) + Q(s,1) \tag{6}$$

where $\hat{x}$ is the mean of the estimation of the hidden state variable and P is the variance of the hidden state variable.

10.3.1.1.2 Single-scale layers

From scale m' in the multi-scale layer, propagate the distributions to the single-scale layers using Eq. (7) (i.e., adapted dynamic equation for use between layers instead of between time-steps) for the mean and Eq. (8) (i.e., to the compute variance from one layer to the other) for variance.

$$\hat{x}(s,l+1) = F(s,l+1)\hat{x}(s,l) + G(s,l+1) \tag{7}$$
$$P(s,l+1) = F(s,l+1)P(s,l)F^T(s,l+1) + Q(s,l+1) \tag{8}$$

Figure 10.4 shows the unconditional distributions being propagated from the root to the leaves. Once m' ($m' = 1$ in Figure 10.7) is reached, the propagation between layers starts.

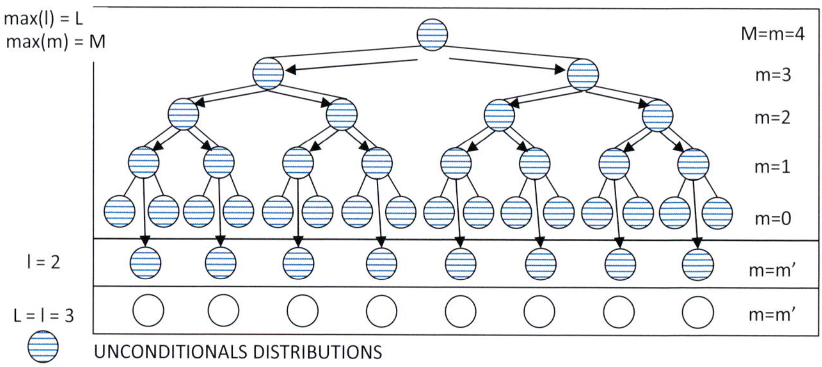

Fig. 10.4. From $m = m'$ propagate unconditional distributions to single-scale layers.

10.3.1.2 *Filtering single-scale layers*

The single-scale soil layers are visited from bottom to top. The bottom layer does not have a prediction from below, so, *a priori* prediction is used. For each layer, the next steps are accomplished as follows:

- An updated estimation is obtained from the prediction, the observations and the use of the Kalman filter.
- A prediction is generated for the next (upper) layer, from the updated estimation and the adapted dynamic equation.

10.3.1.2.1 *A Priori* Prediction

This process assigns values to the deepest layer only, which is a single-scale.

The prediction of this layer must be computed beforehand. Thus, it is equivalent to the unconditional distribution (see Fig. 10.5).

10.3.1.2.2 Kalman filter

The Kalman filter requires 2 inputs:

(1) Predictions of the current layer given the observations in the lower layers:
$\hat{x}(s, l|Y(s, l+))$: mean of estimation of x in the cell s, layer l, given all the observations in cell s for layers below l.
Or, shortened as $\hat{x}_{PL}(s, l)$: Mean of prediction of x from the lower layer.
$P(s, l|Y(s, l+))$: variance of estimation of x in the cell s, layer l, given all the observations in cell s for layers below l.

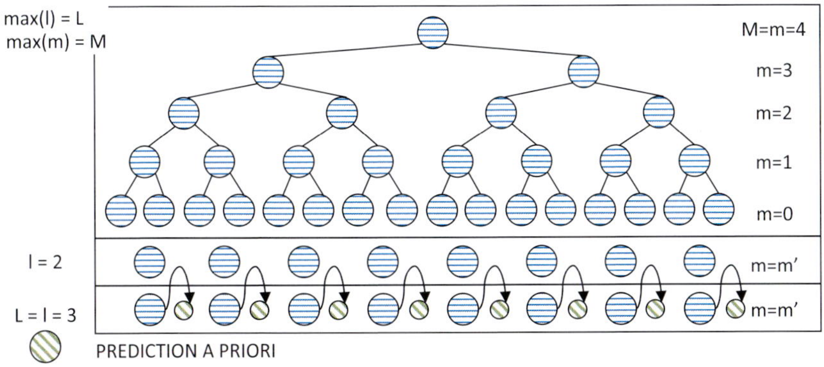

Fig. 10.5. The *a priori* prediction is initialized with the unconditional distribution.

Or, shortened as $P_{PL}(s,l)$: Variance of prediction of x from the lower layer.

(2) Observations in the current layer:
$y(s,l)$: Observation.
$R(s,l)$: Variances of the observation.

To compute the KF, the next steps are required:

(1) Using the variances in the predictions and observations, a Kalman gain matrix K will be computed from Eq. (9) as follows:

$$K(s,l) = \frac{P_{PL}(s,l)C^T(s,l)}{C(s,l)P_{PL}(s,l)C^T(s,l) + R(s,l)} \tag{9}$$

The gain matrix K represents how much the new observation can modify the prediction. The higher the variance in the observation (compared to the prediction variance), the less it will be affected. Each value of this matrix, has a value in the range from 0 to 1. A value of zero indicates that the observation should be ignored, while a value of one indicates that the observation can be used alone to estimate the hidden variable.

(2) Compute the innovation which represents the change in the estimation due to the observation IF the gain were 1.

To compute the innovation, the prediction $\hat{x}_{PL}(s,l)$ is used as a parameter in the observation equation to obtain an estimation of the observation for the case of a perfect prediction, the observation estimated from the prediction would be

$$\hat{y}(s,l) = C(s,l)\hat{x}_{PL}(s,l) + D(s,l) \tag{10}$$

Thus, the difference between the estimated observation and the actual observation is computed to obtain the innovation:

$$IN(s,l) = [y(s,l) - \hat{y}(s,l)] \tag{11}$$

The output of the KF is the updated estimation. It is the prediction innovated by the observation in the proportion indicated by the gain (see Fig. 10.8) below:

$$\hat{x}_U(s,l) = \hat{x}_{PL}(s,l) + K(s,l)IN(s,l) \tag{12}$$

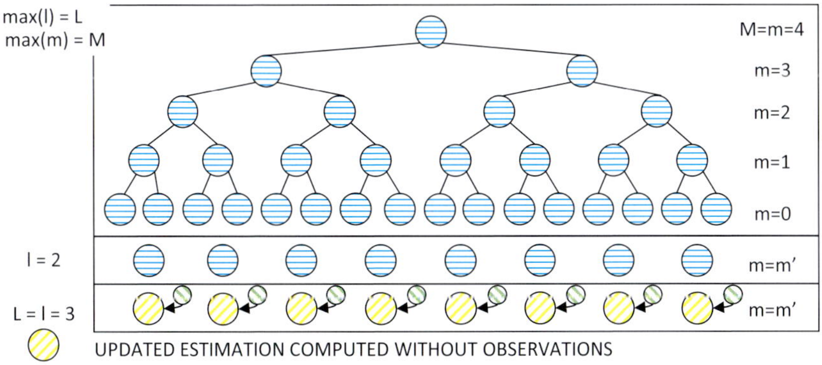

Fig. 10.6. The updated estimation for the hidden state is the prediction when no observations available.

Equation (12) updates estimation when observation becomes available. Equation (13) computes the variance of the updated estimation.

$$P_U(s,l) = [I - K(s,l)C(s,l)]P_{PL}(s,l) \tag{13}$$

Or, if no observation is available (see Fig. 10.6), we update the estimation by:

$$\hat{x}_U(s,l) = \hat{x}_{PL}(s,l) \tag{14}$$

where $\hat{x}_U(s,l) = \hat{x}(s,l|Y(s,l))$ is the estimation of x given the observations for the current layer and the layers below. Or shortened, x updated for layer l in the cell s.

10.3.1.2.3 Prediction

Compute the prediction for the next layer (see Fig. 10.7) using the updated estimation for the current layer and the adapted dynamic equation (Eq.(15)). This version of the dynamic equation is an inverse of the original (Eq.(7)). So, the parameters (F, G, Q) require to be adjusted (Eqs. (17)–(19)).

Equation (15) is an adapted dynamic equation which estimates the mean of the prediction

$$\hat{x}_{PL}(s,l) = \tilde{F}(s,l+1)\hat{x}_U(s,l+1) + \tilde{G}(s,l+1) \tag{15}$$

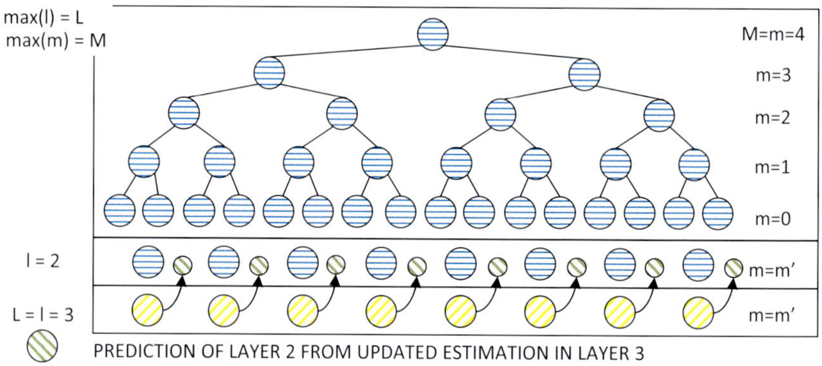

Fig. 10.7. Use the updated estimation to predict the next upper layer.

Equation (16) estimates the variance of the prediction

$$P_{PL}(s,l) = \tilde{F}(s,l+1)P_U(s,l+1)\tilde{F}^T(s,l+1) + \tilde{Q}(s,l+1) \tag{16}$$

Equation (17) adjusts F for the inverse dynamic equation

$$\tilde{F}(s,l) = F^{-1}(s,l)[I - Q(s,l)P^{-1}(s,l)] \tag{17}$$

Equation (18) adjusts G for the inverse dynamic equation

$$\tilde{G}(s,l) = -G(s,l) \tag{18}$$

Equation (19) adjusts Q for the inverse dynamic equation

$$\tilde{Q}(s,l) = [I - Q(s,l)P^{-1}(s,l)]Q(s,l) \tag{19}$$

Once the prediction for the next layer is done, the updated estimation for this one can be computed (see Fig. 10.8).

Once the single-scale layers have been visited and the multi-scale layer is reached, some special steps must be done.

First, the prediction sent by the latest single-scale layer must be received in the same scale (m') in the multi-scale tree (see Fig. 10.10). If m' is the finest scale, the process will continue to predict and update upward (see Fig. 10.9, left panel). But if m' is not the finest scale, the predictions have to be stored to be used later. The filtering will start again in the leaves of the scale tree and, once m' is reached, the predicted stored values from the single-scale layers are fused with the predictions from the tree (see Fig. 10.9, right panel).

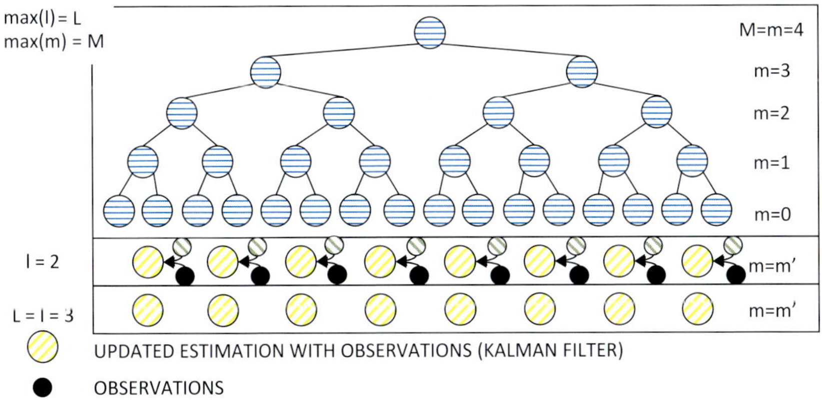

Fig. 10.8. The updated estimation for the hidden state is the result of a KF between the upcoming prediction and observation in the current layer.

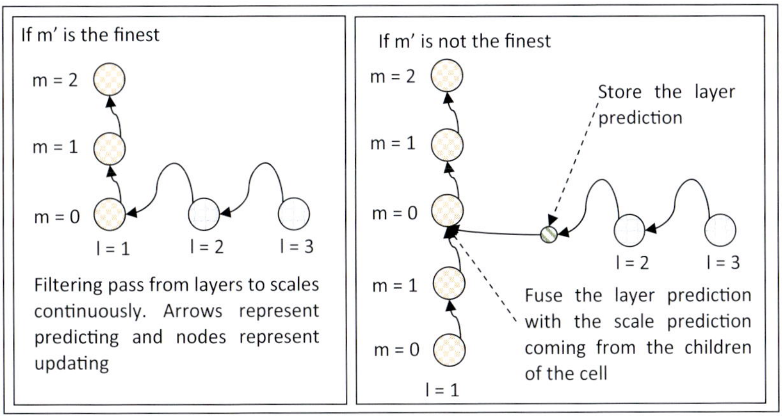

Fig. 10.9. Passing from filtering layers to filtering scales.

10.3.1.3 *Filtering multi-scale layer*

In this process, the scales of the multi-scale layer are visited. For each one, the incoming prediction and the observations are used together with a KF to obtain an updated estimation of the hidden state variable. The incoming prediction is one of the following types:

- *A Priori* of prediction: Obtained from the unconditional distribution through initialization (Sec. 10.3.1.1).

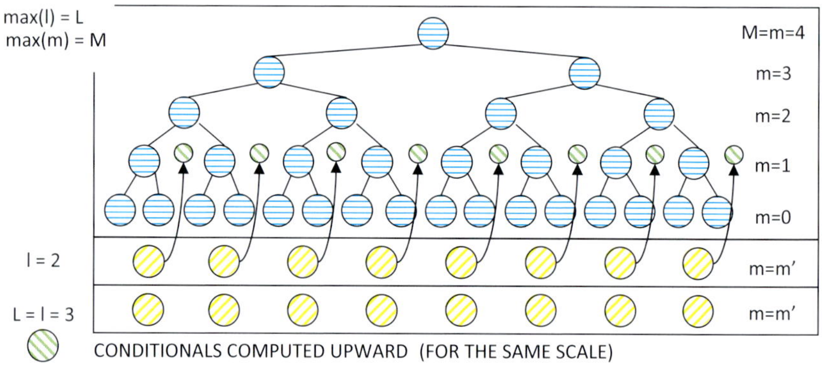

Fig. 10.10. Pass from single-scale to multi-scale through the scale m'.

- Prediction from a single-scale layer which is in the same scale (m') (see Sec. 10.3.1.2.3).
- Prediction from a lower scale in the multi-scale layer. The prediction from the lower scale has been made and the children of each cell have been fused to generate only one prediction for each cell (see Sec. 10.3.1.3.4).
- Prediction from a single-scale layer and from a lower scale in the multi-scale layer. The prediction from the lower scale has been made and the children of each cell have been fused, together with the prediction from the single-scale layer, to generate a single prediction for each cell (see Eqs. (32) and (33)).

10.3.1.3.1 Initialize conditional distributions

If the finest scale in the tree is m', use the predictions obtained from the previous layer (see Figs. 10.9 and 10.11[1]). If not, use the unconditional distributions (see Figs. 10.9 and 10.12). But, once the algorithm reaches the scale m' (see Fig. 10.9, right panel), use Eqs. (32) and (33) to include the prediction coming from the single-scale as another input to the fusion (see Sec. 10.3.1.3.4).

10.3.1.3.2 Kalman filter

This step is analogous to Sec. 10.3.1.2.2. If observations exist for the current scale, then use KF to improve the prediction for the current scale. If not, the estimation for the current scale is the prediction (see Figs. 10.13 and 10.14).

[1] A configuration (Fig. 10.11) with one fewer scale is used from now on in the procedure explanation.

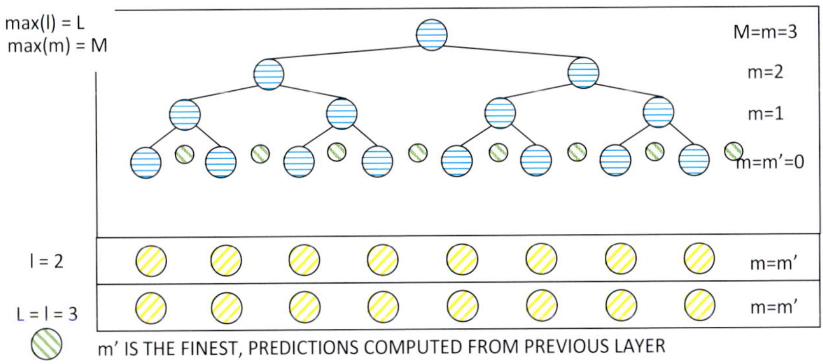

Fig. 10.11. Case when predictions are available from the single-scale layers in the finest scale.

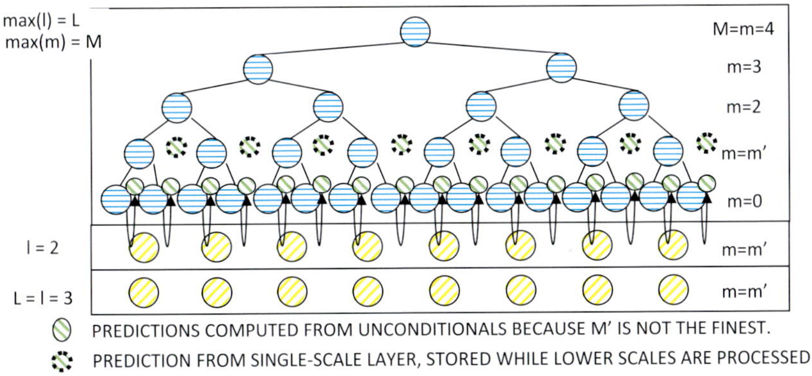

Fig. 10.12. Starting upward sweep in the tree from leaves. Use unconditionals except if upcoming prediction from single-scale layers is available in the initial, finest scale.

KF requires two inputs:

(1) Predictions of the current scale given observations in the finer scales. $\hat{x}(s, l = 1|Y(s-, l = 1))$: Mean of estimation of x in the cell s, layer $l = 1$, given all the observations in cell s for scales finer than the scale of s.

Or, shortened as $\hat{x}_{PS}(s, 1)$: Mean of prediction of x from the finer scales in layer 1.

$P(s, l = 1|Y(s-, l = 1))$: Variance of estimation of x in the cell s, layer $l = 1$, given all the observations in cell s for scales finer than the scale of s.

Or, shortened as $P_{PS}(s, 1)$: Variance of prediction of x from the finer scales in layer 1.

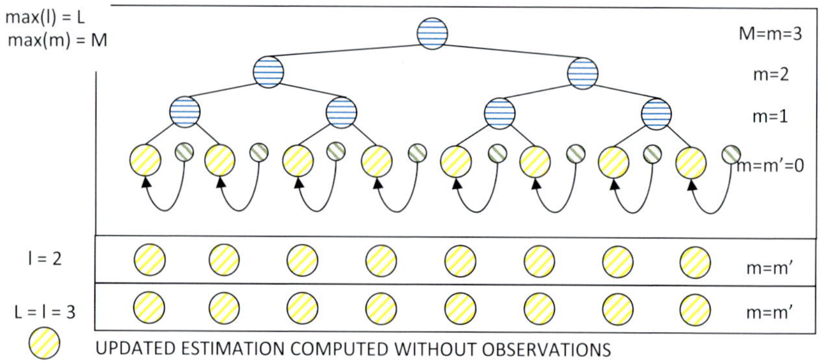

Fig. 10.13. Obtaining updated estimation from prediction only, no observations available.

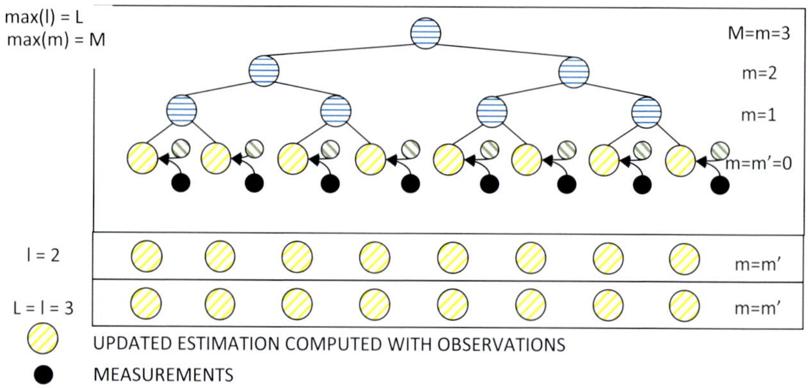

Fig. 10.14. Case when observations available. Computing updated estimation from predictions and observations with a KF.

(2) Observations in the current scale in layer 1.
$y(s,1)$: Observation.
$R(s,1)$: Variance of the observation.

To compute the KF, the next steps are required:

(1) Using the variances in the predictions and observations, a Kalman gain matrix K will be computed:

$$K(s,1) = \frac{P_{PS}(s,1)C^T(s,1)}{C(s,1)P_{PS}(s,1)C^T(s,1) + R(s,1)} \tag{20}$$

For an explanation of the gain matrix K , see Sec. 10.3.1.2.2.

(2) Compute the innovation.
To compute the innovation, the prediction $\hat{x}_{PS}(s,1)$ is used as a parameter in the observation equation to obtain an estimation of the observation for the case of a perfect prediction, then the observation estimated from the prediction is given by

$$\hat{y}(s,1) = C(s,1)\hat{x}_{PS}(s,1) + D(s,1) \tag{21}$$

Then, the difference between the estimated observation and the actual observation is computed to obtain the innovation:

$$IN(s,1) = [y(s,1) - \hat{y}(s,1)] \tag{22}$$

The output of the KF is the updated estimation. It is the prediction innovated by the observation, in the proportion indicated by the gain (see Fig. 10.14):

$$\hat{x}_U(s,1) = \hat{x}_{PS}(s,1) + K(s,1)IN(s,1) \tag{23}$$

Equation (23) computes the updated estimation when observations are available. Equation (24) computes the variance of the updated estimation

$$P_U(s,1) = [I - K(s,1)C(s,1)]P_{PS}(s,1) \tag{24}$$

Or, if the observations are not available (see Fig. 10.13), then the updated estimation is computed as follows,

$$\hat{x}_U(s,1) = \hat{x}_{PS}(s,1) \tag{25}$$

where $\hat{x}_U(s,1) = \hat{x}(s,1|Y(s,1))$ is the estimation of x given the observations for the current scale and the scales below. Or shortened, x updated for scale of s in layer 1.

10.3.1.3.3 Prediction

This step is analogous to Sec. 10.3.1.2.3. Using the updated estimation of the current scale (m) obtained from the previous step, one can compute the prediction of the next coarser scale. An adapted version of the dynamic equation (see Eq. (26)), for predicting between scales is used. It is the inverse of the originally defined in Eq. (5), and thus the parameters need to be adjusted (Eqs. (28) and (29)). Equation (26) also shows that the unconditional mean is subtracted from the updated estimation of the child, and, at the end, it is summed up again. This is because the equation was originally designed for hidden state variables with unconditional zero mean, and thus, this transformation must be done. Equation (26) provides the adapted dynamic equation for scales

$$\hat{x}_{PhS}(s,1) = [\tilde{A}(\alpha_h s,1)(\hat{x}_U(\alpha_h s,1) - \hat{x}(\alpha_h s,1))] + \hat{x}(\alpha_h s,1) \tag{26}$$

where $\hat{x}_{PhS}(s,1)$ is the prediction of x in the cell s from the child h, in the first layer; $\tilde{A}(\alpha_h s,1)$ is the parameter A for the child h adjusted; and $\hat{x}_U(\alpha_h s,1)$ is the updated estimation of x in the child h.

Equation (27) computes the variance of the prediction from the child h

$$P_{PhS}(s,1) = \tilde{A}(\alpha_h s,1)P_U(\alpha_h s,1)\tilde{A}^T(\alpha_h s,1) + \tilde{Q}(\alpha_h s,1) \tag{27}$$

where $P_{PhS}(s,1)$ is the variance of the predicted x from child h; $\tilde{A}(\alpha_h s,1)$ is the parameter A for the child h adjusted; $P_U(\alpha_h s,1)$ is the variance of the updated estimation in the child h; and $\tilde{Q}(\alpha_h s,1)$ is the variance of predictions in the scale of the child h.

Parameter A is adjusted by

$$\tilde{A}(s,1) = \frac{I - \frac{Q(s,1)}{P(s,1)}}{A(s,1)} \tag{28}$$

And parameter Q is adjusted by

$$\tilde{Q}(s,1) = \left[I - \frac{Q(s,1)}{P(s,1)}\right] Q(s,1) \tag{29}$$

Each cell will receive a prediction from each of its children (see Fig. 10.15). The predictions need to be fused in terms of providing only one prediction for each parent cell.

10.3.1.3.4 Fusion of predictions

Given that the iterations are going up in the scale tree, each cell will have one prediction from each descendant. Thus, there will be multiple predictions for each cell. These predictions must be combined to generate only one value. To fuse the predictions, subtract the unconditional estimation from each of the predictions, and then perform a weighted average among these differences using the inverse of the variances as the weighting factors. The result is a predicted difference from the unconditional estimations. Add the predicted difference to the unconditional mean to obtain the fused prediction. If there are predictions available coming from other layers (the

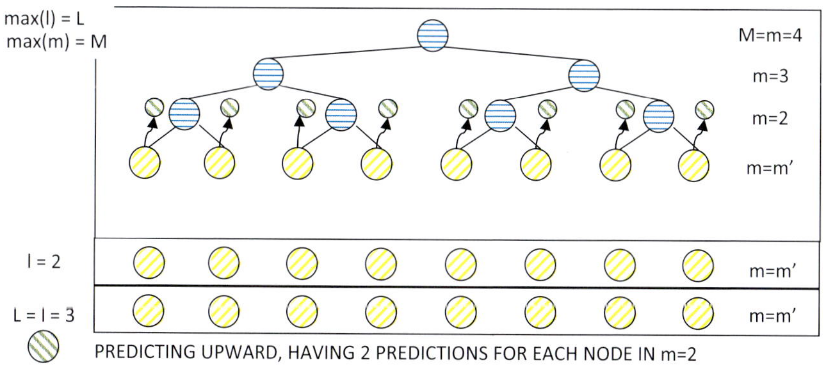

Fig. 10.15. Process of predicting.

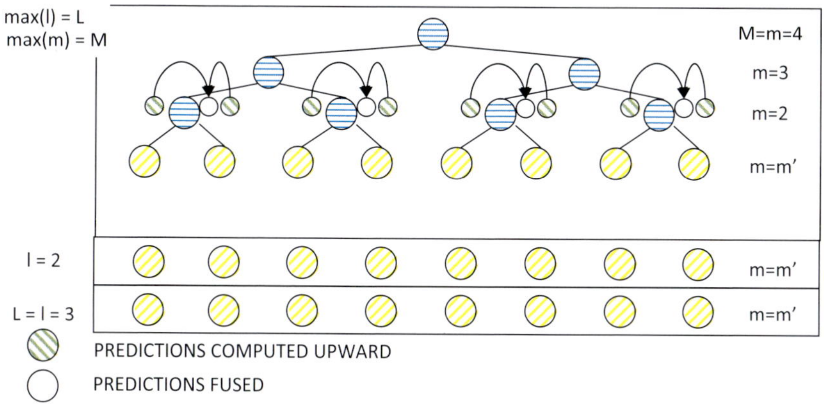

Fig. 10.16. Getting only one prediction from all children (fusing).

single-scale layers) in the same scale, include them in the weighted average as well (see Fig. 10.9).

If $m \neq m'$ (only children will be fused, no predictions from other soil layers would be available, see Fig. 10.16), the fused mean is computed by

$$\hat{x}_{PS}(s,1) = \hat{x}(s,1) + P_{PS}(s,1) \sum_{h=1}^{H(s)} \frac{\hat{x}_{PhS}(s,1) - \hat{x}(s,1)}{P_{PhS}(s,1)} \tag{30}$$

In equation (30), $H(s)$ represents the number of children of cells s. The fused variance is computed as shown below. The inverse of the fused variance is the sum of the inverse variances from each child. But first, all of the inverse variances standardized by subtracting the unconditional inverse variance.

$$\frac{1}{P_{PS}(s,1)} - \frac{1}{P(s,1)} = \sum_{h=1}^{H(s)} \left[\frac{1}{P_{PhS}(s,1)} - \frac{1}{P(s,1)}\right] \Rightarrow \frac{1}{P_{PS}(s,1)}$$

$$= \frac{1}{P(s,1)} + \sum_{h=1}^{H(s)} \left[\frac{1}{P_{PhS}}(s,1)\right] - \sum_{h=1}^{H(s)} \left[\frac{1}{P(s,1)}\right]$$

$$\Rightarrow \frac{1}{P_{PS}(s,1)} = \frac{1}{P(s,1)} + \sum_{h=1}^{H(s)} \left[\frac{1}{P_{PhS}(s,1)}\right] - \frac{H(s)}{P(s,1)}$$

$$\Rightarrow \frac{1}{P_{PS}(s,1)} = \frac{1-H(s)}{P(s,1)} + \sum_{h=1}^{H(s)} \left[\frac{1}{P_{PhS}(s,1)}\right]$$

$$P_{PS}(s,1) = \left[\frac{1-H(s)}{P(s,1)} + \sum_{h=1}^{H(s)} \frac{1}{P_{PhS}(s,1)}\right]^{-1} \tag{31}$$

It can be seen in Eq. (30) that the unconditional estimation is subtracted from the prediction of each child, and then summed up as a weighted average. The weight of each of the differences is the inverse of the variance of the corresponding child. The result of the sum is then divided by the inverse of the fused variance.

If $m = m'$ (i.e., the predictions from all children and from the other layers are available for the fusion process, see Fig. 10.9), then the fused mean including the layer prediction is computed as,

$$\hat{x}_{PS}(s,1) = \hat{x}(s,1) + P_{PS}(s,1) \left\{ \frac{\hat{x}_{PL}(s,1) - \hat{x}(s,1)}{P_{PL}(s,1)} \right.$$
$$\left. + \sum_{h=1}^{H(s)} \frac{\hat{x}_{PhS}(s,1) - \hat{x}(s,1)}{P_{PhS}(s,1)} \right\} \tag{32}$$

And the fused variance including the layer prediction is computed as,

$$P_{PS}(s,1) = \left[-\frac{H(s)}{P(s,1)} + \frac{1}{P_{PL}(s,1)} + \sum_{h=1}^{H(s)} \frac{1}{P_{PhS}(s,1)}\right]^{-1} \tag{33}$$

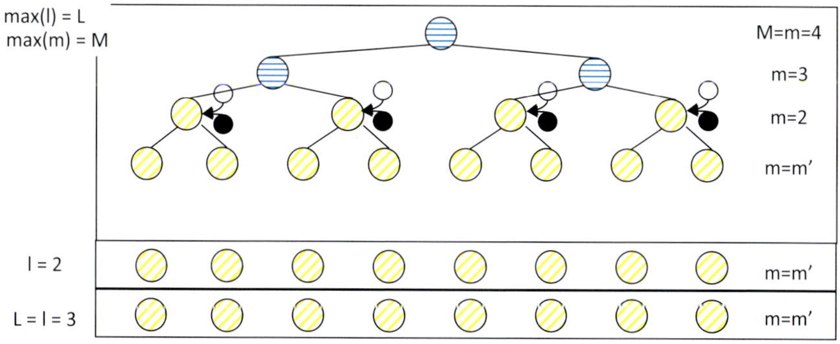

Fig. 10.17. Updating estimation.

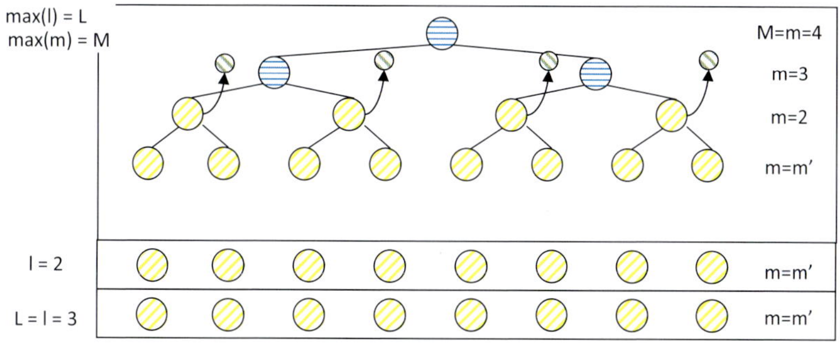

Fig. 10.18. A schematic showing the process of predicting.

The updating, predicting, and fusing steps that complete one scale of the upward sweep of the multi-scale tree are summarized from Figs. 10.17 to 10.19.

At the end of the upward sweep, all of the cells in the multi-scale and single-scale layers are updated. Therefore, for every cell s, the upward sweep produces an estimation of the hidden state variable that includes the influence of every available observation in: (a) cells in lower layers located directly under s, and (b) cells in finer scales which are descendants of s in the multi-scale tree (i.e., either at finer scales and/or deeper layers). Now it is necessary to pass the effect of all of the observations downward to all the cells in such a way that each cell has an estimation of the hidden state variable given the full set of observations (i.e., below and above the cell of interest). This is the objective of the downward sweep.

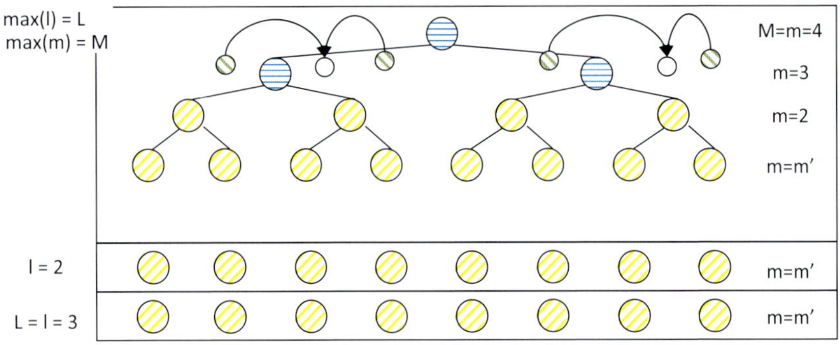

Fig. 10.19. A schematic showing the process of fusing predictions.

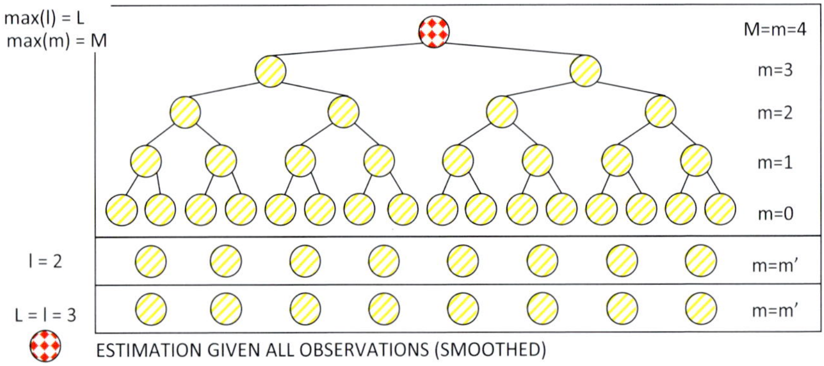

Fig. 10.20. Once the upward sweep is done, the root has the estimated hidden state variable given all the observations.

10.3.2 *Downward sweep*

At this point, the root of the tree has an estimation of the hidden state variable x, given all the available observations (see Fig. 10.20). The downward sweep visits all nodes (or cells) scale-by-scale (and later layer-by-layer) starting from the scale just below the root node. It computes the estimation of the hidden state variable (in each node) given all the available observations.

10.3.2.1 *Smoothing multi-scale layer*

For each cell, compute an estimated value of the hidden state variable given all the available observations. This is called a smoothed estimation.

The smoothing process is equivalent to obtaining the filtering estimations forward and backward, and, weighting the solutions on their inverse variances (Rothrock and Thomas, 1992). The smoother gain matrix J (see Eq. (34)) represents how the backward filter affects the forward filtered solution which has already been computed. The innovation here (see Eq. (35)) is the difference between the smoothed and the predicted estimations in the upper cell. It represents the difference between the forward and the backward filters. The smoother gain matrix is given by

$$J(s,1) = P_U(s,1)\frac{A^T(s,1)}{P_{PS}(\gamma s,1)} \tag{34}$$

where $J(s,1)$ is the smoother gain matrix; $P_U(s,1)$ is the variance of the updated estimation of the current cell; $\tilde{A}^T(s,1)$ is the parameter A corrected for the inversed adapted dynamic equation; and $P_{PS}(\gamma s,1)$ is the variance of the predicted estimation of the parent cell.

The innovation of the smoother is expressed as,

$$IS(s,1) = \hat{x}_S(\gamma s,1) - \hat{x}_{PS}(\gamma s,1) \tag{35}$$

where $IS(s,1)$ is the innovation of the current cell associated with the smoother process; $\hat{x}_S(\gamma s,1)$ is the mean of the smoothed estimation of the parent cell. Also written as $\hat{x}(\gamma s, l=1|Y(s=\mathit{root}, l=1))$; and $\hat{x}_{PS}(\gamma s,1)$ is the mean of the predicted estimation of the parent cell.

The smoothed estimation is equal to the updated estimation plus a correction given by Eq. (36), which is proportional to the gain and to the difference between the backward and the forward filter as follows:

$$\hat{x}_S(s,1) = \hat{x}_U(s,1) + J(s,1)IS(s,1) \tag{36}$$

The variance of the smoothed estimation is expressed as,

$$P_S(s,1) = P_U(s,1) + J(s,1)[P_S(\gamma s,l) - P_{PS}(\gamma s,l)]J^T(s,1) \tag{37}$$

where $P_S(s,1)$ is the variance of the smoothed estimation and $P_S(\gamma s,l)$ is the variance of the smoothed estimation of the parent cell.

In the multi-scale tree, the smoothing process is computed downward until the leaves are reached (see Figs. 10.21 and 10.22).

For each node, a cross covariance between the node and its parent node is computed by Eq. (38). This information is needed for the EM algorithm

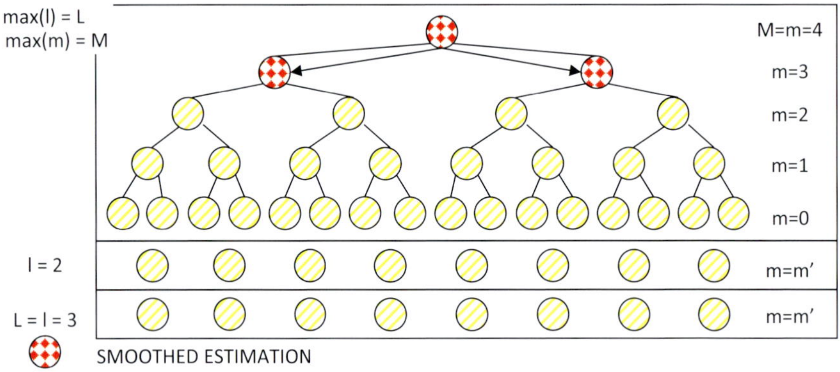

Fig. 10.21. Smoothing from root to next scale.

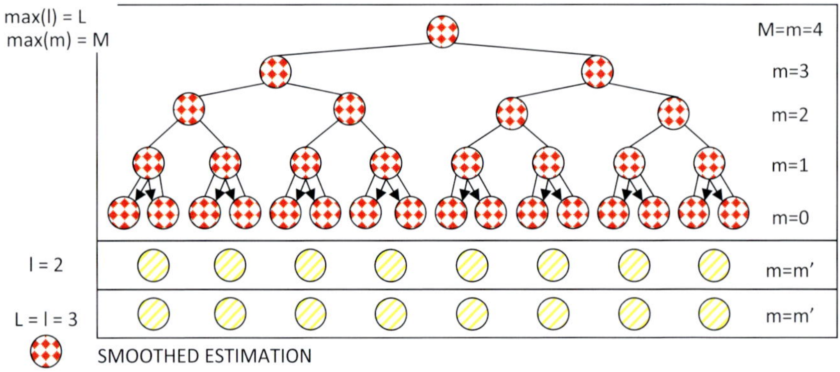

Fig. 10.22. Smoothing from parents to children.

(see Sec. 10.5).

$$P_S(\{s, \gamma s\}, 1) = J(s, 1) P_S(\gamma s, l) \tag{38}$$

10.3.2.2 *Smoothing single-scale layers*

This process is analogous to the smoothing steps described earlier. The objective is to compute the value of the hidden state variable given all the available observations. Thus, it is started with the estimation of a hidden state variable given all the observations in the deeper and/or the same soil layers (i.e., the updated estimation). Then, the difference between

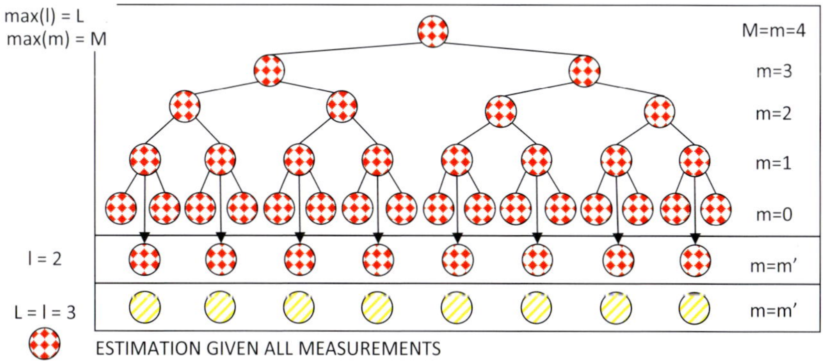

Fig. 10.23. Smoothing from multi-scale to single-scale through m'.

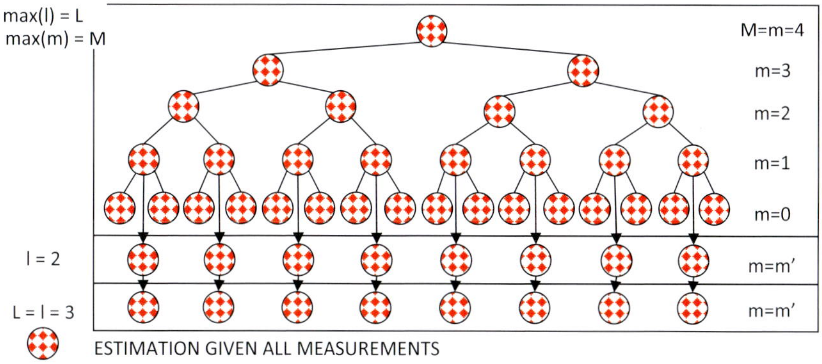

Fig. 10.24. Smoothing from upper to lower layers.

the predicted and smoothed values associated with the shallower cell is multiplied by a gain matrix and is added to the updated estimation.

In this example, the process starts in soil layer 2, based on the cells from soil layer 1 (i.e., the upper layer) that are in the multi-scale layer, i.e., in the scale m' (see Fig. 10.23). The process continues until the deepest soil layer is reached (see Fig. 10.24). The smoothed estimation of the mean is expressed as,

$$\hat{x}_S(s,l) = \hat{x}_U(s,l) + J(s,l)[\hat{x}_S(s,l-1) - \hat{x}_{PL}(s,l-1)] \tag{39}$$

The smoothed estimation of the variance is expressed as,

$$P_S(s,l) = P_U(s,l) + J(s,l)[P_S(s,l-1) - P_{PL}(s,l-1)]J^T(s,l) \tag{40}$$

The gain matrix is

$$J(s,l) = \frac{P_U(s,l)\tilde{F}^T(s,l)}{P_{PL}(s,l-1)} \tag{41}$$

For each node, the cross covariance between the current and the upper layers is also computed by Eq. (42), which is needed for the parameter estimations in the EM algorithm.

$$P_S(s,\{l-1,l\}) = J(s,l-1)P_S(s,l) \tag{42}$$

10.4 EM Algorithm for Parameter Estimation

The expectation maximinzation (EM) algorithm is used to estimate the parameters required in the MKS-based framework for data assimilation. Assuming stationarity or homogeneity of the parameters, the set of parameters can be summarized as:

$$\Theta = \begin{Bmatrix} \hat{x}(root,1), \\ P(root,1), \\ Q(m,l) for \{0 \leq m < M; 1 \leq l < L\}, \\ R(m,l) \end{Bmatrix}$$

The first two parameters represent the distribution of the hidden state variable in the root of the multi-scale tree. The third represents the covariance matrices for the noise term in the dynamic equation, and the fourth is the covariance matrix for the noise term in the observation equation. For the third and fourth parameters, the homogeneity assumption implies that the noises have spatially homogeneous covariances. This is the reason to use $Q(m,l)$ and $R(m,l)$ instead of $Q(s,l)$ and $R(s,l)$, but one can relax this requirement if it is needed.

The EM algorithm modifies the parameters to maximize the true likelihood of the data. It has two steps:

- The E step: Consist of computing conditional expectations required for the M step. These expectations are known as sufficient statistics.
- The M step: Update the parameter values.

The EM algorithm must be executed at the end of each MKS iteration to update the parameters for the next iteration.

10.4.1 *E step*

To compute the statistics defined in the E step, it is required that a full iteration of the MKS algorithm have been completed. In this way, all the required inputs become available. The E step must be done at the end of each full MKS iteration.

First, define the operator

$$\langle\cdot\rangle = E[\cdot|Y(s = root, l = 1)]$$

In other words, it is the expected value given all the available observations.

The sufficient statistics required are:

(1) $\langle x(s,l)\rangle = \hat{x}(s,l|Y(s = root, l = 1)) = \hat{x}_S(s,l)$
(2) $\langle x(s,l)x^T(s,l)\rangle = P_S(s,l) + \hat{x}_S(s,l)\hat{x}_S^T(s,l)$
(3) $\langle x(s,1)x^T(\gamma s,1)\rangle = P_S(\{s,\gamma s\},1) + \hat{x}_S(s,1)\hat{x}_S^T(\gamma s,1)$
For $s \neq root$
(4) $\langle x(s,l-1)x^T(s,l)\rangle = P_S(s,\{l-1,l\}) + \hat{x}_S(s,l-1)\hat{x}_S^T(s,l)$
For $s \in s'$ and $2 \leq l \leq L$. $s' = \{s|s \in m'\}$

10.4.2 *M Step*

Use Eqs. (43)–(45) to update parameters $P(root,1)$, $Q(m,l)$, and $R(m,l)$. Note that the parameter $\hat{x}(root,1)$ will not be updated; it remains at the same value after the initialization step in which all of the observations is available at the current time-step are employed to estimate it. Equation (43) computes the unconditional variance of the root as,

$$P(root,1) = \langle x(root,1)x^T(root,1)\rangle - \hat{x}(root,1)\hat{x}^T(root,1) \tag{43}$$

Equation (44) computes the variance of the observations as,

$$\begin{aligned} R(m,l) = \frac{1}{n_{m,l}}\sum_s\{&[y(s,l) - C(s,l)\langle x(s,l)\rangle - D(s,l)]^T[y(s,l) \\ &- C(s,l)\langle x(s,l)\rangle - D(s,l)]\} \end{aligned} \tag{44}$$

where $n_{m,l}$ is the number of observations in scale m, layer l. Equation (45) computes the variance of the dynamic equation as,

$$Q(m,l=1) = \frac{1}{N_{m,l=1}}\sum_{\text{s in } m}\left\{\begin{array}{l}\langle x(s,l=1)x^T(s,l=1)\rangle \\ -\langle x(s,l=1)x^T(\gamma s,l=1)\rangle A^T(s,l=1)\end{array}\right\} \tag{45}$$

where $N_{m,l=1}$ is the number of cells (or nodes) associated with a hidden state, in the scale m, layer $l = 1$.

Once the parameters are updated, the MKS algorithm is executed again with the updated values of the parameters.

10.5 Application of the MKS-Based Framework with EM Method for Data Assimilation

The MKS-based framework together with the EM algorithm (Parada and Liang, 2004) can be widely used to improve predictions and estimations of hydrologic variables in areas where observations are available from instrument(s) working at different spatial scales, and models which work at different spatial scales through data assimilation (e.g., Parada and Liang, 2008). This framework can also be applied to conduct data fusion (e.g., Wang *et al.*, 2011). The parameters needed and the procedures of employing the MKS-based framework are briefly summarized below.

Parameters:

(1) $\hat{x}(root, 1)$: Average of the averages of all observations and/or predictions (i.e., mean of the means) at different scales, or, received as an input.
(2) $P(root, 1)$: Required as an input.
(3) $Q(m, l)$: Required as an input.
(4) $R(m, l)$: Required as an input.

To start, the parameters must be specified using data from observations or be estimated *a priori*. Then, the MKS-based framework runs first using the initial parameters. At the end of the MKS-based framework, the EM algorithm is run to compute sufficient statistics and then, to update the initial parameters. The likelihood of the estimations for the true hidden variables is computed. The MKS-based framework runs again with a newly updated set of parameters and the process is repeated until the likelihood does not change or meets a given criterion. The estimated hidden state variables at this point are considered the optimal ones.

10.5.1 *Algorithm complexity*

The MKS-EM algorithm is designed to be executed for every time-step that has observations. Thus, it is worth analyzing the computational complexity of the algorithm and estimating the importance of the cost added to each time-step. First, a list of some required variables is as follows:

M: Number of scales.
L: Number of layers.
m': Scale for the model.
modelside: Number of cells in the grid for the model scale. It equals to $2^{m'}$.
maxside: Number of cells in the grid for the finest scale. It equals to 2^M.
maxbound: Higher bound for the **number of elements** required to be walked over in one MKS iteration. It is a pessimistic approach, but it is good enough to show which operations are critical and which are not. It is equal to the number of cells in a grid ($maxside^2$) times the number of layers times the number of scales per layer. Using the previously defined symbols, the expression can be written as follows:

$$(max\ side)^2 \cdot L \cdot M.$$

iterMKS: Number of MKS iterations until EM converges.

Analysis of the complexity of each of the steps:

MKS: Upward sweep:

Initialize unconditionals: This step starts from the root going downward to visit all the multi-scales and layers. The computational complexity is *maxbound*.
Initialize conditional prediction: $maxside^2$.
Kalman filter (infer improved estimation of the hidden variable):

K: $maxside^2$
Innovation: $maxside^2$
Update estimation: $maxside^2$.

Result for Kalman filter: 3 $maxside^2$.
Prediction: $maxside^2$.
Fusion of predictions: $maxside^2$.

Result for upward sweep: $5 maxside^2 * M * L = 5.\ maxbound.$
Downward sweep: J matrix, smoothing: $2 maxside^2$

$2\ maxside^2 * M * L = 2\ maxbound.$

Result for MKS: 7 *maxbound.*
Expectation maximization:

E Step: *maxbound*

M Step: Compute R: *maxbound*

Compute Q: *maxbound*.

Result for Expectation maximization: 3 *maxbound*.
Total result: *iterMKS* * 10 *maxbound*

$= iterMKS * 10\ (maxside^{2*} L^* M)$.

If the hierarchical structure is given, L and M are constant. *iterMKS* have had values from 3 to 8 in our experiments (4 on average). Using the computational complexity operator, one has

$\Rightarrow \mathrm{O}(iterMKS * 10\ (maxside^2 * L * M))$

$= \mathrm{O}(maxside^2)$

$= \mathrm{O}(\mathrm{side}^2)$.

This implies that the computational cost added to the process for time steps having observations is proportional to the square of the number of cells for each side of the grid, or, linearly proportional to the total number of cells. In our experience this computational cost added is pretty small. And, because the complexity is only squared (even linear with respect to the number of cells), we can assume that the cost will not increase significantly.

10.6 Example

In this section, a step-by-step illustration of the MKS-based framework is included through a hypothetical example to help the readers not only understand the concepts and methodology of the our extended MKS-based framework described in this chapter, but also see how this framework works step-by-step.

10.6.1 *Configuration*

In this example the MKS-based framework is run for a case with three soil layers of which the top layer is a multi-scale layer with four spatial scales. The scale for the single-scale layers is $m' = 0$. It is assumed that there are satellite observations available for soil layer $l = 1$ in the scale $m = 1$. The VIC-3L model is running in scale $m = 0$ associated with the three soil layers. The VIC-3L results are also treated as observations in the scale $m = 0$ of the top soil layer. For this example there are also observations available in scale $m = 0$ for layers 2 and 3. In the finest scale, the study area is divided into a grid with 8×8 cells. For each cell in each layer and scale, the Gaussian probability distributions are being determined by the mean and variance of the hidden state variable.

10.6.2 *VIC-3L model results (observations)*

Table 10.1 shows the soil moisture results of the VIC-3L model, for the current time step, over the top soil layer and the observations for the other layers. The darker cells represent higher moisture and the light ones represent lower moisture. Color scale for the soil moisture values is showed in Table 10.2.

10.6.3 *Satellite observations*

It is assumed that satellite observations are available for soil layer $l = 1$ at scale $m = 1$ (Table 10.3).

10.6.4 *Estimation of unconditional mean for the root node $\hat{x}$ (root, 1)*

First, the average is computed for each scale. Then, the average of the averages is computed. The result is the estimation of the unconditional mean for the root node of the multi-scale layer.

10.6.5 *Parameters*

The EM algorithm's results are sensitive to the initial values of P, Q, and R parameters. Thus, it would be more effective if these parameters are selected having some criteria in mind. For example, if there is previous knowledge of the variances for the observations and model, a range can be defined for each Q and R. Thus, Q and R will be limited to those specified ranges throughout the EM iterations. Additionally, Q can be assigned with a greater value than R if it is known that the model is more reliable than the observations, otherwise, R can be assigned to be greater than Q. If there is no information about Q and R, the algorithm can be started using the same value for both. In this way, the EM algorithm will find an optimal set of values for them without an initial preference. If there is not any prior information at all, just prescribe these parameters using the same values included in this example.

The range used in this example is [0.1;3].

Table 10.1. Scale $m = 0$ observations. (Observations in the 3 soil-layers)

	0.380	0.350	0.390	0.410	0.420	0.420	0.430	0.430	0.306	0.250	0.321	0.329	0.365	0.330	0.351	0.339	0.234	0.154	0.242	0.232	0.267	0.248	0.278	0.284
	0.350	0.360	0.380	0.410	0.410	0.430	0.420	0.430	0.292	0.295	0.323	0.350	0.352	0.365	0.345	0.374	0.217	0.224	0.240	0.267	0.270	0.314	0.256	0.289
	0.360	0.370	0.390	0.400	0.410	0.440	0.420	0.420	0.285	0.294	0.329	0.348	0.340	0.375	0.357	0.333	0.212	0.201	0.230	0.288	0.251	0.283	0.289	0.258
m=0	0.380	0.380	0.390	0.410	0.420	0.440	0.430	0.440	0.325	0.284	0.328	0.329	0.340	0.342	0.339	0.349	0.236	0.185	0.268	0.235	0.287	0.283	0.260	0.250
m'=0	0.370	0.370	0.380	0.410	0.400	0.410	0.390	0.420	0.306	0.302	0.326	0.336	0.326	0.341	0.340	0.353	0.217	0.244	0.250	0.283	0.274	0.258	0.271	0.262
	0.360	0.360	0.390	0.380	0.390	0.360	0.380	0.400	0.273	0.275	0.303	0.293	0.334	0.265	0.311	0.340	0.175	0.188	0.208	0.216	0.268	0.198	0.228	0.279
	0.350	0.370	0.370	0.360	0.360	0.340	0.360	0.380	0.286	0.306	0.305	0.287	0.292	0.267	0.294	0.321	0.224	0.245	0.252	0.219	0.216	0.184	0.216	0.258
	0.330	0.340	0.350	0.350	0.340	0.320	0.340	0.360	0.257	0.284	0.268	0.279	0.265	0.249	0.271	0.302	0.205	0.231	0.186	0.218	0.209	0.186	0.200	0.217
	LAYER 1								LAYER 2								LAYER 3							

Table 10.2. Moisture color scale.

0.100	0.150	0.200	0.250	0.300	0.350	0.400	0.450

Table 10.3. Satellite observations.

m=1	0.307	0.332	0.436	0.438
	0.331	0.338	0.422	0.422
	0.269	0.291	0.351	0.367
	0.295	0.265	0.376	0.346

Table 10.4. Estimation of the unconditional mean.

m=1	0.307	0.332	0.436	0.438	
	0.331	0.338	0.422	0.422	AVERAGE 0.349
	0.269	0.291	0.351	0.367	
	0.295	0.265	0.376	0.346	

m=0 m'=0	0.38	0.35	0.39	0.41	0.42	0.42	0.43	0.43	
	0.35	0.36	0.38	0.41	0.41	0.43	0.42	0.43	
	0.36	0.37	0.39	0.4	0.41	0.44	0.42	0.42	
	0.38	0.38	0.39	0.41	0.42	0.44	0.43	0.44	AVERAGE 0.386
	0.37	0.37	0.38	0.41	0.4	0.41	0.39	0.42	
	0.36	0.36	0.39	0.38	0.39	0.36	0.38	0.4	
	0.35	0.37	0.37	0.36	0.36	0.34	0.36	0.38	
	0.33	0.34	0.35	0.35	0.34	0.32	0.34	0.36	

AVERAGE OF SCALES AVERAGES
0.368

Parameter C for the observation equation is set to be 1 (since there is no expansion or reduction in mapping the soil moisture to the measurement) and parameter D is used to account for the bias of each layer to the unconditional mean.

Parameter A is set to 1 to keep mass conservation between scales. Therefore, the unconditional mean is the same for all scales and it is computed as the average of the layers' average, including only layers with observations (see Table 10.4). The unconditional variance starts with Q in the coarser scale and, for the next (finer) scales the following formula applies:

$$P(m, 1) = AP(m + 1, 1)A + Q(m, 1)$$

Table 10.6 shows the setup of these parameters for the multi-scale layer. Parameters F are set to have a value of 1. Parameter G is computed for the corresponding variables in the VIC-3L model (see Eq. (4)).

The unconditional mean starts in layer 1 with the value from the multi-scale parameters. For the next layers, the adapted dynamic equation (i.e., Eq. (1)) is used. The unconditional variance is computed with the formula: $P(m', l) = F(m', l)P(m', l-1)F(m', l) + Q(m', l)$. $C = 1$, and $D = 0$ as discussed earlier. $H(s) = 4$, indicating four children per parent.

10.6.6 *Unconditional distributions*

The values for the unconditional mean and variance for the multi-scale layer come from Table 10.5. The values for the unconditional mean and variance for the single-scale layers, come from Table 10.6.

Table 10.5. Multi-scale layer parameters.

	LAYER 1 MULTI-SCALE					
	SCALE	Q	R	D	UNCONDITIONAL MEAN	UNCONDITIONAL VARIANCE
ROOT	3	1.5	1.5	0	0.368	1.5
	2	1.5	1.5	0	0.368	3
	1	1.5	1.5	-0.01848	0.368	4.5
LEAVES	0	1.5	1.5	0.01848	0.368	6

A	1

Table 10.6. Single-scale layers parameters.

	SINGLE-SCALE LAYERS						
	LAYER	F	G	Q	R	UNCONDITIONAL MEAN	UNCONDITIONAL VARIANCE
TOP	1					0.368	6
	2	1	-0.07	1.5	1.5	0.298	7.5
BOTTOM	3	1	-0.08	1.5	1.5	0.218	9

C	1
D	0

H(s)	4

	LAYER 1 MEAN	L 1 VAR	L 2 MEAN	L 2 VAR	L 3 MEAN	L 3 VAR
m=3	0.368	1.5				
m=2	0.368	3				
m=1	0.368	4.5				
m=0 m'=0	0.368	6	0.298	7.5	0.218	9

By now, we have finished the estimations of the unconditional distributions for each cell as shown above.

10.6.7 *Layer3_Scale0: prediction a priori (from unconditionals)*

The results for the first prediction are shown in Table 10.7.

10.6.8 *Layer3_Scale0: K gain matrix*

Equation (9) is used to compute the gain matrix. For example, for the first cell in the scale = 0 layer = 3:

$$K(s,l) = \frac{P_{PL}(s,l)C^T(s,l)}{C(s,l)P_{PL}(s,l)C^T(s,l) + R(s,l)}$$
$$= \frac{9 \cdot 1}{1 \cdot 9 \cdot 1 + 1.5} = 0.857$$

10.6.9 *Layer3_scale0: innovation*

To obtain the observation estimated from the prediction, use Eq. (10):

$$\hat{y}(s,l) = C(s,l)\hat{x}_{PL}(s,l) + D(s,l) = 1 \cdot 0.218 + 0 = 0.218$$

To calculate the innovation, use Eq. (11). For the first cell in layer 3 scale 0:

$$IN(s,l) = [y(s,l) - \hat{y}(s,l)] = 0.2340 - 0.218 = 0.016$$

Table 10.7. First prediction is made from the unconditional distributions.

LAYER 3
MEAN
0.218
VARIANCE
9

LAYER 3							
INNOVATION							
0.016	-0.064	0.025	0.014	0.049	0.031	0.061	0.067
-0.001	0.007	0.023	0.049	0.052	0.096	0.039	0.072
-0.006	-0.017	0.013	0.070	0.033	0.065	0.071	0.040
0.018	-0.033	0.050	0.018	0.070	0.065	0.042	0.033
-0.001	0.027	0.033	0.065	0.056	0.040	0.053	0.044
-0.043	-0.030	-0.010	-0.001	0.050	-0.019	0.010	0.061
0.007	0.027	0.034	0.002	-0.002	-0.034	-0.002	0.040
-0.012	0.013	-0.032	0.000	-0.008	-0.032	-0.018	0.000

10.6.10 *Layer3_scale0: update*

Use Eqs. (12) and (13). For the first cell:

$$\hat{x}_U(s,l) = \hat{x}_{PL}(s,l) + K(s,l)IN(s,l) = 0.218 + 0.857 \cdot 0.016 = 0.232$$

	L3_S0 OBSERVATION								L3_S0 PREDICTION	L3_S0 UPDATED							
MEAN	0.234	0.154	0.242	0.232	0.267	0.248	0.278	0.284		0.232	0.163	0.239	0.230	0.260	0.244	0.270	0.275
	0.217	0.224	0.240	0.267	0.270	0.314	0.256	0.289		0.217	0.223	0.237	0.260	0.263	0.301	0.251	0.279
	0.212	0.201	0.230	0.288	0.251	0.283	0.289	0.258		0.213	0.203	0.228	0.278	0.246	0.274	0.279	0.252
	0.236	0.185	0.268	0.235	0.287	0.283	0.260	0.250	0.218	0.234	0.189	0.261	0.233	0.278	0.274	0.254	0.246
	0.217	0.244	0.250	0.283	0.274	0.258	0.271	0.262		0.217	0.241	0.246	0.274	0.266	0.252	0.263	0.256
	0.175	0.188	0.208	0.216	0.268	0.198	0.228	0.279		0.181	0.192	0.209	0.217	0.261	0.201	0.227	0.271
	0.224	0.245	0.252	0.219	0.216	0.184	0.216	0.258		0.223	0.241	0.247	0.219	0.216	0.189	0.216	0.252
	0.205	0.231	0.186	0.218	0.209	0.186	0.200	0.217		0.207	0.229	0.190	0.218	0.210	0.190	0.202	0.217
VARIANCE	1.5								9	1.286							

By visually comparing the prediction, the observation and the updated estimation, it can be seen that the updated one is an intermediate value between the predicted and the observed.

10.6.11 *Layer2_scale0: prediction*

LAYER 2							
MEAN PREDICTED							
0.273	0.216	0.279	0.272	0.297	0.283	0.305	0.309
0.261	0.266	0.277	0.296	0.299	0.330	0.289	0.312
0.257	0.249	0.270	0.312	0.285	0.308	0.312	0.290
0.275	0.238	0.297	0.274	0.311	0.308	0.291	0.285
0.261	0.280	0.285	0.308	0.302	0.290	0.299	0.293
0.231	0.240	0.254	0.260	0.297	0.248	0.269	0.305
0.266	0.281	0.286	0.263	0.260	0.237	0.260	0.290
0.253	0.271	0.239	0.262	0.255	0.239	0.249	0.261

$$\hat{x}_{PL}(s,2) = \tilde{F}(s,l+1)\hat{x}_U(s,l+1) + \tilde{G}(s,l+1) = \tilde{F}(s,3)\hat{x}_U(s,3) + \tilde{G}(s,3)$$

$$\tilde{F}(s,3) = F^{-1}(s,3)[I - Q(s,3)P^{-1}(s,3)] = \frac{1 - \frac{1.5}{9}}{1} = 0.833$$

$$\tilde{G}(s,3) = -G(s,3) = 0.08$$

$$\hat{x}_{PL}(s,2) = \tilde{F}(s,l+1)\hat{x}_U(s,l+1) + \tilde{G}(s,l+1) = \tilde{F}(s,3)\hat{x}_U(s,3) + \tilde{G}(s,3)$$
$$= 0.833 \cdot 0.232 + 0.08 = 0.273$$

10.6.12 *Layer2_scale0: gain matrix and innovation*

LAYER 2							
INNOVATION							
0.032	0.034	0.042	0.057	0.069	0.046	0.046	0.030
0.032	0.029	0.046	0.054	0.053	0.035	0.056	0.062
0.027	0.044	0.059	0.036	0.055	0.067	0.045	0.043
0.051	0.046	0.030	0.055	0.029	0.034	0.048	0.064
0.045	0.022	0.041	0.028	0.025	0.051	0.041	0.060
0.043	0.035	0.049	0.033	0.037	0.017	0.042	0.035
0.020	0.025	0.019	0.025	0.032	0.030	0.034	0.031
0.004	0.013	0.030	0.017	0.010	0.010	0.022	0.041

10.6.13 *Layer2_scale0: update*

LAYER 2							
MEAN UPDATED							
0.292	0.236	0.303	0.305	0.337	0.311	0.332	0.327
0.279	0.283	0.304	0.328	0.330	0.351	0.322	0.349
0.273	0.275	0.305	0.333	0.317	0.348	0.338	0.315
0.304	0.265	0.315	0.307	0.328	0.328	0.319	0.322
0.287	0.293	0.309	0.324	0.316	0.320	0.323	0.328
0.256	0.261	0.283	0.280	0.319	0.258	0.293	0.326
0.277	0.296	0.297	0.277	0.279	0.255	0.280	0.308
0.255	0.278	0.256	0.272	0.261	0.245	0.262	0.285

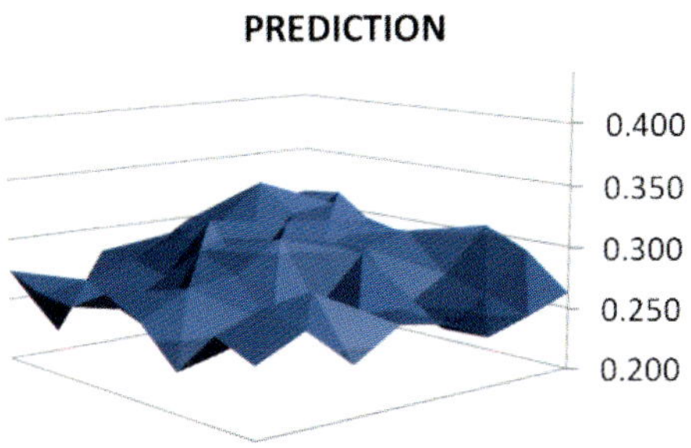

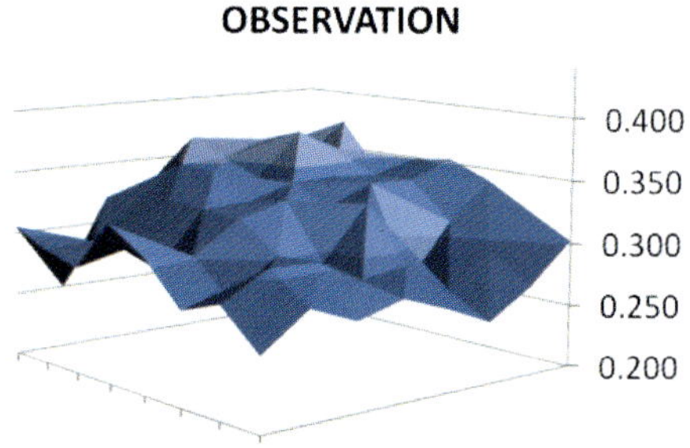

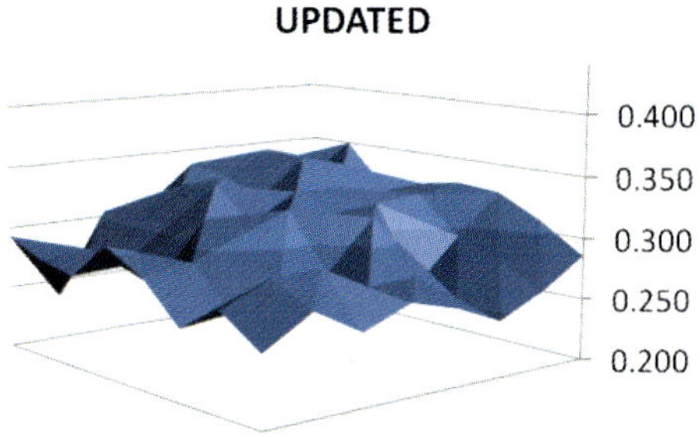

10.6.14 *Layer1_scale0: predict*

If m' is the finest scale in the tree, predictions from the single-scale can be used in the multi-scale tree.

LAYER 1							
MEAN PREDICTED							
0.304	0.259	0.313	0.314	0.339	0.318	0.335	0.331
0.293	0.297	0.313	0.332	0.334	0.351	0.328	0.349
0.289	0.290	0.314	0.336	0.324	0.348	0.341	0.322
0.313	0.282	0.322	0.315	0.333	0.332	0.326	0.328
0.300	0.305	0.317	0.329	0.323	0.326	0.329	0.333
0.275	0.279	0.297	0.294	0.325	0.276	0.305	0.331
0.292	0.307	0.308	0.292	0.293	0.274	0.294	0.316
0.274	0.293	0.275	0.287	0.279	0.266	0.279	0.298

10.6.15 *Layer1_scale0: gain matrix and innovation*

Use Eq. (20) for the gain matrix. Use Eqs. (21) and (22) for the innovation. For the first cell, we have

$$K(s,1) = \frac{P_{PS}(s,1)C^T(s,1)}{C(s,1)P_{PS}(s,1)C^T(s,1) + R(s,1)}$$

$$= \frac{1.765 \cdot 1}{1 \cdot 1.765 \cdot 1 + 1.5} = 0.541$$

$$\hat{y}(s,1) = C(s,1)\hat{x}_{PS}(s,1) + D(s,1) = 1 \cdot 0.304 + 0.01848 = 0.322$$

$$IN(s,1) = [y(s,1) - \hat{y}(s,1)] = 0.380 - 0.322 = 0.058$$

LAYER 1							
INNOVATION							
0.058	0.073	0.059	0.077	0.062	0.083	0.076	0.080
0.038	0.045	0.048	0.059	0.058	0.061	0.074	0.063
0.053	0.061	0.057	0.045	0.068	0.073	0.061	0.079
0.048	0.080	0.049	0.076	0.069	0.089	0.086	0.094
0.052	0.047	0.044	0.062	0.059	0.065	0.043	0.069
0.067	0.063	0.075	0.068	0.046	0.065	0.057	0.051
0.040	0.045	0.044	0.050	0.049	0.048	0.047	0.045
0.037	0.029	0.057	0.044	0.043	0.036	0.042	0.043

10.6.16 *Layer1_scale0: update*

Use Eqs. (23) and (24) to update the estimation. For the first cell:

$$\hat{x}_U(s,1) = \hat{x}_{PS}(s,1) + K(s,1)IN(s,1) = 0.304 + 0.541 \cdot 0.058 = 0.335$$

LAYER 1							
MEAN UPDATED							
0.335	0.298	0.345	0.356	0.373	0.363	0.377	0.375
0.314	0.321	0.339	0.364	0.365	0.384	0.368	0.383
0.317	0.323	0.345	0.361	0.360	0.388	0.374	0.365
0.339	0.325	0.349	0.356	0.370	0.381	0.372	0.379
0.328	0.330	0.341	0.363	0.355	0.361	0.352	0.370
0.311	0.313	0.337	0.330	0.350	0.311	0.335	0.358
0.313	0.331	0.331	0.319	0.319	0.300	0.320	0.341
0.294	0.308	0.306	0.311	0.302	0.285	0.302	0.322

10.6.17 *Layer1_scale1: prediction*

Use Eqs. (26) to (29) to compute the probability distribution of the prediction and the modified parameters.

$$\hat{x}_{PhS}(s,1) = [\tilde{A}(\alpha_h s,1)(\hat{x}_U(\alpha_h s,1) - \hat{x}(\alpha_h s,1))] + \hat{x}(\alpha_h s,1)$$

$$\tilde{A}(\alpha_h s,1) = \frac{I - \frac{Q(\alpha_h s,1)}{P(\alpha_h s,1)}}{A(\alpha_h s,1)} = \frac{1 - \frac{1.5}{6}}{1} = 0.75$$

$$\hat{x}_{PhS}(s,1) = [0.75 \cdot (0.335 - 0.368)] + 0.368 = 0.343$$

LAYER 1							
MEAN PREDICTED							
0.343	0.315	0.350	0.359	0.372	0.364	0.374	0.373
0.327	0.333	0.346	0.365	0.366	0.380	0.368	0.379
0.330	0.334	0.351	0.362	0.362	0.383	0.372	0.366
0.346	0.336	0.354	0.359	0.369	0.377	0.371	0.376
0.338	0.339	0.348	0.364	0.358	0.363	0.356	0.369
0.325	0.326	0.345	0.340	0.355	0.326	0.343	0.360
0.327	0.340	0.340	0.331	0.331	0.317	0.332	0.348
0.313	0.323	0.321	0.325	0.318	0.306	0.319	0.333

10.6.18 *Layer1_scale1: fusion*

First, use Eq. (31) to compute the fused variance.

$$P_{PS}(s,1) = \left[\frac{1-H(s)}{P(s,1)} + \sum_{h=1}^{H(s)} \frac{1}{P_{PhS}(s,1)}\right]^{-1}$$

$$= \left[\frac{1-4}{4.5} + \frac{1}{1.581} + \frac{1}{1.581} + \frac{1}{1.581} + \frac{1}{1.581}\right]^{-1} = 0.537$$

Then, use Eq. (30) to compute the fused mean.

$$\hat{x}_{PS}(s,1) = \hat{x}(s,1) + P_{PS}(s,1) \times \sum_{h=1}^{H(s)} \frac{\hat{x}_{PhS}(s,1) - \hat{x}(s,1)}{P_{PhS}(s,1)}$$

$$= 0.368 + 0.537 \cdot \left(\frac{0.343-0.368}{1.581} + \frac{0.315-0.368}{1.581} + \frac{0.327-0.368}{1.581} + \frac{0.333-0.368}{1.581}\right)$$

$$= 0.316$$

	LAYER 1							
	MEAN FUSED				VARIANCE FUSED			
m=1	0.316	0.351	0.371	0.376	0.537	0.537	0.537	0.537
	0.326	0.353	0.375	0.372	0.537	0.537	0.537	0.537
	0.319	0.342	0.344	0.354	0.537	0.537	0.537	0.537
	0.311	0.316	0.300	0.320	0.537	0.537	0.537	0.537

10.6.19 *Layer1_scale1: gain matrix and innovation*

	LAYER 1 SCALE 1 INNOVATION			
m=1	0.009	0.000	0.084	0.081
	0.023	0.004	0.065	0.068
	-0.032	-0.033	0.025	0.031
	0.002	-0.032	0.094	0.045

10.6.20 *Layer1_scale1: update*

	LAYER 1 SCALE 1							
	MEAN UPDATED				VARIANCE UPDATED			
m=1	0.319	0.351	0.393	0.397	0.395	0.395	0.395	0.395
	0.332	0.354	0.392	0.390	0.395	0.395	0.395	0.395
	0.311	0.334	0.351	0.362	0.395	0.395	0.395	0.395
	0.311	0.307	0.325	0.332	0.395	0.395	0.395	0.395

10.6.21 *Layer1_scale2: predict*

	LAYER 1 MEAN PREDICTED				LAYER 1 VARIANCE PREDICTED			
m=2	0.335	0.356	0.385	0.387	1.176	1.176	1.176	1.176
	0.344	0.358	0.384	0.383	1.176	1.176	1.176	1.176
	0.330	0.345	0.356	0.364	1.176	1.176	1.176	1.176
	0.330	0.327	0.339	0.344	1.176	1.176	1.176	1.176

10.6.22 *Layer1_scale2: fusion and update*

Results of this fusion can be used as the results for updating, because there are no observations at this scale.

	LAYER 1 SCALE 2			
	MEAN UPDATED		VARIANCE UPDATED	
m=2	0.340	0.392	0.416	0.416
	0.319	0.344	0.416	0.416

10.6.23 *Layer1_scale3: predict*

	LAYER 1			
	MEAN PREDICTED		VARIANCE PRED.	
m=3	0.354	0.380	0.854	0.854
	0.343	0.356	0.854	0.854

10.6.24 *Layer1_scale3: fusion and update*

	LAYER 1 SCALE 3	
	MEAN UPDATED	VARIANCE UPDATED
m=3	0.351	0.373

10.6.25 *Layer1_scale3: smoothing*

It is meaningless to smooth the root node of the multiscale tree because the last update is already an estimation of the hidden variable, given all the available observations.

	LAYER 1 SCALE 3	
	MEAN SMOOTHED	VARIANCE SMOOTHED
m=3	0.351	0.373

10.6.26 *Layer1_scale2: smoothing gain matrix and innovation*

Use Eq. (34) to compute the gain matrix and Eq. (35) to compute the innovation. For the first cell, we have

$$J(s,1) = P_U(s,1)\frac{\tilde{A}^T(s,1)}{P_{PS}(\gamma s,1)} = 0.416\frac{0.5}{0.854} = 0.244$$

$$\tilde{A}(s,1) = \frac{I - \frac{Q(s,1)}{P(s,1)}}{A(s,1)} = \frac{1-\frac{1.5}{3}}{1} = 0.5$$

$$IS(s,1) = \hat{x}_S(\gamma s,1) - \hat{x}_{PS}(\gamma s,1) = 0.351 - 0.354 = -0.003$$

	LAYER 1 SCALE 2 GAIN MATRIX		LAYER 1 SCALE 2 INNOVATION	
m=2	0.244	0.244	-0.003	-0.029
	0.244	0.244	0.008	-0.005

10.6.27 *Layer1_scale2: smoothing*

Use Eq. (36) to obtain the smoothed estimation. For the first cell, we have

$$\begin{aligned}\hat{x}_S(s,1) &= \hat{x}_U(s,1) + J(s,1)IS(s,1)\\ &= 0.340 + 0.244 \cdot (-0.003) = 0.3396\end{aligned}$$

	LAYER 1 SCALE 2			
	MEAN SMOOTHED		VARIANCE SMOOTHED	
m=2	0.3396	0.385	0.388	0.388
	0.321	0.343	0.388	0.388

10.6.28 *Layer1_scale2: covariance with parents*

The covariance with parent nodes will be used in the second part calculation of the framework (i.e., the EM part). Use Eq. (38). For example, for the first cell, we have

$$P_S(\{s,\gamma s\},1) = J(s,1)P_S(\gamma s,l) = 0.244 \cdot 0.373 = 0.091$$

LAYER 1 SCALE 2 COV. WITH PARENTS	
0.091	0.091
0.091	0.091

10.6.29 *Layer1_scale1: smoothing*

	LAYER 1 SCALE 1							
	MEAN SMOOTHED				VARIANCE SMOOTHED			
m=1	0.320	0.347	0.393	0.396	0.356	0.356	0.356	0.356
	0.331	0.349	0.392	0.391	0.356	0.356	0.356	0.356
	0.309	0.328	0.348	0.357	0.356	0.356	0.356	0.356
	0.309	0.306	0.326	0.332	0.356	0.356	0.356	0.356

10.6.30 *Layer1_scale1: covariance with parents*

LAYER 1 SCALE 1	COVARIANCE WITH PARENTS	0.087

10.6.31 *Layer1_scale0: smoothing*

	LAYER 1 SCALE 0							
	MEAN SMOOTHED							
m=0 m'=0	0.326	0.300	0.343	0.351	0.381	0.374	0.385	0.384
	0.311	0.316	0.340	0.357	0.376	0.389	0.379	0.389
	0.318	0.322	0.345	0.356	0.372	0.391	0.381	0.375
	0.333	0.323	0.347	0.353	0.379	0.386	0.380	0.384
	0.317	0.318	0.334	0.349	0.351	0.356	0.352	0.365
	0.305	0.306	0.331	0.326	0.348	0.320	0.341	0.357
	0.307	0.319	0.318	0.309	0.317	0.303	0.320	0.335
	0.293	0.303	0.300	0.304	0.305	0.293	0.307	0.321

10.6.32 *Layer1_scale0: covariance with parents*

LAYER 1 SCALE 0	COVARIANCE WITH PARENTS	0.137

10.6.33 *Layer2_scale0: smoothing*

For smoothing between the layers, the equations are very similar. First, the gain matrix is obtained from Eq. (41). Then, the innovation and the smoothed estimation are obtained with Eq. (39). For the first cell, we have

$$\tilde{F}(s,l) = \frac{I - \frac{Q(s,l)}{P(s,l)}}{F(s,l)} = \frac{1 - \frac{1.5}{7.5}}{1} = 0.8$$

$$J(s,l) = \frac{P_U(s,l)\tilde{F}^T(s,l)}{P_{PL}(s,l-1)} = \frac{0.882 \cdot 0.8}{1.765} = 0.400$$

$$\begin{aligned}\hat{x}_S(s,l) &= \hat{x}_U(s,l) + J(s,l)[\hat{x}_S(s,l-1) - \hat{x}_{PL}(s,l-1)] \\ &= 0.292 + 0.400 \cdot [0.326 - 0.304] = 0.301\end{aligned}$$

LAYER 2 SCALE 0							
MEAN SMOOTHED							
0.301	0.252	0.316	0.320	0.354	0.333	0.352	0.348
0.286	0.291	0.315	0.338	0.347	0.366	0.342	0.365
0.285	0.288	0.317	0.341	0.336	0.365	0.354	0.336
0.312	0.281	0.325	0.322	0.347	0.349	0.341	0.345
0.294	0.299	0.316	0.332	0.327	0.332	0.333	0.341
0.268	0.272	0.297	0.293	0.328	0.275	0.308	0.336
0.283	0.301	0.301	0.284	0.288	0.267	0.290	0.315
0.263	0.283	0.266	0.278	0.272	0.255	0.273	0.294

10.6.34 *Layer2_scale0: covariance with upper layer*

For the covariance with the upper layer, use Eq. (42):

$$P_S(s, \{l-1, l\}) = J(s, l-1)P_S(s, l) = 0.400 \cdot 0.630 = 0.252$$

LAYER 2 SCALE 0	**COVARIANCE WITH PARENTS**	0.252

10.6.35 *Layer3_scale0: smoothing*

LAYER 3 SCALE 0							
MEAN SMOOTHED							
0.246	0.181	0.257	0.254	0.288	0.269	0.293	0.294
0.230	0.236	0.256	0.280	0.286	0.318	0.278	0.305
0.227	0.223	0.252	0.292	0.272	0.302	0.300	0.275
0.252	0.211	0.275	0.257	0.295	0.294	0.278	0.276
0.233	0.250	0.261	0.286	0.279	0.273	0.280	0.280
0.199	0.208	0.230	0.233	0.276	0.215	0.246	0.286
0.232	0.251	0.255	0.230	0.230	0.203	0.231	0.265
0.212	0.235	0.204	0.226	0.219	0.199	0.215	0.234

10.6.36 *Layer3_scale0: covariance with upper layer*

LAYER 3 SCALE 0	**COVARIANCE WITH UPPER LAYER**	0.350

10.6.37 *E step — part 1*

Compute some statistics that are required for the estimation of the parameters. $\langle x(s,l)\rangle = \hat{x}(s,l|Y(s = root, l = 1)) = \hat{x}_S(s,l) = 0.351$. This first statistic is equal to the smoothed mean for all the layers and scales. No additional calculations are required.

MEAN SMOOTH LAYER 1

m=3				
	0.351			

m=2		
	0.340	0.385
	0.321	0.343

m=1				
	0.320	0.347	0.393	0.396
	0.331	0.349	0.392	0.391
	0.309	0.328	0.348	0.357
	0.309	0.306	0.326	0.332

m=0, m'=0								
	0.326	0.300	0.343	0.351	0.381	0.374	0.385	0.384
	0.311	0.316	0.340	0.357	0.376	0.389	0.379	0.389
	0.318	0.322	0.345	0.356	0.372	0.391	0.381	0.375
	0.333	0.323	0.347	0.353	0.379	0.386	0.380	0.384
	0.317	0.318	0.334	0.349	0.351	0.356	0.352	0.365
	0.305	0.306	0.331	0.326	0.348	0.320	0.341	0.357
	0.307	0.319	0.318	0.309	0.317	0.303	0.320	0.335
	0.293	0.303	0.300	0.304	0.305	0.293	0.307	0.321

MEAN SMOOTH LAYER 2

m=0, m'=0								
	0.301	0.252	0.316	0.320	0.354	0.333	0.352	0.348
	0.286	0.291	0.315	0.338	0.347	0.366	0.342	0.365
	0.285	0.288	0.317	0.341	0.336	0.365	0.354	0.336
	0.312	0.281	0.325	0.322	0.347	0.349	0.341	0.345
	0.294	0.299	0.316	0.332	0.327	0.332	0.333	0.341
	0.268	0.272	0.297	0.293	0.328	0.275	0.308	0.336
	0.283	0.301	0.301	0.284	0.288	0.267	0.290	0.315
	0.263	0.283	0.266	0.278	0.272	0.255	0.273	0.294

MEAN SMOOTH LAYER 3

m=0, m'=0								
	0.246	0.181	0.257	0.254	0.288	0.269	0.293	0.294
	0.230	0.236	0.256	0.280	0.286	0.318	0.278	0.305
	0.227	0.223	0.252	0.292	0.272	0.302	0.300	0.275
	0.252	0.211	0.275	0.257	0.295	0.294	0.278	0.276
	0.233	0.250	0.261	0.286	0.279	0.273	0.280	0.280
	0.199	0.208	0.230	0.233	0.276	0.215	0.246	0.286
	0.232	0.251	0.255	0.230	0.230	0.203	0.231	0.265
	0.212	0.235	0.204	0.226	0.219	0.199	0.215	0.234

10.6.38 *E step — part 2*

$$\langle x(s,l)x^T(s,l)\rangle = P_S(s,l) + \hat{x}_S(s,l)\hat{x}_S^T(s,l) = 0.373 + 0.351 \cdot 0.351 = 0.496$$

This calculation is made for every cell in the calculation space.

E STEP PART 1 LAYER 1

m=3	0.496							
m=2	0.503				0.536			
	0.490				0.505			
m=1	0.458		0.476		0.510		0.513	
	0.465		0.478		0.510		0.508	
	0.451		0.464		0.476		0.483	
	0.451		0.449		0.462		0.466	
m=0 m'=0	0.736	0.719	0.747	0.753	0.775	0.770	0.778	0.777
	0.726	0.729	0.745	0.757	0.771	0.781	0.773	0.781
	0.730	0.733	0.748	0.756	0.768	0.783	0.775	0.770
	0.741	0.734	0.750	0.754	0.773	0.779	0.774	0.777
	0.730	0.731	0.741	0.751	0.752	0.756	0.754	0.763
	0.722	0.723	0.739	0.736	0.750	0.732	0.746	0.757
	0.724	0.731	0.731	0.725	0.730	0.721	0.732	0.742
	0.715	0.721	0.719	0.722	0.722	0.715	0.724	0.733

E STEP PART 1 LAYER 2

m=0 m'=0	0.791	0.764	0.800	0.803	0.826	0.812	0.824	0.822
	0.783	0.785	0.800	0.815	0.821	0.835	0.818	0.834
	0.782	0.784	0.801	0.817	0.814	0.834	0.826	0.814
	0.798	0.780	0.806	0.804	0.821	0.823	0.817	0.820
	0.787	0.790	0.800	0.811	0.808	0.811	0.811	0.817
	0.772	0.775	0.789	0.786	0.808	0.776	0.795	0.814
	0.781	0.791	0.791	0.781	0.784	0.772	0.785	0.800
	0.770	0.781	0.772	0.778	0.774	0.766	0.775	0.787

E STEP PART 1 LAYER 3

m=0 m'=0	0.986	0.958	0.991	0.990	1.008	0.997	1.011	1.012
	0.978	0.981	0.991	1.004	1.007	1.026	1.002	1.018
	0.977	0.975	0.989	1.011	0.999	1.016	1.015	1.001
	0.989	0.970	1.001	0.991	1.012	1.012	1.003	1.001
	0.980	0.987	0.993	1.007	1.003	1.000	1.003	1.003
	0.965	0.968	0.978	0.979	1.001	0.971	0.986	1.007
	0.979	0.988	0.990	0.978	0.978	0.967	0.979	0.995
	0.970	0.980	0.967	0.976	0.973	0.965	0.971	0.980

10.6.39 *E step — part 3*

$$\begin{aligned}\langle x(s,1)x^T(\gamma s,1)\rangle &= P_S(\{s,\gamma s\},1) + \hat{x}_S(s,1)\hat{x}_S^T(\gamma s,1)\\ &= 0.091 + 0.3396 \cdot 0.351 = 0.210\end{aligned}$$

This calculation is made for every cell in the calculation space that has a parent node.

LAYER 1

m=3								

m=2		
	0.210	0.091
	0.091	0.091

m=1				
	0.195	0.087	0.238	0.087
	0.087	0.087	0.087	0.087
	0.186	0.087	0.206	0.087
	0.087	0.087	0.087	0.087

m=0, m'=0								
	0.014	0.000	0.016	0.000	0.021	0.000	0.021	0.000
	0.000	0.000	0.000	0.000	0.000	0.000	0.000	0.000
	0.014	0.000	0.016	0.000	0.020	0.000	0.020	0.000
	0.000	0.000	0.000	0.000	0.000	0.000	0.000	0.000
	0.013	0.000	0.015	0.000	0.017	0.000	0.017	0.000
	0.000	0.000	0.000	0.000	0.000	0.000	0.000	0.000
	0.013	0.000	0.013	0.000	0.014	0.000	0.015	0.000
	0.000	0.000	0.000	0.000	0.000	0.000	0.000	0.000

10.6.40 *E step — part 4*

$$\begin{aligned}\langle x(s,l-1)x^T(s,l)\rangle &= P_S(s,\{l-1,l\}) + \hat{x}_S(s,l-1)\hat{x}_S^T(s,l)\\ &= 0.252 + 0.301 \cdot 0.326 = 0.350\end{aligned}$$

This calculation is made for every cell in the calculation space that has a soil layer above.

LAYER 2								LAYER 3							
0.350	0.327	0.360	0.364	0.387	0.377	0.387	0.385	0.424	0.396	0.432	0.432	0.452	0.440	0.453	0.453
0.341	0.344	0.359	0.373	0.382	0.394	0.381	0.394	0.416	0.419	0.431	0.445	0.450	0.467	0.445	0.462
0.342	0.345	0.361	0.373	0.377	0.395	0.387	0.378	0.415	0.414	0.430	0.450	0.442	0.461	0.457	0.443
0.356	0.343	0.365	0.365	0.383	0.387	0.381	0.384	0.429	0.410	0.440	0.433	0.453	0.453	0.445	0.446
0.345	0.347	0.357	0.368	0.367	0.370	0.369	0.377	0.419	0.425	0.433	0.445	0.442	0.441	0.443	0.446
0.333	0.335	0.350	0.347	0.366	0.340	0.357	0.372	0.404	0.407	0.419	0.418	0.441	0.410	0.426	0.446
0.339	0.348	0.348	0.340	0.343	0.333	0.345	0.357	0.416	0.426	0.427	0.416	0.417	0.405	0.417	0.434
0.329	0.337	0.332	0.336	0.335	0.327	0.336	0.346	0.406	0.417	0.405	0.413	0.410	0.401	0.409	0.419

10.6.41 *M step*

In the M step, the configuration parameters of the MKS are recomputed. For the unconditional variance of the root, use Eq. (43):

$$\begin{aligned} P(root,1) &= \langle x(root,1)x^T(root,1)\rangle - \hat{x}(root,1)\hat{x}^T(root,1) \\ &= 0.496 - 0.368 \cdot 0.368 = 0.361 \end{aligned}$$

For the variance of observations, use Eq. (44). For example, layer 1 of scale 1 has 16 satellite observations. For the first cell, we have

$$\begin{aligned} &[y(s,l) - C(s,l)\langle x(s,l)\rangle - D(s,l)]^T \cdot \\ &[y(s,l) - C(s,l)\langle x(s,l)\rangle - D(s,l)] \\ &\quad = [0.307 - 1 \cdot 0.3196 - (-0.01848)]^T \cdot \\ &[0.307 - 1 \cdot 0.3196 - (-0.01848)] = 0.0000349 \end{aligned}$$

LAYER 1 SCALE 1			
ESTIMATED OBSERVATION VARIANCE			
0.0000349	0.0000149	0.0037828	0.0036157
0.0003321	0.0000538	0.0023080	0.0024884
0.0004694	0.0003416	0.0004647	0.0007801
0.0000152	0.0004942	0.0047206	0.0010993
		SUM	0.0210

Using these values for the layer 1 of scale 1, we have

$$\begin{aligned} R(m,l) &= \frac{1}{n_{m,l}} \sum_s \left\{ \begin{array}{l} [y(s,l) - C(s,l)\langle x(s,l)\rangle - D(s,l)]^T \cdot \\ [y(s,l) - C(s,l)\langle x(s,l)\rangle - D(s,l)] \end{array} \right\} \\ &= \frac{1}{16}[0.021] = 0.00131 \end{aligned}$$

The new parameters are:

	LAYER 1 MULTI-SCALE						
	SCALE	Q	R	D	UNC. MEAN	UNC. VAR.	n_{ml}
ROOT	3	1.5	1.5	0	0.3676	0.361	1
	2	0.388	1.5	0	0.3676	0.749	4
	1	0.359	0.0013	-0.0185	0.3676	1.108	16
LEAVES	0	0.742	0.0008	0.0185	0.3676	1.850	64

A	1

	SINGLE-SCALE LAYERS							
	LAYER	F	G	Q	R	UNC. MEAN	UNC. VAR.	N_{ml}
TOP	1					0.36761	1.85028	64
	2	1	-0.07	0.462	0.000040	0.29761	2.31210	64
BOTTOM	3	1	-0.08	0.580	0.000242	0.21761	2.89257	64

C	1
D	0

H(s)	4

10.6.42 *Final iteration*

After five iterations, the EM algorithm does not significantly change the estimated parameters. The final converged parameters are shown below:

	LAYER 1 MULTI-SCALE						
	SCALE	Q	R	D	UNC. MEAN	UNC. VAR.	n_{ml}
ROOT	3	1.5	1.5	0	0.3676	0.006	1
	2	0.121	1.5	0	0.3676	0.127	4
	1	0.115	0.0010	-0.0185	0.3676	0.242	16
LEAVES	0	0.165	0.0007	0.0185	0.3676	0.407	64

A	1

	SINGLE-SCALE LAYERS							
	LAYER	F	G	Q	R	UNC. MEAN	UNC. VAR.	N_{ml}
TOP	1					0.36761	0.40734	64
	2	1	-0.07	0.041	0.000036	0.29761	0.44842	64
BOTTOM	3	1	-0.08	0.044	0.000227	0.21761	0.49201	64

C	1
D	0

H(s)	4

And the final estimations of the soil moisture are shown below:

MEAN SMOOTH LAYER 1

m=3

0.363

m=2

0.344	0.393
0.323	0.349

m=1

0.323	0.349	0.401	0.404
0.335	0.352	0.399	0.397
0.309	0.328	0.352	0.361
0.311	0.306	0.335	0.337

m=0, m'=0

0.329	0.298	0.344	0.354	0.382	0.374	0.386	0.385
0.312	0.317	0.340	0.361	0.376	0.391	0.379	0.391
0.318	0.323	0.346	0.359	0.371	0.394	0.382	0.375
0.336	0.324	0.349	0.355	0.379	0.388	0.381	0.386
0.319	0.321	0.336	0.354	0.354	0.359	0.354	0.369
0.305	0.307	0.332	0.327	0.350	0.317	0.340	0.359
0.308	0.323	0.322	0.311	0.319	0.303	0.320	0.338
0.292	0.304	0.300	0.305	0.305	0.290	0.306	0.322

MEAN SMOOTH LAYER 2

m=0, m'=0

0.302	0.251	0.316	0.321	0.354	0.332	0.352	0.348
0.286	0.291	0.315	0.339	0.346	0.367	0.342	0.365
0.285	0.288	0.317	0.341	0.336	0.366	0.354	0.336
0.313	0.281	0.325	0.322	0.347	0.350	0.341	0.345
0.295	0.299	0.316	0.334	0.328	0.333	0.333	0.343
0.268	0.272	0.297	0.293	0.328	0.274	0.307	0.337
0.283	0.302	0.302	0.284	0.289	0.266	0.290	0.316
0.262	0.282	0.266	0.278	0.271	0.254	0.272	0.294

MEAN SMOOTH LAYER 3

m=0, m'=0

0.246	0.180	0.257	0.254	0.288	0.268	0.293	0.294
0.229	0.235	0.255	0.280	0.286	0.318	0.277	0.305
0.226	0.222	0.251	0.292	0.271	0.302	0.299	0.274
0.252	0.211	0.274	0.256	0.295	0.294	0.278	0.275
0.233	0.249	0.261	0.286	0.279	0.273	0.279	0.280
0.199	0.207	0.230	0.232	0.276	0.214	0.245	0.285
0.231	0.251	0.254	0.229	0.230	0.203	0.231	0.265
0.211	0.234	0.203	0.226	0.218	0.198	0.213	0.233

10.7 Symbols

x : Hidden state variable
F, G : Parameters for the linear transformation of the hidden state variable, from one layer to the next one.
w : White noise with variance Q'
Q' : Variance of the noise in the time-dynamic equation (1)
$\hat{x}$: Estimation of the hidden state variable
y : Observation
$\hat{y}$: Estimation of the observation
C, D : Parameters for the linear transformation from the hidden state variable to observations.
v : White noise with variance R'
R' : Variance of the noise in the observation equation (2) associated with time
P : Variance of the hidden state variable
L : Number of soil-layers in the model
l : One of the soil-layers
M : Number of scales
m : One of the scales
m' : Scale of the single-scale layers
A : Parameter for transforming the hidden variable from coarser to finer scales
Q : Variance of the noise in the multi-scale dynamic equation (5)
R : Variance of the noise in the observation associated with multi-scale
θ_i^t : Soil moisture at time t in the layer of i
z_i : Deepest edge of the layer i
h : Pressure head
$\hat{x}_{PL}(s, l)$: Mean of prediction of x from the lower layer
$\hat{x}_{PS}(s, 1)$: Mean of prediction of x from the coarser scale
$P_{PL}(s, l)$: Variance of prediction of x from the lower layer
$P_{PS}(s, 1)$: Variance of prediction of x from the coarser scale
K : Kalman gain
$IN(s, l)$: Innovation
$\hat{x}_U$: Updated estimation of the hidden state variable
$P_U(s, l)$: Variance of the updated estimation
$\hat{x}_{PhS}$: Prediction of the hidden state from the coarser layer from the descendant h

$J(s,1)$: Smoother gain matrix
$IS(s,1)$: Innovation in the smoother
$\hat{x}_S(\gamma s,1)$: Smoothed estimation of the hidden variable
$P_S(s,1)$: Variance of the smoothed estimation of the hidden variable
$N_{m,l=1}$: The number of cells (or nodes) associated with a hidden state
any var(s,l) : A variable defined in cell s located in the soil-layer l. "s" can be located in any level of a multi-scale tree or simply in a single-scale grid.
$H(s)$: Number of children of the cell s

$$Y(s,l) = \{y(\varsigma,\kappa) : [(\varsigma = s) \vee (\varsigma : \textit{descendant of } (s))] \wedge [\kappa \geq l]\}$$
$$Y(s-,\cdot) = \{y(\varsigma,\cdot) : (\varsigma : \textit{descendant of } (s))\}$$
$$Y(\cdot,l+) = \{y(\cdot,\kappa) : \kappa > l\}$$

Acknowledgements

This work was partially supported by the NOAA grant of NA06DAR 4310130 to the University of Pittsburgh.

References

Boulet, G., Kerr, Y. and Chehbouni, A. (2002). Deriving catchment-scale water and energy balance parameters using data assimilation based on extended Kalman filtering. *Hydrological Sciences Journal*, **47**(3), 449–467.

Bowling, *et al.* (2003). Simulation of high latitude hydrological processes in the Torne-Kalix basin: PILPS Phase 2(e), Part 1: Experiment description and summary intercomparisons. *Journal of Global and Planetary Change*, **38**(1–2), 1–30.

Chen, T., Henderson-Sellers, A., Milly, P. *et al.* (1997). Experimental results from the project for intercomparison of land-surface parameterization schemes. *Journal of Climate*, **10**(6), 1194–1215.

Cherkauer, K. A. and Lettenmaier, D. P. (1999). Hydrologic effects of frozen soils in the upper Mississippi River basin. *Journal of Geophysical Research*, **104**, 19599–19610.

Cherkauer, K. A. and Lettenmaier, D. P. (2003). Simulation of spatial variability in snow and frozen soil. *Journal Geophysical Research*, **108**(D22), Art. No. 8858.

Chou, K. C., Willsky, A. S. and Benveniste, A. (1994). Multiscale recursive estimation, data fusion, and regularization. *IEEE Transactions on Automatic Control*, **39**(3), 479–492.

Crow, W. T. and Wood, E. F. (2003). The assimilation of remotely sensed soil brightness temperature imagery into a land surface model using Ensemble

Kalman filtering: A case study based on ESTAR measurements during SGP97. *Advances in Water Resources*, **26**, 137–149.

Fieguth, P. W., Karl, W. C., Willsky, A. S. and Wunsch, C. (1995). Multiresolution optimal interpolation and statistical analysis of TOPEX/POSEIDON satellite altimetry. *IEEE Transactions on Geosciences and Remote Sensing*, **33**(2), 280–292.

Galantowicz, J. F., Entekhabi, D. and Njoku, E. G. (1999). Tests of sequential data assimilation for retrieving profile soil moisture and temperature from observed L-band radio brightness. *IEEE Transactions on Geosciences and Remote Sensing*, **37**(4), 1860–1870.

Hoeben, R. and Troch, P. A. (2000). Assimilation of active microwave observation data for soil moisture profile estimation. *Water Resources Research*, **36**(10), 2805–2819.

Houser, P. R., Shuttleworth, W. J., Famiglietti, J. S., Gupta, H. V., Syed, K. H. and Goodrich, D. C. (1998). Integration of soil moisture remote sensing and hydrologic modeling using data assimilation. *Water Resources Research*, **34**(12), 3405–3420.

Kalman, R. E. (1960). A new approach to linear filtering and prediction problems. *Journal of Basic Engineering*, **82**(1), 35–45.

Kannan, A., Ostendorf, M., Karl, W. C., Castanon, D. A. and Fish, R. K. (2000). ML parameter estimation of a multiscale stochastic process using the EM algorithm. *IEEE Transactions on Signal Processing*, **48**(6), 1836–1840.

Kumar, P. (1999). A multiple scale state-space model for characterizing subgrid scale variability of near-surface soil moisture. *IEEE Transactions on Geosciences and Remote Sensing*, **37**(1), 182–197.

Kumar, P. and Kaleita, A. L. (2003). Assimilation of near-surface temperature using extended Kalman filter. *Advances in Water Resources*, **26**, 79–93.

Lannoy, De, G. J. M., Reichle, R. H., Houser, P. R. and Arsenault, K. R. (2010). Satellite-scale snow water equivalent assimilation into a high-resolution land surface model. *Journal of Hydrometeorology*, **11**(2), 352–369, doi:10.1175/2009JHM1192.1.

Leung, L. R., Hamlet, A. F., Lettenmaier, D. P. and Kumar, A. (1999). Simulations of the ENSO hydroclimate signals in the Pacific Northwest Columbia River basin. *Bulletin of the American Meteorological Society*, **80**(11), 2313–2329.

Liang, X., Lettenmaier, D. P., Wood, E. F. and Burges, S. J. (1994). A simple hydrologically based model of land surface water and energy fluxes for general circulation models. *Journal of Geophysical Research*, **99**(D7), 14415–14428.

Liang, X., Lettenmaier, D. P. and Wood, E. F. (1996a). Surface soil moisture parameterization of the VIC-2L model: Evaluation and modification. *Global and Planetary Change*, **13**, 195–206.

Liang, X., Lettenmaier, D. P. and Wood, E. F. (1996b). One-dimensional statistical dynamic representation of subgrid spatial variability of precipitation in the two-layer variable infiltration capacity model. *Journal of Geophysical Research: Atmospheres*, **101**(D16), 21403–21422.

Liang, X., Wood, E., Lettenmaier, D. *et al.* (1998). The project for intercomparison of land-surface parameterization schemes (PILPS) Phase 2(c) Red-Arkansas River basin experiment, Part 2: Spatial and temporal analysis of energy fluxes. *Global and Planetary Change*, **19**, 137–159.

Liang, X., Wood, E. F. and Lettenmaier, D. P. (1999). Modeling ground heat flux in land surface parameterization schemes. *Journal of Geophysical Research*, **104**(D8), 9581–9560.

Liang, X. and Xie, Z. (2001). A new surface runoff parameterization with subgrid-scale soil heterogeneity for land surface models. *Advances in Water Resources*, **24**, 1173–1193.

Liang, X., Xie, Z. and Huang, M. (2003). A new parameterization for surface and groundwater interactions and its impact on water budgets with the variable infiltration capacity (VIC) land surface model. *Journal of Geophysical Research*, **108**(D16), 8613, doi:10.1029/2002JD003090.

Lohmann, D., Lettenmaier, D., Liang, X. *et al.* (1998). The project for intercomparison of land-surface parameterization schemes (PILPS) Phase 2(c) Red-Arkansas River basin experiment, Part 3: Spatial and temporal analysis of water fluxes. *Global and Planetary Change*, **19**, 161–179.

Luettgen, M. R. and Willsky, A. S. (1995). Multiscale smoothing error models, *IEEE Transactions on Automatic Control*, **40**(1), 173–175.

Mao, D. Z. and Cherkauer, K. A. (2009). Impacts of land-use change on hydrologic responses in the Great Lakes region. *Journal of Hydrology*, **374**(1–2), 71–82.

Margulis, S. A., Entekhabi, D,. McLaughlin, D. and Dunne, S. (2003). Toward the real-time observation of hydrologic land surface fluxes from space: Data assimilation using the ensemble Kalman filter. In *17th Conference on Hydrology, American Meteorological Society*, February, 9–13.

Montaldo, N. and Albertson, J. D. (2003). Multi-scale assimilation of surface soil moisture data for robust root zone soil moisture predictions. *Advances in Water Resources*, **26**, 33–44.

Nijssen, B., Lettenmaier, D. P., Liang, X., Wetzel, S. W. and Wood, E. F. (1997). Streamflow simulation for continental-scale river basins, *Water Resources Research*, **33**(4), 711–724.

Nijssen, B., O'Donnell, G. M., Hamlet, A. F. and Lettenmaier, D. P. (2001a). Hydrologic sensitivity of global rivers to climate change. *Climatic Change*, **50**(1–2), 143–175.

Nijssen, B., O'Donnell, G. M., Lettenmaier, D. P., Lohmann, D. and Wood, E. F. (2001b). Predicting the discharge of global rivers. *Journal of Climate*, **14**(15), 3307–3323.

Ni-Meister, W., Walker, J. P. and Houser, P. R. (2005). Soil moisture initialization for climate prediction: Characterization of model and observation errors. *Journal of Geophysical Research*, **110**, D13111, doi:10.1029/2004JD005745.

Pan, M., Wood, E. F. McLaughlin, D. B., Entekhabi, D. and Luo, L. (2009). A multiscale ensemble filtering system for hydrologic data assimilation, Part I: Implementation and synthetic experiment. *Journal of Hydrometeorology*, **10**, 794–806.

Parada, L. M. and Liang, X. (2003a). A downscaling framework for L-band radio brightness temperature imagery. *Journal of Geophysical Research*, **108**(D22), 8863, doi:10.1029/2003JD003637.

Parada, L. M. and Liang, X. (2003b). A stochastic modeling approach for characterizing the spatial structure of L-band radio brightness temperature imagery. *Journal of Geophysical Research*, **108**(D22), 8862, doi:10.1029/2003JD003567.

Parada, L. M. and Liang, X. (2004). Optimal multiscale Kalman filter for assimilation of near-surface soil moisture for land surface models. *Journal of Geophysical Research*, **109**, D24109, doi:10.1029/2004JD004745.

Parada, L. M. and Liang, X. (2008). Impacts of spatial resolutions and data quality on soil moisture data assimilation. *Journal of Geophysical Research*, **113**, D10101, doi:10.1029/2007JD009037.

Pauwels, V. R. N., Hoeben, R., Verhoest, N. E. C. and De Troch, F. P. (2001). The importance of the spatial patterns of remotely sensed soil moisture in the improvement of discharge predictions for small-scale basins through data assimilation. *Journal of Hydrology*, **251**, 88–102.

Pauwels, V. R. N., Hoeben, R., Verhoest, N. E. C., De Troch, F. P. and Troch, P. A. (2002). Improvement of TOPLATS-based discharge predictions through assimilation of ERS-based remotely sensed soil moisture values. *Hydrological Processes*, **16**(5), 995–1013.

Reichle, R. H., McLaughlin, D. B. and Entekhabi, D. (2002a). Hydrologic data assimilation with the ensemble Kalman filter. *Monthly Weather Review*, **130**, 103–130.

Reichle, R. H., Walker, J. P., Koster, R. D. and Houser, P. R. (2002b). Extended versus ensemble Kalman filtering for land data assimilation. *Journal of Hydrometeorology*, **3**, 728–740.

Rothrock, D. A. and Thomas, D. R. (1992). Ice modeling and data assimilation with the Kalman smoother. *Carsey FD (ed) microwave remote sensing of sea ice, AGU Geophysical Monograph American Geophysical Union Washington D.C.*, 68, 405–418.

Sun, C., Walker, J. P. and Houser, P. R. (2004). A methodology for snow data assimilation in a land surface model. *Journal of Geophysical Research*, **109**(D8), D08108.

Walker, J. P. and Houser, P. R. (2001). A methodology for initializing soil moisture in a global climate model: Assimilation of near-surface soil moisture observations. *Journal of Geophysical Research*, **106**(D11), 11761–11774.

Wang, S., Liang, X. and Nan, Z. (2011). How much improvement can precipitation data fusion achieve with a multiscale Kalman Smoother-based framework? *Water Resources Research*, **47**(3), W00H12, doi:10.1029/2010WR009953.

Wood, E. F., Lettenmaier, D. P., Liang, X., Nijssen, B. and Wetzel, S. W. (1997). Hydrological modeling of continental scale basins. *Annual Review of Earth and Planetary Sciences*, **25**, 279–300.

Wood, E. F. *et al.* (1998). The project for intercomparison of land-surface parameterization schemes (PILPS) Phase 2(c) Red-Arkansas River basin experiment, Part 1: Experiment description and summary intercomparisons, *Global and Planetary Change*, **19**(1–4), 115–135.

Part 4

Application

Chapter 11

Overview of the North American Land Data Assimilation System (NLDAS)

Youlong Xia*,†, Brian A. Cosgrove‡, Michael B. Ek*, Justin Sheffield§, Lifeng Luo¶, Eric F. Wood§, Kingtse Mo‖ and the NLDAS team**

**NOAA/NCEP Environmental Modeling Center, Camp Springs, MD, USA*

†IMSG/NCEP Environmental Modeling Center, Camp Springs, MD, USA

‡NOAA/NWS Office of Hydrologic Development, Silver Spring, MD, USA

§Department of Environmental and Civil Engineering, Princeton University, Princeton, NJ, USA

¶Department of Geography, Michigan State University, East Lansing, MI, USA

‖NOAA/NCEP Climate Prediction Center, Camp Springs, MD, USA

The National Oceanic and Atmospheric Administration (NOAA) National Centers for Environmental Prediction (NCEP) Environmental Modeling Center (EMC), together with its NOAA Climate Program Office (CPO) Climate Prediction Program of the Americas (CPPA) partners, have established a North American Land Data Assimilation System (NLDAS). The system runs multiple land surface models (LSMs) over the Continental United States (CONUS) to generate long-term hourly, 1/8th degree hydrological and meteorological products. NLDAS was initiated in 1998 as a collaborative project between

**The NLDAS team comprises all scientists in NLDAS-1 (Kenneth Mitchell, Dag Lohmann, Paul Houser, Eric Wood, John Schaake, Alan Robock, Brian Cosgrove, Justin Sheffield, Qingyun Duan, Lifeng Luo, R. Wayne Higgins, Rachel Pinker, J. Dan Tarpley, Dennis Lettenmaier, Curtis Marshall, Jared Entin, Ming Pan, Wei Shi, Victor Koren, Jesse Meng, Bruce Ramsay, and Andrew Bailey) as well as new additions in NLDAS-2 (Charles Alonge, Jiarui Dong, Yun Fan, Kintse Mo, Ben Livneh, David Mocko, Helin Wei, Andy Wood, and Youlong Xia).

S. Liang, X. Li & X. Xie (eds.), *Land Surface Observation, Modeling and Data Assimilation*, 335–376.

NOAA, NASA, and several universities to improve the generation of initial land surface conditions for numerical weather prediction models. The first phase of NLDAS (NLDAS-1, 1998–2005) centered on the construction of the overall NLDAS system and on the assessment of the ability of the four NLDAS LSMs to accurately simulate water fluxes, energy fluxes, and state variables. These LSMs included the Noah, Mosaic, Sacramento Soil Moisture Accounting (SAC-SMA), and Variable Infiltration Capacity (VIC) models. Building on the results of NLDAS-1, the project entered into a second phase (NLDAS-2, 2006–present) which has included upgraded forcing data and LSMs, model intercomparison studies, real-time monitoring of extreme weather events, and seasonal hydrologic forecasts. NLDAS-1 and NLDAS-2 have also spurred and supported other modeling activities, including high-resolution 1 km land surface modeling and the establishment of regional and global land data assimilation systems. NLDAS-2 operates on both a real-time monitoring mode and an ensemble seasonal hydrologic forecast mode. In the monitoring mode, land states (soil moisture and snow water equivalent) and water fluxes (evaporation, total runoff, and streamflow) from real-time LSM executions are depicted as anomalies and percentiles with respect to their own model-based climatology. One key application of the real-time updates is for drought monitoring over the CONUS, and NLDAS supports both NOAA Climate Prediction Center (CPC) and US National Integrated Drought Information System (NIDIS) drought monitoring activities.

The uncoupled ensemble seasonal forecast mode generates downscaled ensemble seasonal forecasts of surface forcing based on a climatological Ensemble Streamflow Prediction (ESP) type approach, a method utilizing CPC Official Seasonal Climate Outlooks, and a third approach using NCEP Climate Forecast System (CFS) ensemble dynamical model predictions. The three sets of forcing ensembles are then used to drive a chosen LSM (currently VIC) in seasonal forecast mode over 14 large river basins that together span the CONUS domain. One- to six-month ensemble seasonal forecast products such as air temperature, precipitation, soil moisture, snowpack, total runoff, evaporation, and streamflow are derived using each forecasting approach. The anomalies and percentiles of the predicted products and the drought probability forecast based on the predicted total column soil moisture for each forcing approach can be used for the purpose of drought prediction over the CONUS, and provide key support for NIDIS and CPC drought forecast efforts.

11.1 Introduction

Over the past decade, real-time and retrospective Land Data Assimilation Systems (LDAS) have emerged as important sources of drought monitoring and forecast data, initial land surface states for numerical weather prediction models, and forcing data for end-user applications in areas including hydrology, meteorology, biology, and public safety. This chapter will detail the emergence and evolution of such systems and will focus, in particular, on the first two phases of the collaborative North American LDAS (NLDAS) project. Details will be provided on the structure and operations of NLDAS, on the validation of land surface output and forcing fields, and on the system's monitoring and forecast capabilities.

11.1.1 *Background of LDAS*

Global atmospheric four-dimensional data assimilation systems have been a mainstay of operational numerical weather prediction centers since the 1970s. Real-time four-dimensional ocean data assimilation on large-scale ocean basins followed in the middle to late 1980s (Ji *et al.*, 1994) on the heels of the Tropical Ocean Global Atmosphere (TOGA) program. However, it was not until the mid-1990s that real-time continental and global LDAS came into existence. The first viable examples of such systems were the coupled land–atmosphere four-dimensional data assimilation systems at major numerical weather prediction centers such as the National Oceanic and Atmospheric Administration's (NOAA) National Centers for Environmental Prediction (NCEP) (Kalnay *et al.*, 1996) and the European Centre for Medium-Range Weather Forecasts (ECMWF) (Gibson *et al.*, 1997). Such coupled land–atmosphere data assimilation systems are often characterized by sizable errors and drift in soil moisture and temperature states and surface energy and water fluxes. These errors stem in part from substantial biases in the surface forcing — in particular, precipitation and solar insolation — from the parent atmospheric model. To reduce these significant biases in the surface forcing, the Global Soil Wetness Project (GSWP) used observation-based precipitation analysis and satellite-based surface insolation fields combined with surface fields from coupled land–atmosphere data assimilation systems to drive uncoupled LSMs over a global domain (Dirmeyer *et al.*, 1999).

The GSWP approach greatly reduced the aforementioned large errors and drift in soil moisture/temperature states and surface energy/water fluxes. This finding inspired the further evolution of LDAS from coupled land atmosphere systems into uncoupled land-only systems. However, the monthly satellite retrievals of precipitation and insolation used in GSWP are not conductive to driving the hourly-to-daily updates of land states needed to initialize operational weather prediction models. Hence, the NLDAS project was initiated in 1998 as a multi-institution collaboration led by NCEP (Mitchell *et al.*, 1999), with the goal of producing land surface states and fluxes via an uncoupled land-only system that relied as much as possible on hourly-to-daily observation-based forcing fields.

Following in the steps of the NLDAS project, a Global Land Data Assimilation System (GLDAS) was developed jointly by scientists at the

National Aeronautics and Space Administration's (NASA) Goddard Space Flight Center (GSFC) and NOAA/NCEP. The main aim of this system was, and continues to be, to produce global land states and fluxes in support of global medium- to seasonal-range weather and climate forecasts.

In addition to using observation-based forcing data to avoid known biases in model-based forcing data, observations of land states (i.e., soil moisture and snowpack) can be used in LDAS-type systems to correct unrealistic LSM states by employing data assimilation techniques. These systems have the potential to provide reliable and accurate initial states for numerical prediction models and so improve weather and seasonal climate predictions and can also provide accurate high-resolution estimates of land surface water fluxes, energy fluxes, and state variables. They are also valuable for simulating and predicting biological and agricultural productivity, for monitoring and forecasting extreme events (i.e., floods and droughts), and for performing a wide array of studies in meteorology, climatology, hydrology, public safety, ecology, geophysics, biophysics, and biochemistry.

11.1.2 *NOAA-NASA-University collaborations and the development of NLDAS*

The NLDAS is a long-term multi-institution project developed jointly by scientists at NOAA/NCEP, NASA GSFC, the NOAA National Weather Service (NWS) Office of Hydrologic Development (OHD), the NOAA Environmental Satellite, Data, and Information Service (NESDIS), the NOAA Climate Prediction Center (CPC), and several universities including Princeton University (PU), the University of Washington (UW), the University of Maryland (UM), and Rutgers University (RU). The study region for NLDAS encompasses the CONUS along with parts of Mexico and Canada (Fig. 11.1). Four NLDAS partner LSMs are executed in the system, and include the Noah (Ek *et al.*, 2003), Mosaic (Koster and Saurez, 1994), Sacramento Soil Moisture Accounting (SAC-SMA, Burnash *et al.*, 1973; Burnash, 1995), and Variable Infiltration Capacity (VIC) (Liang *et al.*, 1994) models. These models are run at a 0.125° spatial resolution and an hourly temporal resolution. NLDAS simulations are executed in real-time at NCEP and, respectively, at NASA GSFC, PU, and UW. NLDAS includes two phases: Phases 1 and 2.

The first phase (NLDAS-1) was initiated in 1998 and centered on retrospective executions of the four NLDAS models using three years of

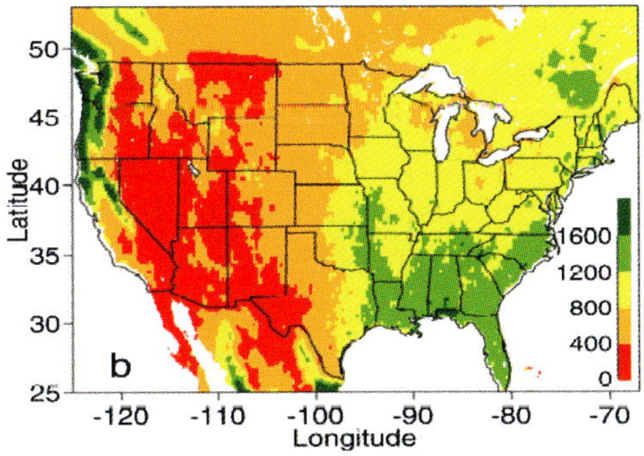

Fig. 11.1. Depiction of domain used for NLDAS-1 and NLDAS-2. Mean annual NLDAS-1 precipitation (mm) for October 1, 1997 to September 30, 1999, is shown. Adapted with permission from Mitchell *et al.* (2004). Copyright 2004, John Wiley & Sons, Inc.

surface forcing data (October 1, 1996 to September 30, 1999) to output land states and fluxes. The first year of each run was used as a spin-up period and the last two years of model output were used for analysis and validation. Following the completion of the retrospective simulations, a real-time continuation of these four model runs was implemented at NOAA/NCEP.

Work in NLDAS-1 involved the construction of the overall NLDAS system, including parameter data sets, software, and analysis tools, and the assessment of the ability of the four NLDAS LSMs to accurately simulate water fluxes, energy fluxes (except SAC-SMA) and state variables. However, the length of the NLDAS-1 output data set was too short to provide the long-term study period and climatology necessary to support regional weather and climate simulations, and the operational analysis and monitoring of droughts and floods. The latter was a critical shortcoming, as NLDAS sought to support activities such as the US Drought Monitor and NCEP's Climate Prediction Center drought monitor/outlook. In addition, weaknesses were found both in the surface forcing data and in the models used in NLDAS-1 that needed to be corrected if model output were to improve in accuracy. Responding to these needs, the NLDAS team initiated the second phase of the project (NLDAS-2) at the first Climate Prediction Program of the Americas (CPPA) Principal Investigator (PI) meeting held in Tucson, Arizona during August 14–16, 2006.

The modeling system developed for NLDAS-2 operates in two modes: an analysis/monitoring mode based on a 29-year NLDAS retrospective run augmented with a real-time extension (from January 1, 2008–present), and a prediction mode based on a seasonal hydrological forecast system

(Ek and NLDAS team, 2011a). The data used to force the long-term NLDAS retrospective run includes input from several ground- and space-based observation platforms. The precipitation forcing is underpinned by a daily gauge-only precipitation analysis over the CONUS domain that includes a Parameter-elevation Regression on Independent Slopes Model (PRISM) (Daly *et al.*, 1994) correction to account for the impacts of topography on precipitation. This daily precipitation analysis is temporally disaggregated to hourly precipitation amounts using NCEP Stage II hourly radar/gauge precipitation (Lin and Mitchell, 2005). The non-precipitation surface forcing is derived from the NCEP retrospective and real-time North American Regional Reanalysis (NARR) (Mesinger *et al.*, 2006). The NARR-based surface downward solar radiation is bias-corrected using seven years (1997–2004) of Geostationary Operational Environmental Satellite (GOES) satellite-derived solar radiation retrievals (Pinker *et al.*, 2003; Alonge and Cosgrove, 2008), in order to overcome the significant positive shortwave biases which characterize the NARR.

The NLDAS prediction mode, consisting of an uncoupled ensemble seasonal hydrological forecast system, was developed jointly by scientists at PU and the UW. The system utilizes three independent approaches for generating downscaled ensemble seasonal forecasts of surface forcing: (1) a climatological ESP-type approach (Ensemble Streamflow Prediction) (Day, 1985), (2) a method utilizing CPC's Official Seasonal Climate Outlook (Wood and Lettenmaier, 2006), and (3) an approach using NCEP's Climate Forecast System (CFS) ensemble dynamical model predictions (Luo and Wood, 2007). For each of these three approaches, 20 ensemble members (forcing realizations) are generated using a Bayesian merging algorithm developed by PU (Luo *et al.*, 2007; Luo and Wood, 2008). The three forcing methods are then used to drive the VIC model over 14 large river basins that together span the CONUS domain.

From these simulations emerge one-to-nine month ensemble seasonal prediction products, including air temperature, precipitation, soil moisture, snowpack, total runoff, evaporation, and streamflow. A primary use of the anomalies and percentiles of these products is for drought prediction. In support of this, in September 2008, the NLDAS-2 prediction mode was transitioned at NCEP/EMC from a research to a quasi-operational seasonal hydrological forecast system. This system is run at the beginning of each month to generate monthly drought forecast products that extend out for six months, which support operational drought monitoring and forecast activities.

11.1.3 *Other LDAS activities around the world*

Since its establishment, NLDAS has had a long and successful history in producing land surface state variables and fluxes (Mitchell *et al.*, 2004), and output from the system has been widely used by meteorologists, hydrologists, climatologists, ecologists, agronomists, policy makers, and educators. Spurred by the success of NLDAS and GLDAS, other continental and national LDAS systems have rapidly developed in recent years. Examples of continental LDAS systems include the European LDAS (Van Den Hurk, 2002), South American LDAS (de Goncalves *et al.*, 2006, 2009), and Middle East and North-African LDAS (Bolten *et al.*, 2010). National LDAS systems include the Canadian LDAS (Balsamo *et al.*, 2007), the Chinese LDAS (Li *et al.*, 2007), and the Korean LDAS (Lim *et al.*, 2008). The success of NLDAS and GLDAS has spurred increased domestic use as well, with surface states from an NCEP-based GLDAS system being used to support the NCEP Climate Forecast System Reanalysis (CFSR) (Saha *et al.*, 2010). As LDAS products find further use in weather and climate predictions and become more useful to climate information services, additional continental/national LDAS systems will most likely be developed.

11.2 NLDAS History

11.2.1 *NLDAS-1*

As mentioned in Sec. 11.1.2, NLDAS-1 was initiated in 1998 as a multi-institution collaboration led by NCEP (Mitchell *et al.*, 1999) with a central mission of generating multi-LSM land surface states and fluxes via an uncoupled land-only system forced with as many ground- and space-based observations as possible. These land surface states were produced for the purpose of providing initial states for regional numerical weather and climate models in an effort to improve weather and seasonal climate simulations. The following sections will focus on the NLDAS-1 model framework, the validation of forcing data and model outputs, and the lessons learned from NLDAS-1.

11.2.1.1 *Framework*

The details of the NLDAS framework were discussed in Mitchell *et al.* (2004), and so only a brief summary is provided here. The NLDAS domain encompasses the continental USA, the northern part of Mexico, and the southern part of Canada (Fig. 11.1). Four LSMs (Noah, Mosaic, SAC-SMA, and VIC) were executed in a common data framework in which they shared a 0.125° regular latitude–longitude grid, land mask, terrain elevation grid, hourly input forcing data, soil texture and vegetation classes; streamflow network and routing model; and GRIB-format output. The elevation at each 0.125° grid cell was derived from the 30 arc-second GTOPO30 digital elevation database of Verdin and Greenlee (1996). Vegetation classifications were derived from a 1-km resolution global dataset of land cover classes that was produced at the UM, which is based on observations from the Advanced Very High Resolution Radiometer (AVHRR) on the NOAA-15 satellite (Hansen *et al.*, 2000).

Adding to these two data sets, the soil texture database over the CONUS was derived from the 1-km State Soil Geographic Database (STATSGO) database of Miller and White (1998), which assigns one of 16 texture classes to each of 11 soil layers over a 2-m depth. Noah, Mosaic, and VIC assume a vertically uniform soil class, the properties of which are drawn from the top layer of the STATSGO database. The one exception to this is Mosaic's soil porosity, which was derived for each Mosaic soil layer using weighted averages from the 11-layer soil textures. Also, while SAC-SMA differs from the other models in that it uses conceptual soil-water storage reservoirs, the STATSGO soil layer information was still utilized, serving as the basis of many of this model's *a priori* parameters. Outside the CONUS, where STATSGO data is unavailable, NLDAS soil texture was derived from the 5-min Food and Agriculture Organization (FAO) global database of Reynolds *et al.* (2000). Although all four NLDAS models used common base fields of vegetation and soil class, they may use different internal model parameter values such as vegetation root depth and density, the number and thickness of soil layers, and the seasonality of vegetation. Further, while each model outputs hourly data, the internal model time steps differed. Mosaic and Noah utilized 15-min time steps, while VIC and SAC-SMA were each executed with an hourly time step.

All NLDAS-1 surface forcing data except for precipitation and downward shortwave radiation were obtained from NCEP's mesoscale Eta model-based Data Assimilation System (EDAS) (Rogers *et al.*, 1995), a continuously cycled North American four-dimension data assimilation system. Two-meter air temperature and specific humidity along with surface pressure and downward longwave radiation were elevation-adjusted using the difference in height between the EDAS and NLDAS topography (Cosgrove *et al.*, 2003a). The NLDAS precipitation forcing over the CONUS was derived through a multi-step process that started with NCEP's 0.25° gauge-only daily precipitation analysis of Higgins *et al.* (2000). After being interpolated to the 1/8th degree NLDAS grid, the daily precipitation was then temporally disaggregated to an hourly time scale using weights derived from the hourly, 4-km, radar/gauge Stage 2 precipitation data set (Cosgrove *et al.*, 2003a). Completing the forcing data set, the downward shortwave radiation fields were derived from GOES-based products (Mitchell *et al.*, 2004) to avoid the significant positive biases present in the EDAS shortwave fields.

As mentioned in Sec. 11.1, four LSMs were executed in NLDAS-1: Noah, Mosaic, VIC, and SAC-SMA. The Noah model was developed as the land component in the NOAA/NCEP mesoscale Eta model (Betts *et al.*, 1997; Chen *et al.*, 1997; Ek *et al.*, 2003). It also serves as the land component of the evolving Weather Research and Forecasting (WRF) regional atmospheric model, NCEP's Global Forecast System (GFS), and the NOAA/NCEP coupled Climate Forecast System (CFS). The model represents the soil freeze-thaw process and its impact on soil heating/cooling and transpiration following Koren *et al.* (1999). The Noah model features four soil layers with spatially invariant thicknesses of 10 cm, 30 cm, 60 cm, and 100 cm. The first three layers form the root zone in non-forested regions, while all layers contribute to the root zone in forested regions.

The Mosaic model was developed for use in NASA's global climate model (Koster and Suarez, 1994, 1996; Koster *et al.*, 2000). It is a Soil–Vegetation–Atmosphere Transfer (SVAT) scheme that accounts for the sub-grid heterogeneity of vegetation and soil moisture using a tiling approach. Up to 10 tiles representing the different types of vegetation within a grid cell can be used in the current configuration of Mosaic. Each vegetation tile features its own energy and water balance, and soil moisture and temperature. Each tile has three soil layers with thicknesses of 10 cm, 30 cm, and 160 cm, with the first two forming the root zone of the model. The water storage in each layer of the soil column is calculated as a weighted average of the water storage from all tiles within the grid cell.

The VIC model was jointly developed by the UW and PU as a macroscale, semi-distributed, grid-based hydrological model (Liang *et al.*, 1994; Wood *et al.*, 1997). VIC can be executed in several modes, with the full water and energy balance version used in NLDAS-1. VIC has three soil layers: the top layer is 10 cm thick, while the bottom two layers feature variable thicknesses which are determined, in part, through calibration. The root zone can span all three layers and depends on the vegetation class and its vertical root distribution. Like the Mosaic model, the VIC model utilizes sub-grid vegetation tiles. The VIC model includes a two-layer energy balance snow model (Cherkauer *et al.,* 2003), which represents snow accumulation and ablation on the ground and overlying forest canopy and uses sub-grid elevation bands to represent the impact of elevation on temperature, precipitation, and snow. VIC has been widely and successfully applied to large river basins in the USA (Nijssen *et al.*, 1997; Lettenmaier *et al.*, 1996; Cherkauer and Lettenmaier, 1999), and elsewhere (Lohman *et al.*, 1998; Lobmeyr *et al.*, 1999), and has participated in national (Maurer *et al.*, 2002) and global scale studies (Sheffield *et al.*, 1996; Nijseen *et al.*, 2001).

The SAC-SMA model was developed as a lumped conceptual hydrological model (Burnash *et al.*, 1973; Burnash, 1995) and is used operationally at National Weather Service (NWS) River Forecast Centers. Seeking to increase the applicability of SAC-SMA over a broad range of scales, the NWS Office of Hydrologic Development developed a distributed version and an accompanying set of nationwide *a priori* model parameters (Koren *et al.*, 2000; 2003). These parameters reduce the need of deriving a calibrated set of parameters and facilitate large scale implementations of the model. While updates to these parameters have been released by OHD, NLDAS makes use of the first version which is based on relationships between vegetation cover and STATSGO soils data, and has not performed further calibration.

SAC-SMA was run together with the distributed version of the Snow-17 model (a temperature indexed-based snow model) (Anderson, 1973) in NLDAS-1 to more fully represent the hydrologic cycle. As is the case with SAC-SMA, *a priori* model parameters have recently been developed for Snow 17. However, as these parameters were not available at the time of the initial NLDAS simulations, constant nationwide values supplied by OHD were used in their place.

The SAC-SMA model represents water storage by dividing the soil column into upper and lower zones which each have tension and free water components. These components represent the active part of the total water storage in a grid. The conceptual water storage components of the SAC-SMA model are not tied to any soil depth or thickness and thus are not well suited for intercomparison with other models or validation against observations. Addressing this issue, as part of a recent frozen ground physics upgrade, SAC-SMA gained the ability to map the conceptual water storages to distinct soil layers [SAC-Heat Transfer (HT) (Koren *et al.*, 2010)].

This upgrade has allowed for the accurate simulation of soil moisture and soil temperature throughout the soil column. SAC-HT has been further upgraded through inclusion of Noah LSM evapotranspiration physics (SAC-HT-Evapotranspiration (ET)), which allow for the internal computation of potential evapotranspiration and for more robust representation of the impact of vegetation on evaporation. While these versions were not used in the base set of NLDAS-1 simulations, they will be considered for use in future simulations. As an interim solution, when soil moisture output at distinct layers was needed in NLDAS, it was computed using post-processing techniques.

SAC-SMA and Snow-17 only use some of the forcing variables that the other NLDAS-1 LSMs require. Specifically, 2-m air temperature and precipitation is used by Snow 17, while potential evapotranspiration, precipitation, and Snow-17 snowmelt are ingested by SAC-SMA. Because SAC-SMA is a conceptual rainfall-runoff water balance model, it does not calculate an energy balance, which causes its list of output variables to differ from that of the other LSMs.

The routing model used with all of the LSMs in NLDAS-1 is detailed in Lohmann *et al.* (1998, 2004). It calculates the timing of the runoff reaching the outlet of a grid cell and transports water via a user-specified river network. The routing model assumes that water can leave a grid cell only in the direction of one of its neighboring grid cells according to a river flow direction mask. Both within-cell and cell-to-cell routing time delays are represented by linear, time-invariant, and causal functions with non-negative impulse response characteristics described in detail in Lohmann *et al.* (1998, 2004). After water is transported out of a grid cell, it is further routed through the river network. Transport within the river is accomplished with the linearized Saint-Venant equation, which includes a dispersion term and an advection term. Two routing modes (distributed and simple lumped approach) were used. The distributed routing mode was used for eight large NLDAS study basins because the runoff reaching these outlets takes several weeks, a timescale more appropriate for distributed routing. The simple lumped routing approach was used for 1145 small NLDAS study basins which feature faster response times and for which Lohmann *et al.* (2004) found good lumped model routing performance.

11.2.1.2 *Validation research*

As outlined in Table 11.1, seven ground- and space-based observation databases were used to validate NLDAS-1 forcing data, energy fluxes, water

Table 11.1. Observation-based databases used in NLDAS-1 for validation.

Database name	Database description
ARM/CART (Atmospheric Radiation Measurement/Cloud and Radiation Testbed)	2-m air temperature, surface pressure, 10-m wind speed, 2-m specific humidity, downward shortwave radiation, downward longwave radiation, upward shortwave radiation, upward longwave radiation, sensible heat flux, latent heat flux, ground heat flux, net radiation at 24 sites over the US Southern Great Plains. The dataset spans July 1, 1997 to September 30, 1999, and features an hourly resolution (Luo *et al.*, 2003).
GOES (Geostationary Operational Environmental Satellite)	Skin temperature, 0.125 spatial resolution, hourly temporal resolution, from October 1, 1996 to September 30, 1999 (Pinker *et al.*, 2003; Mitchell *et al.*, 2004).
Illinois (Soil Moisture Databank)	Monthly soil moisture at 11 soil depths (down to 2 m) at 18 sites in Illinois in the USA (Robock *et al.*, 2000; Schaake *et al.*, 2004), October 1997 to September 1999.
IMS (Ice Mapping System)	Daily snow cover over the Northern Hemisphere, 25 km spatial resolution, from February 1, 1997 to September 30, 1999 (Ramsay, 1998; Sheffield *et al.*, 2003).
Oklahoma Mesonet	Soil moisture and temperature at 4 soil layers (down to 0.7 m) at 72 stations over the Oklahoma Mesonet network in the USA with an hourly resolution (Robock *et al.*, 2003), October 1, 1997 to September 30, 1999.
SNOTEL (Snowpack Telemetry Network, USA)	Daily snow water equivalent at 110 SNOTEL sites over the Western USA (Pan *et al.*, 2003), October 1, 1997 to September 30, 1999.
USGS (US Geological Survey)	Daily streamflow at 1145 small basins and 9 large basins over the CONUS, October 1, 1997 to September 30, 1999.

fluxes, and state variables. Luo *et al.* (2003) used observations from 24 ARM/CART sites to evaluate NLDAS-1 forcing data and concluded that differences between the forcing data and these observations were small at hourly-to-monthly time scales for all variables except precipitation. For precipitation, root mean square error, bias and correlation statistics show that while the agreement was marginal at an hourly timescale, it was much better at daily and longer time scales. And while shortwave radiation compared well against observations, it should be noted that the direct merging of EDAS and GOES downward shortwave radiation fields created large spatial and temporal discontinuities.

Table 11.2. Observation-based databases used in NLDAS-2 for validation.

Database name	Database description
Ameriflux (American Flux Station Network)	Monthly net radiation, sensible heat flux, latent heat flux, and ground heat flux from 21 Ameriflux stations spanning from January 1997 through December 2007 (Mo *et al.*, 2011; Ek *et al.*, 2011b).
ARM/CART (Atmospheric Radiation Measurement/Cloud and Radiation Testbed)	Net radiation, sensible heat flux, latent heat flux, ground heat flux, skin temperature at 24 sites over the US Southern Great Plain. The dataset covers from January 1, 1997 to September 30, 1999, with an hourly resolution (Ek *et al.*, 2011b).
Illinois (Soil Moisture Databank)	Monthly soil moisture at 11 soil depths (down to 2 m) at 18 sites, Illinois, USA (Robock *et al.*, 2000), from January 1985 to December 2004.
Oklahoma Mesonet	Soil moisture and temperature at 4 soil layers (down to 0.7 m) at 72 stations over the Oklahoma Mesonet network in the USA with an hourly resolution (Robock *et al.*, 2003). Soil moisture observations span the period from January 1, 1997 to December 31, 2002, and soil temperature is available from January 1, 1997 to December 31, 1999.
US Soil Temperature	Monthly soil temperature at six soil layers (down to 1 m) at 137 sites over the CONUS, spanning from January 1, 1979 through December 2002 (Hu and Feng, 2003, Xia *et al.*, 2013).
USGS (US Geological Survey)	Daily streamflow at 961 small basins and 8 large basins over the CONUS, from October 1, 1979 to September 30, 2007 (Xia *et al.*, 2012a, 2012b).

Complementing the forcing validation discussed above was the validation of NLDAS-1 model output. The validation of soil temperature showed that the two NLDAS models which output soil temperature were able to produce accurate simulations of soil temperature at depths of 5 cm and 25 cm (Robock *et al.*, 2003). In contrast to these positive results, the validation of model-produced streamflow (Lohmann *et al.*, 2004) revealed poor simulations of streamflow over most of NLDAS-1 study basins. Developed as part of the Model Parameter Estimation Experiment (MOPEX) using a US Geological Survey (USGS) list of unregulated basins along with input from NOAA/NWS/OHD, PU, and UW, this collection of 1145 study basins are 10,000 km^2 or less in size and have no visible signs of reservoir operation. In these basins, Mosaic and SAC-SMA were characterized by large negative biases, while VIC displayed large positive biases, particularly over the Upper Midwest and Southeast USA. Similarly, poor simulation of snow water equivalent was observed at 110 SNOTEL stations, particularly in the western mountainous areas of the US (Pan *et al.*, 2003). This analysis showed that Noah exhibited overly large sublimation and early snowmelt when compared to observations and simulations from the other three models. These results stemmed from the failure to account for the impacts of topography on precipitation in the NLDAS-1 forcing data and from the gauge undercatch of snowfall due to wind (Pan *et al.*, 2003; Lohmann *et al.*, 2004).

NLDAS-1 validation activities also revealed substantial biases and intermodel differences in total runoff and evaporation over the ARM/CART region of the Southern Great Plains. The validation of surface energy fluxes over this same region (Robock *et al.*, 2003) yielded similar findings. These results echo the findings of earlier Project for the Intercomparison of Land Surface Parameterization Schemes (PILPS) studies which found that the partitioning of precipitation into total runoff and evaporation is often an area of model disparity. A second area of model disparity was highlighted by Mitchell *et al.* (2004), who found that the different aerodynamic conductance values used by each model caused large intermodel differences in sensible heat flux and skin temperature. These differences were found to further impact latent heat flux and the partitioning of water fluxes. A final area of model disagreement was found by Robock *et al.* (2003) and Schaake *et al.* (2004) to be soil moisture, once again echoing previous findings from the PILPS series of experiments. Complicating this issue, an NLDAS-1 spin-up study also showed that the one year allotted for model spin-up may not have been long enough to allow soil moisture states to reach equilibrium in some NLDAS grid cells (Cosgrove *et al.*, 2003b).

The findings of intermodel differences emphasize the importance of successfully assimilating remotely sensed land surface temperature, soil moisture, and snowpack into LSMs in order to improve accuracy of land surface states and fluxes. Such improved land states and fluxes are critical for enhancing the prediction ability of numerical weather and seasonal climate models.

11.2.1.3 *Lessons from NLDAS-1*

NLDAS-1 successfully established the NLDAS modeling framework including the general system configuration, model parameters, model makeup, forcing derivation procedure, model output, and validation tools. These hallmark accomplishments culminated in the publishing of 10 NLDAS-1 papers in the *Journal of Geophysical Research* (see Table 1 of Mitchell *et al.*, 2004). For all of its success, NLDAS-1 was characterized by several shortcomings that collectively can be viewed as lessons learned. Chief among these, NLDAS-1 featured a relatively short three-year retrospective analysis spanning October 1996 through September 1999, and a real-time run covering the period from October 1999 to the present. As such, NLDAS-1 lacked a monitoring mode underpinned by a long-term retrospective run and was unable to support a seasonal hydrologic prediction mode. These shortcomings handicapped the application of forcing data and the use of multi-model outputs in the initialization of weather and climate models. It also limited the usefulness of NLDAS-1 to groups such as NIDIS which are involved in the monitoring and prediction of extreme weather and climatic events.

As previously mentioned, NLDAS-1 was also hindered by precipitation and shortwave biases which translated, in part, to deficiencies in the land surface states output by the NLDAS models. Taken together, the shortcomings outlined above highlighted the need for improved NLDAS forcing data and model products, and led to the extension of NLDAS into a second phase, NLDAS-2.

11.2.2 *NLDAS-2*

NLDAS-2 was initiated at the CPPA PI meeting held in Tucson, Arizona, in 2006 as a CPPA "core project". This multi-institution collaborative project led by NOAA/NCEP is currently ongoing. The two largest differences from NLDAS-1 are (1) the production of a new 30-year forcing data set and associated 30-year NLDAS retrospective run, and (2) the addition of a seasonal hydrologic prediction component. These enhancements benefit the land surface modeling community in general but were specifically aimed at supporting drought monitoring and prediction efforts led by NIDIS and NOAA/CPC. All of the NOAA, NASA, and university partners returned to support NLDAS-2.

11.2.2.1 *Improvement of forcing and models*

The non-precipitation NLDAS-2 land surface forcing fields are derived from the analysis fields of the NCEP North American Regional Reanalysis (NARR) (Mesinger *et al.*, 2006). The NARR consists of (1) a 25-year retrospective (1979–2004) model run and (2) a daily real-time update produced at NCEP (2005–present). NARR analysis fields feature a 32-km spatial resolution and a three-hourly temporal resolution. NARR fields that are utilized to generate NLDAS-2 forcing fields are first spatially interpolated to the finer resolution of the NLDAS-2 1/8th degree grid, and then temporally interpolated to the NLDAS-2 hourly frequency. Additionally, surface pressure, surface downward longwave radiation, near-surface air temperature, and near-surface specific humidity are elevation-adjusted to account for the differences between the NARR and NLDAS terrain. This vertical adjustment is based on the traditional vertical lapse rate of 6.5 K/km for air temperature. The spatial interpolation, temporal disaggregation, and elevation correction processes remained the same from NLDAS-1 to NLDAS-2 and are presented in Cosgrove *et al.* (2003a).

Seeking to improve upon the NLDAS-1 precipitation forcing data set, Alonge and Cosgrove (2008) based NLDAS-2 precipitation forcing upon a temporal disaggregation of the 1/8th degree gauge-only CPC analysis of daily precipitation that includes a topographic adjustment based on the widely applied PRISM climatology (Daly *et al.*, 1994). This approach reduced the orographic-related negative precipitation bias present in NLDAS-1, in particular, over western mountainous regions. However, it does not address the issue of gauge undercatch of solid precipitation, an issue that will be addressed through future investigations. NLDAS-2 investigators also sought to improve upon the NLDAS-1 downward shortwave radiation fields: toward this end, Alonge and Cosgrove (2008) employed a ratio-based (Berg *et al.*, 2003) bias correction technique and the UM's Surface Radiation Budget (SRB) data set (Pinker *et al.*, 2003) to correct NARR downward shortwave radiation (also see http://www.emc.ncep.noaa.gov/mmb/nldas/LDAS8th/forcing/forcing_narr.shtml). This treatment mitigated much of the bias present in the NARR downward shortwave radiation fields while maintaining their spatial and temporal continuity. A final forcing-related change from NLDAS-1 to NLDAS-2 centered on the source of potential evaporation data used by SAC-SMA. While SAC-SMA simulations in NLDAS-1 utilized PE values generated by the Noah LSM, SAC-SMA simulations in NLDAS-2 used a climatological monthly PE data set. Operational support experience at NOAA/OHD and research work at NOAA/NCEP (Xia *et al.*, 2012b) have shown this to be a superior approach, and the change was implemented in an effort to reduce the large mean annual negative runoff biases found in NLDAS-1 SAC-SMA simulations.

Improvements made in formulating NLDAS-2 extended beyond forcing data and into the participating LSMs through efforts to tune model parameters (i.e., model calibration) and add or improve physical processes. Wei *et al.* (2013) upgraded Noah LSM model physics and added calibrated model parameter sets to improve Noah's simulation of latent heat, total runoff and skin temperatures for the warm season. In complementary research, Livneh *et al.* (2010) enhanced the Noah snow model by adding representation of cold content/melt-refreeze physics and a snow age albedo algorithm, by increasing the maximum snow albedo, and by constraining aerodynamic conductance in stable boundary layer cases to improve the simulation of SWE, sublimation, snowmelt, and snow cover. VIC also benefited from model upgrades: in particular, Troy *et al.* (2008) applied a parameter transfer approach to calibrate model parameters to observed streamflow from 1130 small basins across the CONUS. Their effort reduced the large bias in total annual runoff which had characterized VIC in NLDAS-1.

NLDAS-2 improvements in forcing data and LSMs were accompanied by improvements in the LSM spin-up process, which addressed the overly short one-year NLDAS-1 spin-up period. A two-step method was used to generate the initial model states in NLDAS-2 for each of the four LSMs. First, climatological land surface states for 00Z on October 1 were generated by averaging together all of the NLDAS-1 00Z October 1 values from 1999 through 2006. These NLDAS-1-based restart files were then used to initialize 15.25-year spin-up simulations for each model from October 1, 1979 to January 1, 1995. The 00Z January 1 states for the last 10 years of this spin-up simulation were averaged together to provide initial states for the official NLDAS-2 retrospective run period of January 1, 1979 to September 1, 2007. Because the streamflow routing code that translates gridded model runoff to stream discharge needs about six months of spin-up time, only 28 years of model output from October 1, 1979 to September 1, 2007, are analyzed in this study.

11.2.2.2 *Evaluation of NLDAS-2 products*

As was the case in NLDAS-1, much of the research within NLDAS-2 has focused on the validation of the project's forcing data and multi-model hydrological output. This is a critical endeavor, as data is most useful to end users when its strengths and weaknesses are known. Validation of model products against *in situ* observations (see Table 11.2) not only can identify forcing data errors, deficiencies in model structure, and model parameter errors but also can assess the quality of the output and its suitability for model initialization, water resource, disaster management, ecological, and agricultural end-user applications.

This first section presents key results from the surface energy-validation of three NLDAS-2 models as detailed in Ek *et al.* (2011b). The outputs from

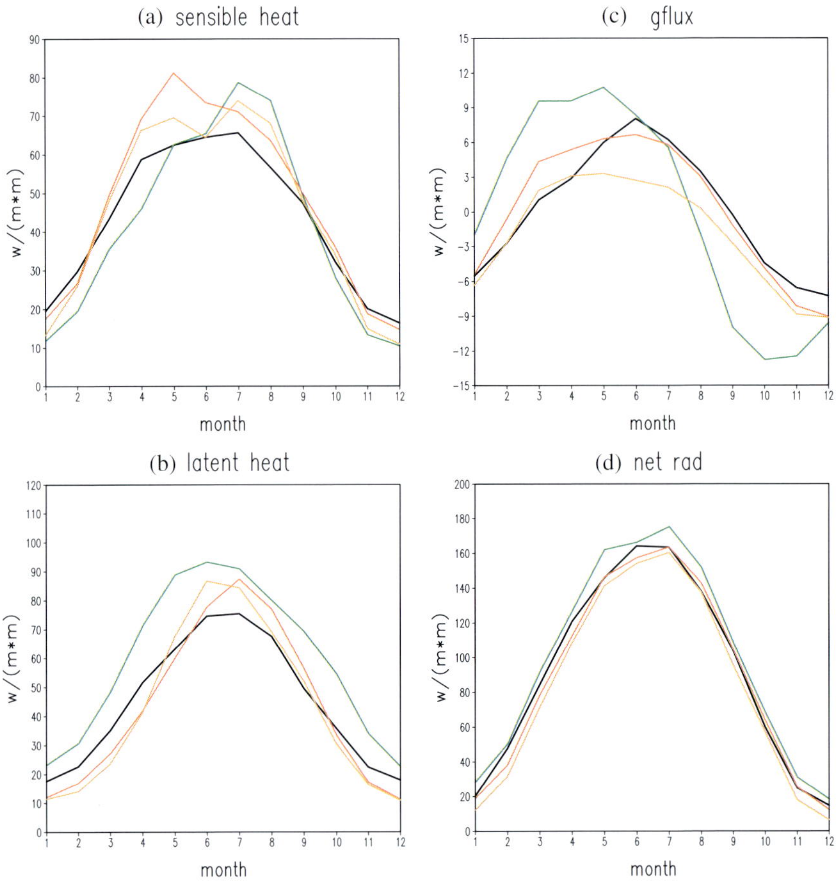

Fig. 11.2. The mean annual cycle of the land surface energy budget: (a) sensible heat, (b) latent heat, (c) ground heat, and (d) net radiation averaged over the USA based on available observations from 21 Ameriflux stations. Traces include observations (black), Noah LSM (red), Mosaic LSM (green), and VIC LSM (orange).

Noah, VIC, and Mosaic are compared with sensible heat flux (Fig. 11.2(a)), latent heat flux (Fig. 11.2 (b)), ground heat flux (Fig. 11.2(c)), and net radiation (Fig. 11.2(d)) observations from the Ameriflux network (part of FLUXNET) (Baldocchi *et al.*, 2001). For each variable, the monthly climatology is computed by averaging over all 21 Ameriflux stations across the CONUS for the period from 1997 to 2007 over which station reports were available.

Overall, Noah and VIC verify closer to observations than does Mosaic. All models generally capture the seasonal cycle of net radiation well in terms of phase and magnitude, although Mosaic features a positive summertime bias. Mosaic also produces overly high latent heat values and overly low sensible values in comparison with other models and observations. In particular, Mosaic greatly overestimates (underestimates) ground heat flux in the winter and spring (summer and fall), indicating that the model directs too much energy into the ground during the summer and fall, and releases too much energy to the atmosphere during winter and spring. VIC also underestimates ground heat flux during the summer and fall due to a significant phase difference when compared with the observations, while the Noah LSM verifies well. Further examining the sensible and latent heat flux analysis, it can be seen that both Noah and VIC capture the seasonal cycle well, although both overestimate flux values during the summertime. Although not depicted in Fig. 11.2, the ensemble mean of the models verifies better than does the output from any individual model (Mo *et al.*, 2011).

Complementing the surface flux studies detailed above, the NLDAS-2 hydrological analysis was conducted over a subset of the 1145 study basins used in NLDAS-1 (Lohmann *et al.*, 2004). For the purposes of the NLDAS-2 hydrological analysis, we selected 961 small basins from the original list of 1145 that featured no missing data over the NLDAS-2 study period. Figure 11.3 shows the relative bias of the mean annual runoff for all four LSMs over these basins. Over the middle section of the country where there is a small mean annual runoff ratio and large evaporation efficiency, it can be seen that Noah and SAC-SMA overestimate the mean annual runoff, Mosaic underestimates the mean annual runoff, and VIC features mixed results. The Mosaic model underestimates the mean annual runoff for all of the basins in the CONUS except for several basins in northern Texas, Oklahoma, and northern Washington, where mean annual runoff is overestimated. It features a large relative bias in the northeast corner of the domain and has small relative biases in the southeast corner. By contrast, the Noah model correctly simulates the mean annual runoff over much of the Eastern USA. This is due to improvements and upgrades made to the Noah model after NLDAS-1 (Livneh *et al.*, 2011; Wei *et al.*, 2013). With the exception of the middle portion of the country, the four-model ensemble mean generally features the smallest annual relative biases.

Figure 11.4 displays the cumulative distribution function of the daily, weekly, and monthly Nash–Sutcliffe Efficiency (NSE) for the four LSMs and their ensemble mean. Focusing on the daily NSE, the Noah and Mosaic

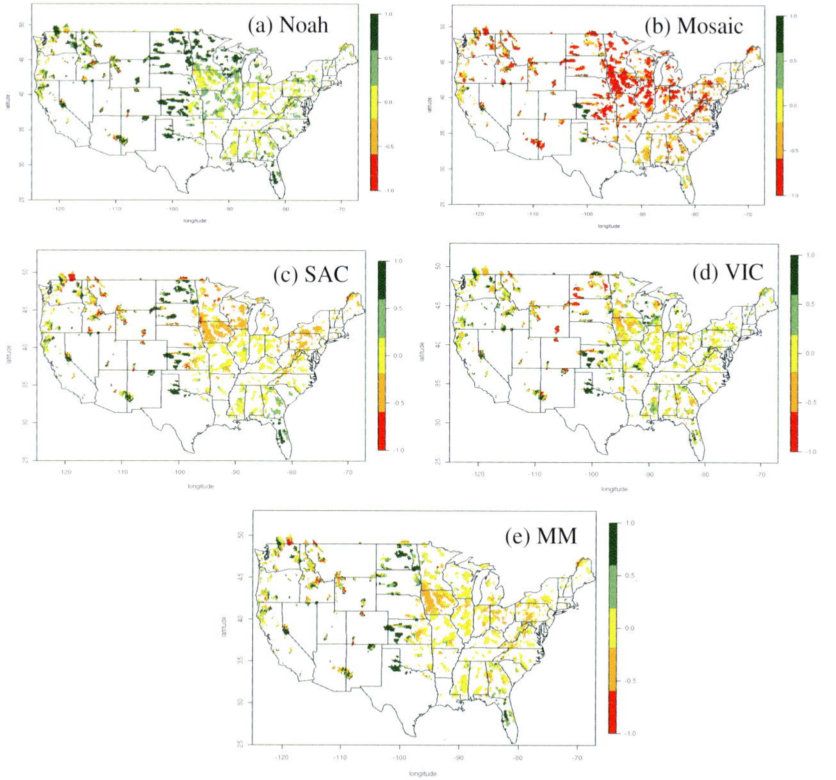

Fig. 11.3. Mean annual runoff relative bias [(model-observed)/observed] for the 961 small study basins for October 1, 1979 to September 30, 2007, from (a) Noah, (b) Mosaic, (c) Sacramento, (d) VIC, and (e) Ensemble Mean (MM). The observed runoff for a given basin is calculated from the basin area and observed basin outlet stream discharge information provided by the USGS.

models feature "useful" prediction skill (NSE $\geqq 0.4$) only over 10% of the study basins, the SAC-SMA and VIC models are useful over about 20–30% of the basins, and the ensemble mean is useful over about 40% of the basins. Performing a similar analysis for the weekly NSE, the percentage of basins over which useful skill is demonstrated increases from 10% to 22% for the Mosaic model, and increases from 10% to 40%, from 30% to 60%, from 20% to 65%, and from 40% to 75% for the Noah model, the SAC-SMA model, the VIC model, and ensemble mean, respectively. This trend of increasing accuracy with increasing analysis period continues for the Noah and VIC models, where an examination of the monthly NSE shows the

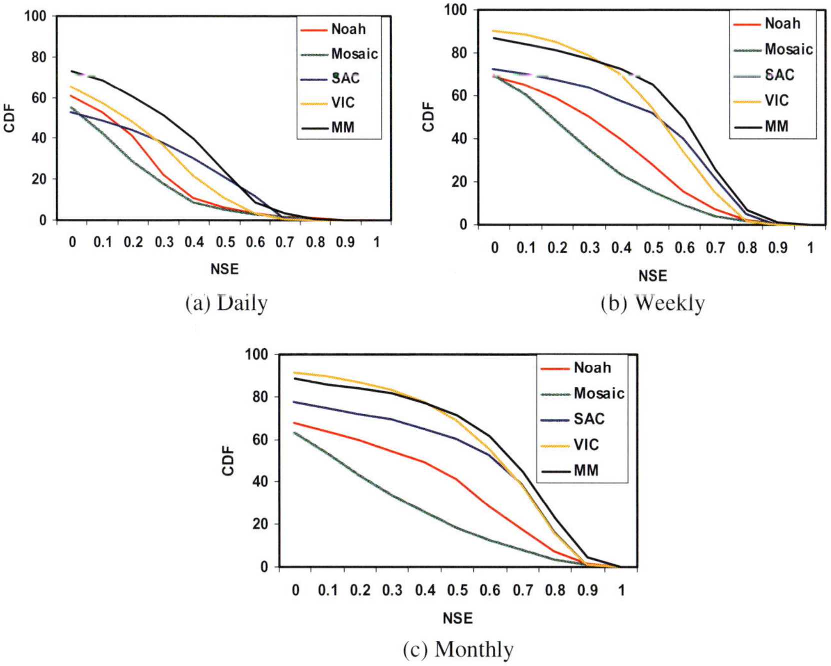

Fig. 11.4. Cumulative distribution function (CDF) of the Nash–Sutcliffe efficiency (NSE) coefficients calculated from the (a) daily, (b) weekly, and (c) monthly mean simulated streamflow of the four NLDAS-2 land surface models (Noah, Mosaic, Sacramento, VIC) and their ensemble mean for the time period — October 1, 1979 to September 30, 2007, over the 961 small study basins (NSE values <0 not shown).

percentage of useful basins increases, respectively, from 40% to 50%, and from 65% to 80%. By contrast, monthly NSE values show little increase over the weekly values when computed for the other two models and the ensemble mean.

Simulation skill can also be assessed from an anomaly correlation perspective as was done by Xia *et al.* (2012b). Judged in this fashion, for all time scales, the simulation skill of all four models and their ensemble mean is generally high over the CONUS — in particular over the eastern region and the west coast where anomaly correlation coefficients are over 0.8. Only a few basins have non-statistically significant correlations at the 95% confidence level between observed and daily, weekly, monthly, and annual mean simulated streamflow anomalies (Xia *et al.*, 2012b). As was the case with the NSE analysis, the SAC-SMA and VIC models display the highest level of skill, and the Noah and Mosaic models feature relatively lower

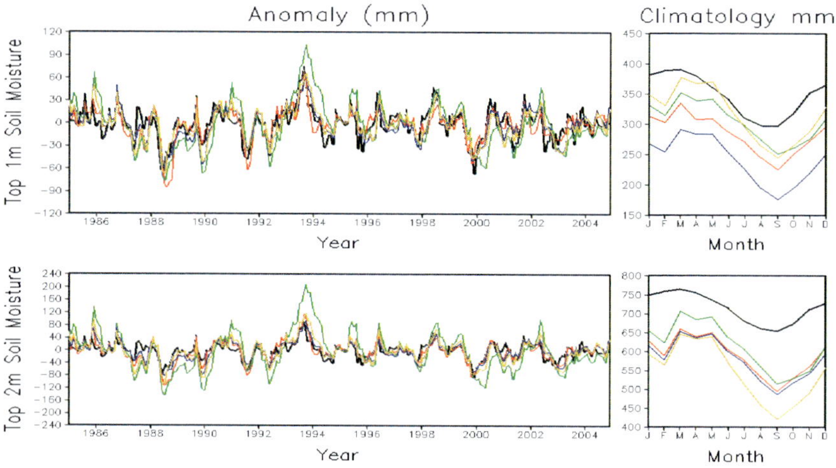

Fig. 11.5. Monthly mean soil moisture anomaly and climatologically averaged seasonal cycle in Illinois for top 1 m and top 2 m soil layers (red line: Noah, green line: Mosaic, blue line: SAC-SMA, orange line: VIC, black line: observation).

simulation skill. However, unlike in the NSE analysis, the ensemble mean does not offer increased seasonal cycle simulation skill over that of the individual LSMs.

Another important indicator of a model's ability to accurately simulate the hydrologic cycle is its depiction of soil moisture anomalies. Figure 11.5 shows the variability of the monthly mean soil moisture anomaly computed over the Illinois soil moisture network for the period from January 1985 to December 2004 for the top 1-m and 2-m soil layers. The results show that all models except for the Mosaic model simulate monthly mean soil moisture anomalies very well, successfully capturing most wet and dry events. However, exceptions exist (e.g., 2002 dry event).

Focusing on the soil moisture anomalies in the top two meters of soil, Noah, Mosaic, SAC-SMA and VIC have anomaly correlations, respectively, of 0.82, 0.63, 0.78, and 0.76. All of the models underestimate the soil moisture in the top 1 m and 2 m of the soil column. In particular, the SAC-SMA model greatly underestimates the soil moisture of the top 1 m, a result that is consistent with findings from NLDAS-1 (Mitchell *et al.*, 2004; Robock *et al.*, 2003; Schaake *et al.*, 2004) while the VIC model greatly underestimates the soil moisture in the top 2 m of the soil column. One contributing factor to this behavior is the lack of inclusion of irrigation and ground water effects in any of the NLDAS-2 models, which can impact soil moisture levels through the redistribution of water. Other factors include deficiencies in the forcing data and in model physics. Further examining each model's performance in turn, the Mosaic model recovers slowly from both positive and negative anomalies, showing stronger anomaly persistence than is seen in either the observations or the other three models. This behavior stems in part from its relatively large field capacity and leads to small anomaly correlation coefficients when compared with the other three models. This poor simulation skill lowered the overall skill of the ensemble mean. As such, the ensemble mean did not offer improved simulation skill versus the majority of the models (i.e., Noah, SAC-SMA, VIC).

Of note is the fact that the NLDAS-2 top 2 m soil moisture anomaly correlation values compare favorably against similar analysis performed using Noah output from other retrospective simulations. In particular, anomaly correlation values for the NCEP–Department of Energy Global Reanalysis 2 (GR2) (Kanamitsu *et al.*, 2002), the NARR (Mesinger *et al.*, 2006), and the Climate Forecast System Reanalysis (CFSR) (Saha *et al.*, 2010) were 0.47, 0.67, and 0.61, respectively. These results indicate that NLDAS-2 serves as a strong source of soil moisture data.

Like soil moisture, soil temperature is also an important LSM product and was a focus of the NLDAS-2 validation effort. We used monthly mean soil temperature observation from 137 NOAA Cooperative Network soil temperature stations across the USA to evaluate Noah LSM simulations at annual and monthly timescales at three soil layers (0–10 cm, 10–40 cm, 40–100 cm) for the period from January 1979 to December 2002 (not shown). The time evolution of the simulated anomalies in all three soil layers follows the observations quite well. However, exceptions exist, and include the 1996 and 1997 warm events, where the simulated soil temperature anomaly is overly cold compared with observations. The phase of the seasonal cycle of soil temperature within the three soil layers is also generally well simulated, but on average values are 2–5 K colder than observations. The anomaly correlations between the observed and simulated soil temperature values are 0.87, 0.77, and 0.69 for the 0–10 cm, 10–40 cm, and 40–100 cm layers, indicating a worsening simulation of soil temperature with depth.

Utilizing data from 72 Oklahoma Mesonet stations (Brock *et al.*, 1995), it was also possible to assess Noah LSM soil temperature simulations at a sub-monthly timescale. Figure 11.6 displays three-year (1997–1999) daily-averaged soil temperature from the Noah LSM. Compared against observations, the Noah LSM does a good job of simulating the near-surface

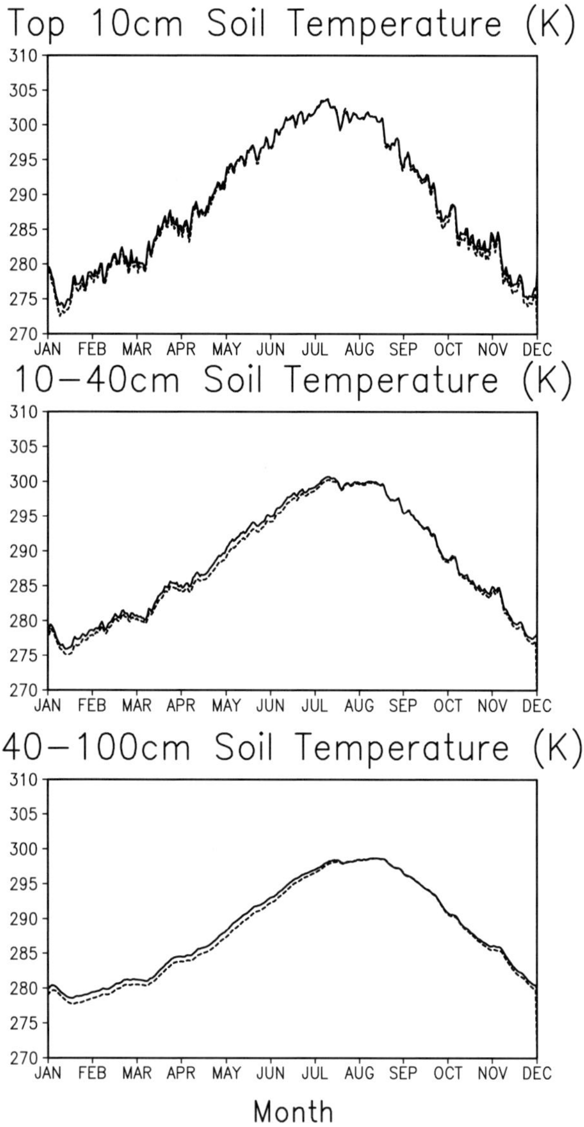

Fig. 11.6. Time series of 3-year (1997–1999) mean modeled soil temperature (dashed line) at the three Noah LSM soil layers (0–10 cm, 10–40 cm, 40–100 cm) compared with observations (solid line) from 5 cm, 25 cm, and 70 cm depths. The soil temperature is spatially averaged from 72 Oklahoma Mesonet stations.

soil temperature. Soil temperature is also well-simulated in the second and third soil layers, with a maximum difference of less than 3 K. This is generally consistent with the results of Robock *et al.* (2003), where a maximum difference of 5 K was found. The smaller maximum difference can be explained by physics updates in the Noah LSM made for NLDAS-2, and by the use of improved forcing data.

Continuing the analysis of model output, Fig. 11.7 shows a comparison between the observation-based radiative skin temperature and that modeled by the Noah, VIC, and Mosaic LSMs in NLDAS-2. Because direct observations of skin temperature were unavailable, the simulated and observed upward longwave radiation fields were converted to equivalent surface skin temperatures by assuming a surface emissivity value of one as used in Robock *et al.* (2003). While in reality the surface emissivity varies and is not exactly one, the three NLDAS models featuring energy balance components (Noah, Mosaic, and VIC) use this assumption in their calculation of upward longwave radiation. The use of this same assumption in the skin temperature comparison therefore ensures consistency and allows for a fair comparison with observations. The plot shows the observed monthly mean diurnal cycle of skin temperature averaged over 22 ARM/CART stations, and the differences between the simulated and observed skin temperature. The observed skin temperature displays a strong diurnal cycle, particularly during winter and summer. The Noah and Mosaic models underestimate night-time skin temperature in the winter and spring and overestimate day-time skin temperature in the summer, although Mosaic has a stronger negative night-time bias during the winter and spring. Differing from these two models, VIC significantly overestimates night-time skin temperature in all four seasons due to the oversimulation of net radiation.

As indicated by Robock *et al.* (2003), model structure and parameter shortcomings may be responsible for the different biases in VIC and Mosaic output. At the same time, a recent Noah LSM update meant to reduce cold season snow sublimation in stable boundary layer cases (Livneh *et al.*, 2011) may be responsible for the large night-time skin temperature biases in the Noah LSM. Future collaboration between the NLDAS team and the land surface modeling community will yield model structure upgrades and refined parameter sets focused on reducing these biases.

11.2.2.3 *Drought monitor and seasonal hydrological forecast system*

Drawing on the land surface outputs from the NLDAS modeling system, the NLDAS drought monitor was originally developed by NASA GSFC

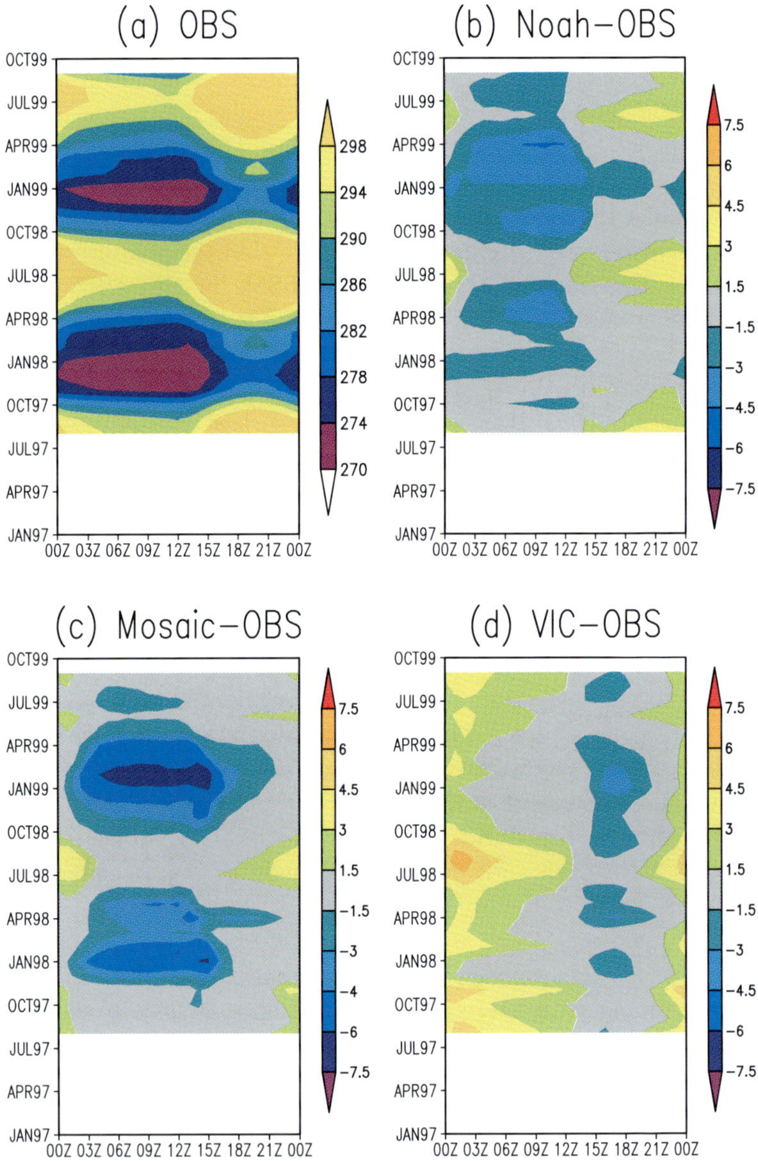

Fig. 11.7. Observed and simulated skin temperature is compared for the period spanning January 1, 1997 to September 30, 1999. Panels are (a) the monthly mean diurnal cycle of the observed skin temperature averaged from 22 ARM/CART stations, (b) the difference of the monthly mean diurnal cycle of skin temperature between the Noah LSM and the observations, (c) the difference between the Mosaic LSM and the observations, and (d) the difference between the VIC LSM and the observations.

scientists in 2006. At that time, it only included output from the Noah and Mosaic models. The monitor at first used 8-year (1997–2005) of NLDAS-1 retrospective and real-time output to generate the top 1 m and total column long-term soil moisture climatologies necessary to characterize current hydrologic conditions but was later revised to use the 28-year (1980–2007) of NLDAS-2 output. It calculated and displayed anomaly and percentile values of soil moisture at daily, weekly, and monthly timescales. In 2008, the drought monitor was transitioned to the NCEP/EMC public server and was subsequently extended to include all four NLDAS-2 models and their ensemble mean. The list of output variables was also expanded, with snow water equivalent, total runoff, streamflow, evaporation and precipitation joining soil moisture which had already been present in the GSFC version of the monitor. Figure 11.8 provides a typical example of NLDAS drought monitor output: the total column soil moisture anomaly on March 8, 2013. In these images, the yellow and red colors indicate the presence of drought conditions in the Southwest, the Great Plains, and the Midwest of the USA.

The inclusion of the four-model ensemble mean in the NLDAS-2 drought monitor is significant, as Diemeyer *et al.* (2006) along with the NLDAS-2 validation covered above demonstrated that a multi-model ensemble typically produces more reliable land surface products than does an individual model. However, there is value in the individual model output as well. The result from each individual model can be used to determine the uncertainty inherent in the ensemble mean and to identify inter-model differences. Moreover, it can also be used to diagnose the strengths and weaknesses of each model.

Complementing the analysis capabilities of the NLDAS drought monitor, the NLDAS-2 seasonal hydrologic forecast system was created through combining the forecast systems of Luo and Wood (2008) and Wood and Lettenmaier (2006). While only the VIC model is currently used in the experimental forecast system, plans exist to incorporate the other three NLDAS-2 models in the near future. Forecasts are produced at the beginning of each month using the initial condition that is closest to the first day of the month. Three hydrologic forecast approaches have been implemented in NLDAS-2, differing in the source of climate forecast data (dynamic or statistical) that is downscaled and bias-corrected to provide the necessary LSM forcing data. The three methods are:

(1) The CFS-based forecast approach which uses seasonal forecasts from the NCEP Climate Forecast System (CFS) dynamical model. In this

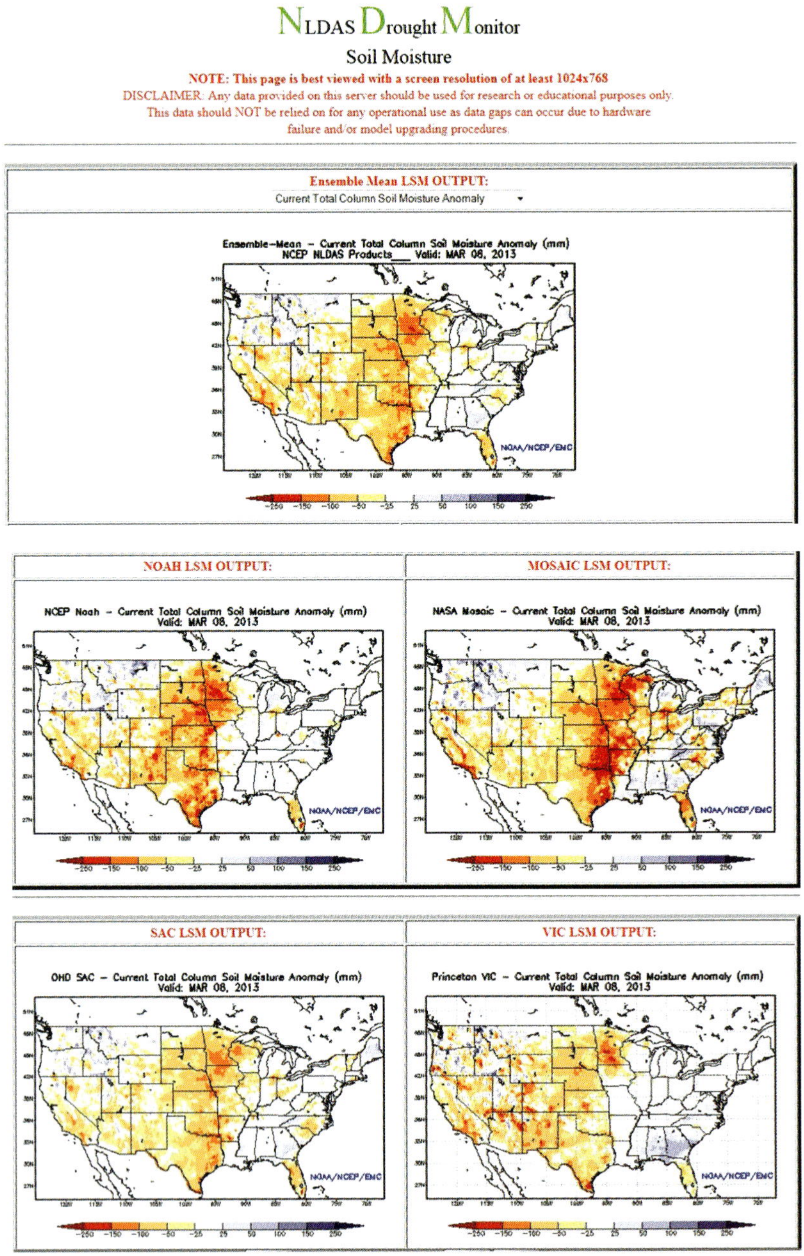

Fig. 11.8. An example page from the NLDAS drought monitor. Page depicts the total column soil moisture anomaly (mm/month) on March 8, 2013, for the four individual NLDAS-2 models and their ensemble mean.

approach, the set of forecasts from the previous month are combined to form a model forecast distribution which is then merged with historic observations using a Bayesian approach developed by Luo *et al.* (2007) and Luo and Wood (2008).

(2) The CPC outlook-based forecast approach which uses the three-month seasonal probability of exceedence (POE) forecasts released by the NCEP Climate Prediction Center (CPC) during the previous month as the forecast distribution for each of the 102 climate divisions of the USA. In this approach, the three-month CPC distributions are converted to monthly distributions and then applied to all NLDAS grid cells within the climate division using a statistical method called the Schaake shuffle. The approach is detailed in Wood and Lettenmaier (2006).

(3) The Ensemble Streamflow Prediction (ESP) approach which uses 20 randomly selected historical atmospheric forcing time series as possible realizations of future conditions.

All three methods described above generate 20 ensemble members and are limited to this value due to the availability of computational resources.

In quasi-operational mode at NCEP/EMC since November 2008, this three-pronged hydrologic forecast system is executed at the beginning of each month by the EMC land-hydrology group. It produces precipitation, streamflow, soil moisture, and evaporation anomalies and percentiles as well as a drought probability forecast based on total column soil moisture. Forecast products are provided to the CPC seasonal drought outlook group and made available to other drought forecast end-users via the NLDAS website at EMC (http://www.emc.ncep.noaa.gov/mmb/nldas/forecast/). This seasonal prediction component of the EMC NLDAS website was transitioned from an earlier version created by PU located at http://hydrology.princeton.edu/forecast/current.php.

NLDAS-2 drought monitor and seasonal forecast products strongly support NIDIS activities including the creation of weekly US Drought Monitor (USDM) analysis and outlooks, the NCEP/CPC monthly drought briefing, and the NCEP/CPC seasonal drought outlook. Figure 11.9 presents a schematic of how NLDAS products are used to support these activities.

NIDIS and other end users of hydrologic data depend on timely and reliable inputs in order to construct their deadline-constrained forecasts and analysis. While the NLDAS-2 drought monitoring and hydrologic forecast system strives to maintain a high degree of up-time, it is still

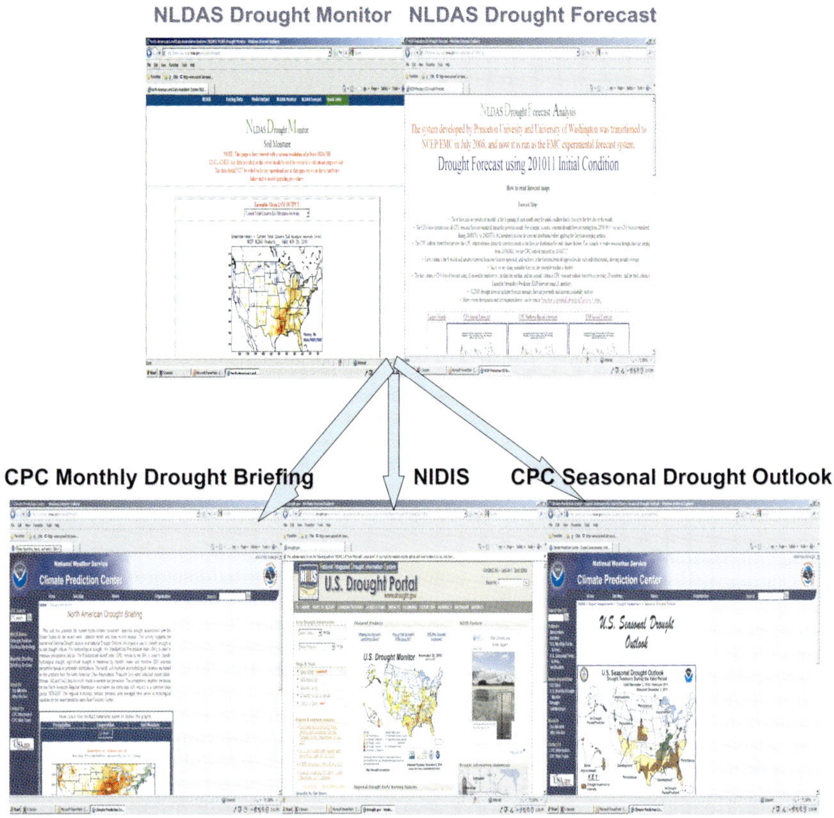

Fig. 11.9. A schematic diagram depicting the support NLDAS products provide for operational CPC and NIDIS drought monitoring and prediction activities (NLDAS drought monitor and forecast website: http://www.emc.ncep.noaa.gov/mmb/nldas; CPC monthly drought briefing website: http://www.cpc.ncep.noaa.gov/products/Drought/; NIDIS drought monitoring website: http://www.drought.gov/drought/; CPC seasonal drought outlook website: http://www.cpc.ncep.noaa.gov/products/expert_assessment/seasonal_drought.html).

a quasi-operational activity and does not feature the stringent requirements and redundant computer/software resources that an operational system would have. Addressing this issue, NCEP/EMC and CPC will transition both NARR and NLDAS-2 into NCEP operations, which will greatly increase the reliability and timeliness of NLDAS-2 data products. Additionally, CONUS-wide validation and assessment of NLDAS-2 drought forecast products will be a major focus of future work, as will the inclusion of additional LSMs in the drought forecast system.

11.2.2.4 *Remaining issues and future directions*

Forcing data were greatly improved in NLDAS-2 over that used in NLDAS-1. In particular, better precipitation forcing in NLDAS-2 led to a reduction in the large negative runoff and positive evaporation biases seen in NLDAS-1 (Xia *et al.*, 2012b). However, several potential areas of improvement remain and provide a direction for future work. For example, the wind-induced undercatch of solid precipitation by gauges was not corrected in NLDAS-2, as there was not a mature correction technique available (although some approaches exist, including Adam and Lettenmaier, 2003; Xia, 2007). In addition, the warm bias present in NARR (Royer and Poirier, 2010), and thus NLDAS-2, 2-m air temperature strongly affected the simulation of snowpack (and thus melt-based streamflow) in all four NLDAS-2 models owing to their use of a 0°C rain/snow determination temperature. With the aforementioned shortcomings in mind, areas of future NLDAS-2 research include a new 2-m temperature observational data set that would avoid the warm temperature bias present in the NARR, as well as a new precipitation data set which includes a correction for gauge undercatch.

Future NLDAS-2 activities extend beyond the improvements in forcing data mentioned above, and into further improvements in model physics and parameters. While the upgraded NLDAS-2 LSMs used in conjunction with the NLDAS-2 surface forcing have produced high-quality land surface products as compared against observed water fluxes, energy fluxes, and state variables (Ek *et al.*, 2011a, 2011b; Xia *et al.*, 2012a, 2012b), inter-model differences are still very large and areas for improvement remain. One such area is in the simulation of streamflow, where NLDAS-2 demonstrated that the SAC-SMA and VIC models, which originally grew out of the hydrological community, performed much better than the Noah and Mosaic models, which have roots in the meteorological modeling community. The former set of models has a closer relationship between precipitation and total runoff anomaly than the latter (Xia *et al.*, 2012c). One way in which this will be leveraged in the future is by implementing portions of the SAC-SMA and/or VIC hydrological modeling schemes into the Noah and Mosaic LSMs. An initial effort to improve streamflow simulations along these lines is underway and detailed in Livneh *et al.* (2012).

Other future research directions include addressing the large negative skin temperature and soil temperature biases exhibited by the Noah and Mosaic models in the cold season, and the negative soil moisture biases generated in the warm season. Further research will also seek to determine why the Noah model generated relatively more total runoff and less evaporation, and the Mosaic model generated relatively less total runoff and more evaporation than the other NLDAS-2 LSMs.

These model cross-pollination efforts align well with the recommendation of the 2004 National Service Integrated Water Science Plan (Carter *et al.*, 2004) to create a unified Hydro-LSM. Featuring strong hydrologic as well as energy balance and land surface state performance, it would seek to increase the efficiency of related research and operations through bridging the divides which have historically existed with the NWS hydrologic and meteorological communities. The intercomparison, model integration, and model improvement activities of NLDAS-2 have brought the community closer to a such an integrated model — one which is capable of providing top-quality energy balance, land surface, and hydrologic fields. Additional work within NLDAS-2 and the broader modeling community will help bring this concept to fruition and will enable enhanced water resource, weather, and climate forecasts and analysis which are becoming increasing vital sources of information.

11.3 Summary and Concluding Remarks

Over the course of two distinct project phases, the NLDAS has successfully produced high-quality surface meteorology and precipitation datasets for use in the Noah, SAC-SMA, VIC, and Mosaic LSMs. These LSMs have generated land surface products including water fluxes (evaporation, runoff/streamflow), energy fluxes (net radiation, sensible heat flux, latent heat flux, ground heat flux), and state products (soil moisture, snow cover, soil moisture, soil temperature, skin temperature). These products have been used in numerous applications by GAPP and then CPPA researchers as well as by end users in other hydrologic research and application communities. The NIDIS organization, in particular, has been a prime focus of NLDAS-2 support. With its growing user-base and maturity level, the NLDAS-2 system will in the future be officially implemented into operations at NCEP, greatly enhancing the timeliness and reliability of the system.

Future activities will also include the continued testing and validation of NLDAS-2 model upgrades. Beyond supporting such validation activities, NLDAS-2 will also assist in developing the next generation of the Noah LSM, the Noah-Multi-Parameterization model (Noah-MP) (Niu *et al.*, 2011; Yang *et al.*, 2011). This model, featuring a multi-layer snow sub-model along with representation of ground water, dynamic vegetation and frozen soil processes, is being tested by several groups including EMC, NCAR, and University of Texas at Austin. Work will also continue on improving forcing data as outlined in Sec. 11.2.2.4.

Notwithstanding the high quality of its products or the range of applications it supports, we recognize that NLDAS-2 system is not an actual data assimilation system *per se*, because there is no assimilation of remotely sensed land-surface states such as soil moisture and snowpack. Seeking to address this issue, the NASA GSFC Hydrologic Sciences Branch is currently collaborating with NCEP/EMC to make NASA's Land Information System (LIS) (Kumar *et al.*, 2006) the software basis of the NLDAS-2 real-time monitoring system. LIS contains multiple data assimilation modules, and the switch from individual non-data assimilation capable model drivers to a unified LIS software framework will facilitate the eventual assimilation of soil moisture, snowpack, brightness temperature, and other remotely sensed variables, along with ground-based measurements such as USGS streamflow, into all four NLDAS-2 LSMs.

Data assimilation will improve the initial land surface conditions ingested by the NLDAS-2 seasonal hydrological forecast system. As detailed above, this system provides one- to six-month anomalies and percentiles of land surface products along with a drought probability forecast based on total column soil moisture. Output from this system is available via the "NLDAS Forecast" tab on the NLDAS website and supports operational drought prediction activities at CPC and NIDIS. However, since the system currently only includes a single model and has not been well-evaluated outside of the Ohio River basin, it is not mature enough to implement operationally at NCEP. Future efforts will thus focus on including additional models in the system and on performing additional validation, allowing for the official implementation of this valuable forecast tool.

Given the wide range of future NLDAS-2 model and forcing improvements as well as the need by the end-user community for a stable product, the NCEP/EMC NLDAS team will maintain two parallel versions of the NLDAS system: a research version and an operational version. Improvements in surface forcing data, model development, and the addition of data assimilation approaches from NLDAS partners will be immediately implemented into the research version for internal evaluation at NCEP/EMC. After upgrades are fully vetted and approved, they will be transferred into the operational version of the system. This will provide a solid pathway through which end-users will continue to benefit from the latest modeling approaches and high-quality output available from the NLDAS-2 system.

Acknowledgments

NLDAS-1 and NLDAS-2 research activities have been supported by the NOAA Office of Global Programs Global Energy and Water Cycle Experiment (GEWEX) Americas Prediction Project, the NOAA Climate Program Office (CPO) Climate Prediction Program of the Americas (CPPA), and the NASA Terrestrial Hydrology Program.

References

Adam, J. C. and Lettenmaier, D. P. (2003). Adjustment of global gridded precipitation for systematic bias. *Journal of Geophysical Research*, **108**, 1–14.

Alonge, C. J. and Cosgrove, B. A. (2008). Application of NARR-based NLDAS ensemble simulations to continental-scale drought monitoring. AGU spring meeting, #H43E-6, Fort Lauderdale, FL, USA.

Anderson, E. A. (1973). National weather service river forecast system: Snow accumulation and ablation model. Technical Memorandium, NWS Hydro-17, National Oceanic and Atmospheric Administration, Silver Spring, MD, USA.

Baldocchi, D. D. *et al.* (2001). FLUXNET: A new tool to study the temporal and spatial variability of ecosystemscale carbon dioxide, water vapor and energy flux densities. *Bulletin of the American Meteorological Society*, **82**, 2415–2434.

Balsamo, G., Mahfouf, J.-F. Bélair, S., and Deblonde, G. (2007). A land data assimilation system for soil moisture and temperature: An information content study. *Journal of Hydrometeorology*, **8**, 1225–1242.

Berg, A. A., Famiglietti, J. S., Walker, J. P., and P. R. Houser (2003). Impact of bias correction to reanalysis products on simulation of North American soil moisture and hydrological fluex. *Journal of Geophysical Research*, **108**, 4490, doi: 10.1029/2002JD003334.

Betts, A., Chen, F., Mitchell, K. and Janjic, Z. (1997). Assessment of the land surface and boundary layer models in two operational versions of the NCEP Eta model using FIFE data. *Monthly Weather Review*, **125**, 2896–2916.

Bolten, J., Rodell, M., Zaitchik, B., Toll, D., Engman, E., Habib, S. and Ozdogan, M. (2010). Development of the Middle Eastern and North African Land Data assimilation system. Geophysical Research Abstracts, 12, EGU2010-5498, EGU General Assembly.

Brock, F. V., Crawford, K. C., Elliott, R. L., Cuperus, G. W., Stadler, S. J., Johnson, H. L. and Eilts, M. D. (1995). The Oklahoma Mesonet: A technical overview. *Journal of Atmospheric and Oceanic Technology*, **12**, 5–19.

Burnash, R. J. C., Ferral, R. L. and McGuire, R. A. (1973). A generalized streamflow simulation system: Conceptual models for digital computer. Technical Report, Joint Federal State River Forecast Center, U.S. National Weather Service and California Department of Water Resource, Sacramento, California.

Burnash, R. J. C. (1995). The NWS river forecast system-catchment modeling. In *Computer Models of Watershed Hydrology*, Singh, V. P. (ed.), pp. 311–366. Littleton, CO: Water Resources Publications.

Carter, G., Seo, D. J., Cline, D., Hartman, R., Mitchell, K. and Zhou, J. (2004). The NWS Integrated Water Science Plan (IWSP), Technical Report, US Department of Commerce, Silver Spring, Maryland, USA.

Chen, F., Janjic, Z. and Mitchell, K. (1997). Impact of atmospheric surface-layer parameterizations in the new land-surface scheme of the NCEP mesoscale Eta model. *Boundary Layer Meteorology*, **85**, 391–421.

Cherkauer, K. A. and Lettenmaier, D. P. (1999). Hydrological effects of frozen soils in the upper Mississippi River basin. *Journal of Geophysical Research*, **104**, 19599–19610.

Cherkauer, K. A., Bowling, L. C. and Lettenmaier, D. P. (2003). Variable Infiltration Capacity (VIC) cold land process model updates. *Global and Planetary Change*, **38**, 151–159.

Cosgrove, B. A., Lohmann, D., Mitchell, K. E. *et al.* (2003a). Real-time and retrospective forcing in the North American Land Data Assimilation System (NLDAS) project. *Journal of Geophysical Research*, **108**(D22), 8842, doi:10.1029/2002JD003118.

Cosgrove, B. A., Lohmann, D., Mitchell, K. E. *et al.* (2003b). Land surface model spin-up behavior in the North American Land Data Assimilation System (NLDAS). *Journal of Geophysical Research*, **108**(D22), 8845, doi:10.1029/2002JD00331603.

Daly, C., Neilson, R. P. and Philips, D. L. (1994). A statistical-topographic model for mapping climatological precipitation over mountainous terrain. *Journal of Applied Meteorology*, **33**, 140–158.

Day, G. N. (1985). Extended streamflow forecasting using NWSRFS. *Journal of Water Resources Planning and Management*, **111**, 157–170.

de Goncalves, L. G. G., Shuttleworth, W. J., Burke, E. J., Houser, P., Toll, D. L., Rodell, M. and Arsenault, K. (2006). Toward a South American Land Data Assimilation System (SALDAS): Aspects of land surface model spin-up using the simplified simple biosphere. *Journal of Geophysical Research*, **111**, D17110, doi:10.1029/2005JD006297.

de Goncalves, L. G. G. *et al.* (2009). The South American Land Data Assimilation System (SALDAS) 5-year retrospective atmospheric forcing datasets. *Journal of Hydrometeorology*, **10**, 999–1010.

Dirmeyer, P., Dolman, A. and Sato, N. (1999). The pilot phase of the global soil wetness project. *Bulletin of the American Meteorological Society*, **80**, 851–878.

Dirmeyer, P. A., Gao, X., Zhao, M., Guo, Z., Oki, T. and Hanasaki, N. (2006). The Second Global Soil Wetness Project (GSWP-2): Multi-model analysis and implications for our perception of the land surface. *Bulletin of the American Meteorological Society*, **87**, 1381–1397.

Ek, M. B., Mitchell, K. E., Lin, Y., Rogers, E., Grunman, P., Koren, V., Gayno, G. and Tarpley, J. D. (2003). Implementation of Noah land surface model advances in the National Centers for Environmental Prediction operational mesoscale Eta model. *Journal of Geophysical Ressearch*, **108**, 8851, doi:10.1029/2002JD003296.

Ek, M. and NLDAS Team (2011a). North American Land Data Assimilation System (NLDAS-2) Phase 2: Development and applications. *GEWEX News*, **21**, 6–7.

Ek, M., Xia, Y., Mo, K. *et al.* (2011b). NLDAS-2 — Long-term high resolution multi-model hydrometeorological data and its application in the US. *Bulletin of the American Meteorological Society* (in preparation).

Gibson, J. K., Kallberg, P., Uppala, S., Hernandez, A., Nomura, A. and Serrano, E. (1997). ERA description. ERA-15 Proj. Rep. Ser. 1, 72 pp., Eur. Cent. for Medium-Range Weather Forecasts, Reading, UK.

Gutman, G. and Ignatov, A. (1998). The derivation of the green vegetation fraction from NOAA/AVHRR data for use in numerical weather prediction models. *Journal of Applied Remote Sensing*, **19**, 1533–1543.

Hansen, M. C., DeFries, R. S., Townshend, J. R. G. and Sohlberg, R. (2000). Global land cover classification at 1 km spatial resolution using a classification tree approach. *International Journal of Remote Sensing*, **21**, 1331–1364.

Henderson-Sellers, A., Pitman, A. J., Love, P. K., Irannejad, P. and Chen, T. H. (1995). The Project for Intercomparison of Land Surface Parameterization Schemes (PILPS): Phases 2 and 3. *Bulletin of the American Meteorological Society*, **76**, 489–503.

Higgins, R. W., Shi, W., Yarosh, E. and Joyce, R. (2000). Improved United States precipitation quality control system and analysis. NCEP/CPC, Atlas 7, Climate Prediction Center, Camp Springs, Maryland, USA.

Hu, Q. and Feng, S. (2003). A daily soil temperature dataset and soil temperature climatology of the contiguous United States. *Journal of Climate*, **42**, 1139–1156.

Ji, M., Kumar, A. and Leetmaa, A. (1994). A multiseason climate forecast system at the National Meteorological Center. *Bulletin of the American Meteorological Society*, **75**, 569–577.

Kalnay, E. *et al.* (1996). NCEP/NCAR 40-year reanalysis project. *Bulletin of the American Meteorological Society*, **77**, 437–472.

Kanamitsu, M., Ebisuzaki, W., Woollen, J., Yang, S.-K., Hnilo, J. J., Fiorino, M. and Potter, G. L. (2002). NCEP-DOE AMIP-II Reanalysis (R-2). *Bulletin of the American Meteorological Society*, **83**, 1631–1643.

Koren, V., Schaake, J., Mitchell, K. E., Duan, Q., Chen, F. and Baker, J. (1999). A paramerization of snowpack and frozen ground intended for NCEP weather and climate models. *Journal of Geophysical Research*, **104**, 19569–19585.

Koren, V., Smith, M., Wang, D. and Zhang, Z. (2000). Use of soil property data in the derivation of conceptual rainfall-runoff model parameters. In *Proceedings of the 15th Conference on Hydrology, American Meterological Society*, Long Beach, CA, pp. 103–116.

Koren, V., Smith, M. and Duan, Q. (2003). Use of *a priori* parameter estimates in the derivation of spatially consistent parameter sets of rainfall-runoff models. In *Advances in the Calibration of Watershed Models*, Duan, Q., Sorooshian, S., Gupta, H., Rosseau, A. and Turcotte, R. (eds.), Vol. 6, pp. 239–254. Water Science and Applications Series, Washington, DC: AGU.

Koren, V., Simith, M., Cui, Z., Cogrove, B., Werner, K. and Zamora, R. (2010). Modification of Sacramento Soil Moisture Accounting Heat Transfer Component (SAC-HT) for Enhanced Evapotranspiration. NOAA (National Oceanic and Atmospheric Administration) Technical Report NWS (National Weather Service) 53, Silver Spring, Maryland, USA, 72pp.

Koster, R. and Suarez, M. (1994). The components of the SVAT scheme and their effects on a GCM's hydrological cycle. *Advances in Water Resources*, **17**, 61–78.

Koster, R., Suarez, M. and Heiser, M. (2000). Variance and predictability of precipitation at seasonal-to-interannual timescales. *Journal of Hydrometeorology*, **1**, 24–46.

Kumar, S. V., Peters-Lidard, C. D., Tian, Y., Houser, P. R., Geiger, J., Olden, S., Lighty, L., Eastman, J. L., Doty, B., Dirmeyer, P., Adams, J., Mitchell, K., Wood, E. F. and Sheffield, J. (2006). Land information system — An interoperable framework for high resolution land surface modeling. *Environmental Modelling and Software*, **21**, 1402–1415.

Lettenmaier, D. P., Abdula, F. A., Wood, E. F. and Smith, J. A. (1996). Application of a macroscale hydrologic model to estimate the water balance of the Arkansas-Red River basin. *Journal of Geophysical Reseatch*, **101**, 7449–7459.

Li X., Huang, C., Che, T., Jin, R., Wang, S., Wang, J., Gao, F., Zhang, S., Qiu, C. and Wang, C. (2007). Development of a Chinese Land Data Assimilation System (CLDAS): Its progress and prospects. *Progress in Natural Sciences*, **17**, 881–892.

Liang, X., Lettenmaier, D. P., Wood, E. F. and Burges, S. J. (1994). A simple hydrologically based model of land surface water and energy fluxes for GCMs. *Journal of Geophysical Research*, **99**, 14415–14428.

Lim, Y.-J., Lee, T. Y. and Byun, K. Y. (2008). Korea Land Data Assimilation System (KLDAS) and its application using WRF. In *22nd Conference on Hydrology*, 88th American Meteorological Society Annual Meeting, New Orleans, LA, January 20–24, 2008, Paper JP1.4.

Lin, Y. and Mitchell, K. E. (2005). The NCEP Stage II/IV hourly precipitation analyses: Development and applications. In *19th Conference on Hydrology*, American Meteorological Society, San Diego, CA, January 9–13, 2005, Paper 1.2.

Livneh, B., Xia, Y., Ek, M. B., Mitchell, K. E. and Lettenmaier, D. (2010). Noah LSM snow model diagnostics and enhancements. *Journal of Hydrometeorology*, **12**, 721–738.

Livneh, B., Restrepo, P. J. and Lettenmaier, D. P. (2012). Development of a unified land model for prediction of surface hydrology and land-atmosphere interactions. *Journal of Hydrometeorology*, **12**, 1299–1320.

Lobmeyr, M., Lohmann, D. and Ruhe, C. (1999). An application of a large scale conceptual hydrological model over the Elbe region. *Hydrology and Earth System Sciences*, **3**, 363–374.

Lohmann, D., Raschke, E. Nijssen, B. and Lettenmaier, D. P. (1998). Regional scale hydrology, Part II: Application of the VIC-2L model to the Weser River, Germany. *Hydrological Sciences Journal*, **43**, 143–158.

Lohmann, D., Mitchell, K. E. Houser, P. R. *et al.* (2004). Streamflow and water balance intercomparisons of four land surface models in the North American Land Data Assimilation System (NLDAS) project. *Journal of Geophysical Research*, **109**, D07S91, doi:10.1029/2003JD003517.

Luo, L., Robock, A., Mitchell, K. E. *et al.* (2003). Validation of the North American Land Data Assimilation System (NLDAS) retrospective forcing over the Southern Great Plains. *Journal of Geophysical Research*, **108**(D22), 8843, doi:10.1029/2002JD003246.

Luo, L., Wood, E. F. and Pan, M. (2007). Bayesian merging of multiple climate model forecasts for seasonal hydrological predictions. *Journal of Geophysical Research*, **112**, D10102, doi:10.1029/2006JD007655.

Maurer, E. P., Wood, A. W., Adam, J. C., Lettenmaier, D. P. and Nijssen, B. (2002). A long-term hydrologically-based data set of land surface fluxes and states for the conterminous United States. *Journal of Climate*, **15**, 3237–3251.

Mesinger, F., DiMego, G., Kalnay, E. *et al.* (2006). North American regional reanalysis. *Bulletin of the American Meteorological Society*, **87,** 343–360.

Miller, D. A. and White, R. A. (1998). A conterminous United States multilayer soil characteristics data set for regional climate and hydrology modeling. *Earth Interactions Journal*, **2**, 1–26.

Mitchell, K. *et al.* (1999). GCIP Land Data Assimilation System (LDAS) project now underway. *GEWEX News*, **9**(4), 3–6.

Mitchell, K. E., Lohmann, D., Houser, P. R. *et al.* (2004). The multi-institution North American Land Data Assimilation System (NLDAS): Utilizing multiple GCIP products and partners in a continental distributed hydrological modeling system. *Journal of Geophysical Research*, **109**, D07S90, doi:10.1029/2003JD003823.

Mo, K., Long, L., Xia, Y., Yang, S. K., Schemm, J. E. and Ek, M. B. (2010). Drought indices based on the Climate Forecast System Reanalysis and ensemble NLDAS. *Journal of Hydrometeorology*, **12**, 181–215.

Nijssen, B., Lettenmaier, D. P., Liang, X., Wetzel, S. W. and Wood, E. F. (1997). Streamflow simulation for continental-scale river basins. *Water Resources Research*, **33**, 711–724.

Nijssen, B., O'Donnell, G. M., Lettenmaier, D. P., Lohmann, D. and Wood, E. F. (2001). Predicting the discharge of global rivers. *Journal of Climate*, **14**, 3307–3323.

Niu, G.-Y. *et al.* (2011). The community Noah land surface model with multiparameterization options (Noah-MP), Part 1: Model description and evaluation with local-scale measurements. *Journal of Geophysical Research*, **116**, D12109, doi:10.1029/2010JD015139.

Pan, M., Ming, P., Sheffield, J. *et al.* (2003). Snow process modeling in the North American Land Data Assimilation System (NLDAS), Part 2: Evaluation of model simulated snow water equivalent. *Journal of Geophysical Research*, **108**(D22), 8850, doi:10.1029/2003JD003994.

Pinker, R. T., Tarpley, J. D., Laszlo, I. *et al.* (2003). Surface radiation budgets in support of the GEWEX Continental-Scale International Project (GCIP) and the GEWEX Americas Prediction Project (GAPP), including the North American Land Data Assimilation System (NLDAS) project. *Journal of Geophysics Research*, **108**(D22), 8844, doi:10.1029/2002JD003301.

Pitman, A. J. and Henderson-Sellers, A. (1998). Recent progress and results from the project for the intercomparison of land surface parameterization schemes. *Journal of Hydrology*, **212**, 128–135.

Ramsay, B. H. (1999). The Interactive Multisensor Snow and Ice Mapping System. *Hydrological Process*, **12**, 1537–1546.

Reynolds, C. A., Jackson, T. J. and Rawls, W. J. (2000). Estimating soil water-holding capacities by linking the food and agriculture organization soil map of the world with global pedon databases and continuous pedotransfer functions. *Water Resources Research*, **36**, 3653–3662.

Robock, A., Vinnikov, K. Y., Srinivasan, G. Entin, J. K., Hollinger, S. E., Speranskaya, N. A., Liu, S., Namkhai, A. (2000). The Global Soil Moisture Data Bank. *Bulletin of the American Meteorological Society*, **81**, 1281–1299.

Robock, A., Luo, L., Wood, E. F. *et al.* (2003). Evaluation of the North American Land Data Assimilation System over the Southern Great Plains during the warm season. *Journal of Geophysical Research*, **108**(D22), 8846, doi:10.1029/2002JD003245.

Rogers, E., Deaven, D. and Dimego, G. (1995). The regional analysis system for the operational "early" Eta model: Original 80-km configuration and recent changes. *Weather Forecasting*, **10**, 810–825.

Rodell, M., Houser, P. R., Jambor, U. *et al.* (2004). The global land data assimilation system. *Bulletin of the American Meteorological Society*, **85**, 381–394.

Royer, A. and Poirier, S. (2010). Surface temperature spatial and temporal variations in North America from homogenized satellite SMMR-SSM/I microwave measurements and reanalysis for 1979–2008. *Journal of Geophysical Research*, **115**, D08110, doi:10.1029/2009JD012760.

Saha, S. *et al.* (2010). The NCEP climate forecast system reanalysis. *Bulletin of the American Meteorological Society*, **91**, 1017–1057.

Schaake, J. C., Duan, Q., Korean, V., Mitchell, K. E., House, P. R., Wood, E. F., Robock, A., Lettenmaier, D. P., Lohmann, D., Cosgrove, B., Sheffield, J., Lou, L., Higgins, R. W., Pinker, R. T. Tarpley, J. D. (2004). An intercomparison of soil moisture fields in the North American Land Data Assimilation System (NLDAS). *Journal of Geophysical Research*, **109**, DO1590, doi: 10.1029/2002JD003309.

Sheffield, J., Pan, M., Wood, E. F. *et al.* (2003). Snow process modeling in the North American Land Data Assimilation System (NLDAS), Part 1: Evaluation of model-simulated snow cover extent. *Journal of Geophysical Research*, **108**(D22), 8849, doi:10.1029/2002JD003274.

Sheffield, J., Goteti, G. and Wood, E. F. (1996). Development of a 50-year high-resolution global dataset of meteorological forcings for land surface modeling. *Journal of Climate*, **19**, 3088–3111.

Troy, T. J., Wood, E. F. and Sheffield, J. (2008). An efficient calibration method for continental-scale land surface modeling. *Water Resources Research*, **44**, W09411, doi:10.1029/2007WR006513.

Van Den Hurk, B. (2002). European LDAS established. *GEWEX News*, **12**, 9.

Verdin, K. L. and Greenlee, S. K. (1994). Development of continental scale digital elevation models and extraction of hydrologic features. Paper presented at *Third International Conference/Workshop on Integrated GIS and Environmental Modeling*, National Center for Geography Information and Analysis, Santa Fe, NM, January 21–26.

Wei, H., Xia, Y., Mitchell, K. E. and Ek, M. B. (2013). Improvement of Noah land surface model for the warm season processes: Assessment of water and energy flux simulation. *Hydrological Processes*, **27**, 297–303.

Wood, E. F., Lettenmaier, D. P., Liang, X., Nijssen, B. and Wetzel, S. W. (1997). Hydrological modeling of continental-scale basins. *Annual Review of Earth and Planetary Sciences*, **25**, 279–300.

Wood, E. *et al.* (1998). The Project for Intercomparison of Land-Surface Parameterization Scheme (PILPS) Phase 2(c) Red-Arkansas River basin experiment, Part 1: Experiment description and summary intercomparisons. *Global and Planetary Change*, **19**, 115–136.

Wood, A. W. and Lettenmaier, D. P. (2006). A testbed for new seasonal hydrologic forecasting approaches in the Western US. *Bulletin of the American Meteorological Society*, **87**, 1699–1712.

Xia, Y. (2007). Calibration of LaD model in the Northeast of the United States using observed annual streamflow. *Journal of Hydrometeorology*, **8**, 1098–1110.

Xia, Y., Mitchell, K. E., Ek, M. B. and NLDAS Team (2012a). Continental water and energy flux analysis and validation for North-American Land Data Assimilation System project, Phase 2, Part 1: Comparison analysis of multi-model results. *JGR-Atmosphere*, **117**, D03109. doi:10.1029/2011JD016048.

Xia, Y., Mitchell, K. E., Ek, M. B. and NLDAS Team (2012b). Continental water and energy flux analysis and validation for North-American Land Data Assimilation System Project, Phase 2, Part 2: Validation of simulated streamflow. *JGR-Atmosphere*, **117**, D03110, doi:1029/2011JD016051.

Xia,Y., Ek, M. B., Wood, A., Wei, H. and Meng, J. (2012c). Comparative analysis of relationships between NLDAS-2 forcing data and model outputs. *Hydrological Processes*, **26**, 467–474.

Xia, Y., Ek, M., Sheffield, J., Livneh, B., Huang, M., Wei, H., Feng, S., Luo, L., Meng, J. and Wood, E. (2013). Validation of Noah — simulated soil temperature in the North American Land Data Assimilation System Phase 2. *Journal of Applied Meteorology and Climatology*, **52**, 455–471.

Yang, Z. L. *et al.* (2011). The community Noah land surface model with multiparameterization options (Noah-MP), Part 2: Evaluation over global river basins. *Journal of Geophysical Research*, **116**, D12110, doi:10.1029/2010JD015140.

Chapter 12

Soil Moisture Data Assimilation for State Initialization of Seasonal Climate Prediction

Wenge Ni-Meister

Department of Geography, Hunter College and Program in Earth and Environmental Sciences, The City University of New York, New York, NY10021, USA

Climate model prediction skill is currently limited in response to poor land surface soil moisture state initialization. However, initial soil moisture state prediction skill can potentially be enhanced by the assimilation of remotely sensed near-surface soil moisture data in offline simulation. This study reviews recent progress on soil moisture data assimilation for state initialization of seasonal climate prediction. Data assimilation is a process of merging observations with a system dynamic model to provide improved estimates of the environmental state variables. Soil moisture data assimilation is relatively new; however, rapid progress has been made in the last few decades with the available remotely sensed soil moisture data and the well-developed data assimilation systems in atmosphere and oceanography.

After briefing the history of soil moisture data assimilation and the basic concepts of data assimilation, the study focuses on the recent progress made in soil moisture data assimilation through a case study of the soil moisture initialization activities for NASA's seasonal and interannual climate prediction. This study is one of the first to evaluate such potential using actual remote sensing data together with field observations to characterize model and observation errors. The example demonstrates that while soil moisture data assimilation has made great progress in the last decade, it is still in its infancy. The accurate estimation of modeling and observation error remains a barrier to the wide use of the soil moisture data assimilation systems. Good quality remotely sensed soil moisture data with accurate uncertainty information at continental and global scale are needed to ensure the success of the operational use of soil moisture data assimilation technique. Further advancement on the current soil moisture data assimilation methods is necessary to be able to assimilate multi-source hydrological remote sensing data into land surface

S. Liang, X. Li & X. Xie (eds.), *Land Surface Observation, Modeling and Data Assimilation*, 377–402.

models (LSMs) for the best use of various remote sensing data sources at continental and global scales for improved seasonal climate prediction.

12.1 Introduction

Soil moisture is a key hydrological state variable which integrates much of the land surface hydrological and biophysical processes. Soil moisture interacts with the overlying atmospheric boundary layer through exchanges of sensible and latent heat fluxes, thus influences the atmospheric circulation. Soil has the ability to store precipitated water from periods of excess for later evaporation and to "remember" the wet or dry weather conditions longer than atmospheric processes (Koster and Suarez, 2000; Dong *et al.*, 2007b). Accurate knowledge of the spatial distribution and temporal variation of soil moisture would provide insight into larger-scale hydrological processes and would serve as good land surface moisture initialization states in fully coupled climate system models for improved seasonal-to-interannual climatological and hydrological prediction (Koster *et al.*, 2004).

Obtaining accurate soil moisture datasets at large spatial scales over a long period is not an easy task. *In situ* soil moisture measurements are generally expensive and no large area soil moisture networks exist to measure soil moisture at high temporal frequency in multiple soil depths. Alternatively, many land surface process models were developed and are available for calculating soil moisture; however, modeled soil moisture data are often prone to error due to insufficient model physics, inaccurate parameterization, and initialization states.

Remote sensing, on the other hand, is ideal for obtaining data at large-scales. Airborne microwave remote sensing data are available to retrieve large-scale soil moisture (Jackson *et al.*, 1995; Jackson, 1997; Jackson *et al.*, 1999). Thanks to satellite remote sensing, global 1/4 degree resolution near-surface soil moisture content has been derived using C-band passive microwave observations from the Nimbus-7 satellite Scanning Multi-frequency Microwave Radiometer (SMMR) for the period of operation from 1979 to 1987 (Owe *et al.*, 2001). Moreover, the Advanced Microwave Scanning Radiometer for the Earth observing system (AMSR-E) on the Aqua satellite launched in 2001 provides global C-band passive microwave soil moisture data at 25 km spatial resolution. While no C-band measurements are available between SMMR and AMSR-E, low-latitude soil moisture has been estimated using Tropical Rainfall Measuring Mission (TRMM) X-band microwave observations since 1998 (Bindlish *et al.*, 2003; Gao *et al.*, 2004,

2006). Further, L-band passive microwave soil moisture data has been launched already, available from 2009 with the launch of the Soil Moisture and Ocean Salinity (SMOS) mission in November 2009. The advantage of satellite data is that they provide a uniform spatial and temporal coverage of the land surface soil moisture data. But one limitation of these microwave-based soil moisture estimates is their confinement to the top few centimeters of soil and spatial and temporal gaps in areas with significant vegetation, soils, and roughness error sources. However, remote sensing data, when integrated with a land surface process model, will be able to provide estimates of soil moisture both at surface and deep in the root-zone with higher spatial and temporal resolution with less error than either remotely sensed data or model simulations alone. Data assimilation offers a means to combine the advantages of modeling with those of remote sensing data.

The use of data assimilation in hydrology is relatively new, however, it has shown promise for improving hydrologic predictions by incorporating remotely sensed soil moisture (e.g. Houser *et al.*, 1998; Walker *et al.*, 2001, 2002; Reichle *et al.*, 2002a,b; Zhang *et al.*, 2005; Reichle and Koster, 2005; Ni-Meister *et al.*, 2006; De Lannoy *et al.*, 2007, 2009; Crow and Reichle, 2008; Crow and Van den Berg, 2010; Huang *et al.*, 2008) and snow (e.g., Dong *et al.*, 2007a; Sun *et al.*, 2004; Rodell and Houser, 2004) into land surface models (LSMs). The focus of this paper is to review the recent progress made in soil moisture data assimilation. Section 12.2 summarizes the history of soil moisture/hydrological data assimilation; Sec. 12.3 introduces the basic concepts and common methods used in soil moisture data assimilation; and Sec. 12.4 presents a case study of soil moisture data assimilation. Sec. 12.5 discusses the future direction of soil moisture data assimilation.

12.2 Brief History of Soil Moisture Data Assimilation

Data assimilation techniques have widely been used in meteorology and oceanography for decades (see Daley, 1991, for detailed description). With substantial enhancement of the remote sensing instruments measuring various atmospheric quantities and the improvements in data assimilation techniques over the last few years, satellite data have been assimilated in Numerical Weather Prediction (NWP) models operationally. This resulted in the improvement of middle range (three days to two weeks) weather predictions to be as good as short range (one day) predictions (Kalnay, 2003). The use of data assimilation in oceanography also resulted in improved prediction of ocean dynamics (Bennett, 1992). Data assimilation methods have been well demonstrated as an effective technique to improve predictions of atmospheric and oceanic dynamics processes.

With the well-developed data assimilation methods from meteorology and oceanography, hydrological data assimilation, particularly soil moisture data assimilation, has made great progress in the last two decades. Jackson *et al.* (1981) were the first to use the direct insertion approach to update soil moisture with remotely sensed surface soil moisture when observations are available. Bernard *et al.* (1981) used synthetic observations to predict surface fluxes based on the direct insertion method. Prevot *et al.* (1984) and Brucker and Witono (1989) used a similar approach, but with real remote sensing data. More recent progress made on soil moisture data assimilation is to test the feasibility of optimal soil moisture data assimilation methods. The research focuses on: (1) the use of Kalman filtering based sequential/real time data assimilation strategy and (2) the non-sequential/retrospective assimilation including the Kalman smoothing or variational data assimilation strategy.

Kalman filter (KF) (Gelb, 1974) is an optimal sequential data assimilation method, which provides the optimal estimate for the current and future states for linear systems when observations are available. To apply to nonlinear hydrological processes, the extended KF (EKF) has been successfully applied in soil moisture data assimilation by linearizing the hydrologic processes (Entekhabi *et al.*,1994; Walker *et al.*, 2001, 2002; Walker and Houser, 2001). It has demonstrated how data assimilation can extend remotely sensed measurements of surface conditions into the underlying soil column.

Ensemble KF (EnKF) was implemented as an alternative to avoid the calculation of the derivative of a tangent linear model to approximate the nonlinear hydrological processes (Reichle *et al.*, 2002a, 2002b; Margulis *et al.*, 2002; Crow and Wood, 2003). During the past decade, the EnKF has been extensively applied to the assimilation of satellite-based surface soil moisture retrievals into LSMs (Reichle and Koster, 2005; Reichle *et al.*, 2007; Bolten *et al.*, 2010; Ni-Meister *et al.*, 2006). Because background error/covariance information in the EnKF is propagated via a Monte Carlo ensemble of model forecasts, many types of error can be represented.

The feasibility of synthesizing distributed fields of soil moisture using the non-sequential data assimilation was first demonstrated by Houser *et al.* (1998), who used a variational (4D) DA approach. A more complicated variational approach has been applied in hydrology (Reichle *et al.*, 2001a,

2001b; Margulis and Entekhabi, 2003). This approach uses observations distributed in space and time with the knowledge of temporal evolution of the model state through the calculation of the adjoint, which is obtained by linearizing the forward model along a trajectory producing the tangent-linear model. However, for the nonlinear hydrologic processes, calculating the adjoint can be difficult. To overcome this problem, Dunne and Entekhabi (2005, 2006) developed an EnKF-based smoothing approach, which uses EnKF to update not only the current and future states but also the states in the past while avoiding calculating the adjoint. They demonstrated that it can be used operationally for soil moisture data assimilation (Dunne and Entekhabi, 2006).

To facilitate the operational application of data assimilation in hydrology for land surface initialization, the Land Data Assimilation Systems (LDAS) conceptual framework has been developed to provide the best estimation of the current state of land surfaces through integration of satellite observations and modeling at a wide range of scales. Several LDAS systems have been implemented in near real time and at high spatial resolutions for North America (Cosgrove *et al.*, 2003; Mitchell *et al.*, 1999), Europe (Van den Hurk, 2002), China (Li *et al.*, 2007; Shi *et al.*, 2011), and the global domain (Rodell *et al.*, 2004; Peters-Lidard *et al.*, 2007; Kumar *et al.*, 2006). These LDASs are forced with real-time numerical weather prediction models outputs with satellite observed surface parameters. Data assimilation methods are being implemented in these systems to incorporate satellite observed state conditions as a constraint to the model dynamics. For example, direct insertion, rule-based approaches, and EnKF were implemented in the current Land Information System (http://lis.gsfc.nasa.gov) to assimilate multiple observations (soil moisture and snow water equivalent (SWE)) into multi-models (Kumar *et al.*, 2008, 2009). This has the potential to improve land surface energy balance and water cycle estimations.

12.3 Basic Concepts of Soil Moisture Data Assimilation

Soil moisture data assimilation is a method of minimizing some of the uncertainties inherent in all LSMs due to their approximation of the complexity in the land surface processes. Observations, if available, from sources such as satellite data can be integrated into the model to update a quantity simulated by the model with the purpose of reducing the error in the model formulation. The correction applied is derived from the respective weightings of the uncertainties of both the model predictions and the observations. There has been much research focused on soil moisture data assimilation. These assimilation studies include the use of variational schemes and variants of the KF schemes. The following reviews the most commonly used optimal data assimilation schemes.

Optimal data assimilation refers to the process varying the model and observational error at each time in a dynamical system to minimize the error between the model predictions and observations. It can be divided into two categories: sequential data assimilation or non-sequential including smoothing and 4D variational data assimilation. In the sequential approach, observations are used to update the state at the current and the future states. In the non-sequential assimilation, observations are used to update not only the current and the future state, but also the past state and it uses observations distributed in space and time to update the knowledge of the temporal revolution of the state.

Sequential and non-sequential methods use different approaches to minimize the mismatch between the model and the observations. The optimal analysis is given by the maximum likelihood *a priori* Bayesian estimate of the system states. This is equivalent to minimizing the square error between the predicted observations by the model and the measured observations, weighted by the inverse of the covariance matrices, over the assimilation window with the cost function (Nichols, 2002),

$$J = \frac{1}{2}(x_0 - x_0^b)^T B_0^{-1}(x_0 - x_0^b) + \frac{1}{2}\sum_{k=0}^{N-1}(h_k(x_k) - y_k)^T R_k^{-1}(h_k(x_k) - y_k) \tag{1}$$

where subscripts 0 and k indicate the system initial state and the system state at time t_k, respectively. $B_0 = E((x - x_0^b)(x - x_0^b)^T)$ is the error covariance of the initial state, x_0^b is the background/model initialization state. This problem can be solved directly to give a sequential assimilation (dynamic observer) scheme, or it can be solved indirectly to give a four-dimensional variational assimilation (direct observer) scheme.

The sequential data assimilation method derives the optimal sequential solution for the nonlinear hydrologic processes system by minimizing the cost function with the background states as the prior/initial state. The optimal solution is given by

$$x_k^a = x_k^b + K_k(h_k(x_k^b) - y_k) \tag{2}$$

where $K_k = B_k H_k^T (H_k B_k H_k^T + R_k)^{-1}$ is the Kalman gain with $H_k = \frac{\partial h_k}{\partial x}|_{x_k^b}$ (the matrix H_k is the Jacobian of the function h_k with respect to x evaluated at x_k^b), $B_k = E((x - x_k^b)(x - x_k^b)^T)$ is the error covariance matrix for the background states x_k^b at time step t_k. Note that K_k is based on the relative magnitudes of the model and observation error covariances (Gelb, 1974).

A linear system has an exact optimal solution, this scheme is called KF (Kalman, 1960). But for nonlinear systems, the calculation of derivatives of linearized equations (Jacobian matrix) is required to propagate the error covariance to approximate the nonlinearities of the hydrologic model. This method is often called EKF. However, the derivatives of the nonlinear LSM used in the EKF sometimes are impossible, which hinders its use with conventional LSM. As an alternative to the EKF for nonlinear problems, the EnKF was developed (Evensen, 1994; Houtekamer and Mitchell, 1998), the model covariances are estimated using a Monte Carlo approach to produce an ensemble of model trajectories. The advantages of EnKF are that it can be used in any forward model and the model does not need to be differentiable. It also offers great flexibility in the specification of model error. The disadvantage is that estimates are conditioned on past measurements only.

The variational data assimilation method finds the model initial conditions that minimize a misfit between model results and observations for the whole analysis period. This approach calculates the derivatives of the cost function J in Eq. (1) with respect to each of the initial vector values, known as the adjoint using the Lagrange multiplier. The analysis period is usually much longer than the model time step (Reichle *et al.*, 2001a, b; Margulis and Entekhabi, 2003). The advantage of variational techniques is that all observations are used in a match mode to obtain the best states. However, variational techniques require that the system is differentiable. Recently De Lannoy *et al.* (2009) combined an EnKF and an adaptive covariance correction method to optimize the variances and retrieve the off-diagonal correlations in *a priori* error covariance matrix. The multi-scale KF (Parada and Liang, 2004; Kumar, 1999; Pan *et al.*, 2009) efficiently considers the spatial dependence and scaling properties of soil moisture.

Data assimilation offers several advantages over traditional methods for retrieving soil moisture from microwave satellite observations and model estimates. First, it provides a framework to best use both model predictions and observations. If the observations are unreliable, the model correction will be very small, and if the observations are reliable, the model correction will be such that the model estimate becomes very close to the observed value. Good sets of remote sensing data are critical to ensure improved estimates of soil moisture. Second, the assimilation procedure estimates not only soil moisture but related energy and mass fluxes. Third, estimates of soil moisture are provided not only near the surface but throughout the soil column. Fourth, it provides continuous soil moisture estimates in time and space rather than updates soil moisture at the specific time and location when/where observations are available. Fifth, the data assimilation procedure can generate estimates at finer resolution than the microwave observations. Sixth, the data assimilation procedure provides information on both the accuracy and uncertainty of soil moisture estimates (Reichle *et al.*, 2001a, 2001b, 2002a; Margulis *et al.*, 2002). It offers the advantage to integrate various amounts of satellite observations and model predictions for improved estimates of soil moisture states at fine spatial and temporal scales. Section 11.4 uses an example to demonstrate some of the advantages listed here.

12.4 Soil Moisture Assimilation — A Case Study

Seasonal climate model prediction accuracy is currently limited due to poor soil moisture state initialization. However, initial soil moisture state prediction accuracy can potentially be enhanced by the assimilation of remotely sensed near-surface soil moisture data in offline simulation. The following uses the current NASA–Global Modeling and Assimilation Office (GMAO)'s land initialization activities as an example to demonstrate the use of KF-based soil moisture data assimilation for improved soil moisture initialization for seasonal to interannual climate prediction.

12.4.1 *Data assimilation algorithm development*

NASA land initialization activities started with developing innovative data assimilation algorithms to merge satellite data and model predictions. To this end, Walker and Houser (2001) included an EKF surface soil moisture data assimilation strategy in the GMAO's catchment-based land surface model (CLSM) (Koster *et al.*, 2000) with the simplified assumption that errors in different catchments are uncorrelated. A one-dimensional KF was used to update the soil moisture prognostic variables of the CLSM because of its computational efficiency.

This assimilation system was refined using a synthetic twin experiment. Walker and Houser (2001) found that the unique catchment model physics are well suited for surface soil moisture assimilation, as the dominant prognostic moisture state variable (catchment deficit) has a significant correlation with surface soil moisture content except in very deep soils. Traditional LSMs generally have a vertical layering structure whose correlation is comparatively modest. Their study has shown that by assimilating near-surface soil moisture observations, errors in forecast soil moisture profiles as a result of poor initialization may be removed and the resulting predictions of runoff and evapotranspiration improved. A later study on soil moisture assimilation into multiple LSMs also demonstrated the advantage of the catchment model for improving root-zone soil moisture estimates through data assimilation due to its strong coupling of surface and root-zone soil moisture (Kumar *et al.*, 2009).

The EnKF has also been implemented in the catchment model (Reichle *et al.*, 2002b), being simply an alternative methodology for propagating the state covariance matrix that does not require model linearization. Reichle *et al.* (2002b) applied the EnKF to the soil moisture estimation problem, found it performed well, and evaluated the relative benefits of the ensemble and EKFs using synthetic soil moisture observations. They found that the ensemble approach does not require the calculation of derivatives of linearized equations (Jacobian matrix) to propagate the error covariance to approximate the nonlinearities of the hydrologic model and the measurement process that are required in the EKF approaches. The error covariance is propagated in the ensemble scheme through a finite ensemble of model trajectories. Furthermore, the ensemble scheme allows non-stationary spatially and temporally correlated representations of the model input and measurement errors and it allows for a wider range of model errors. It is extremely computational expensive to consider the horizontal correlation in the model using the extended scheme. Small ensemble members are required for computational efficiency in the ensemble scheme to match the performance of the EKF.

However, the performance of both filters is dependent on the assumption that model predictions and observations are unbiased, and that reliable estimates of model and observation error are available, which are difficult to estimate without an estimate of the truth for comparison. Therefore characterizing the model and satellite observation error becomes an important part of assimilation studies that use real satellite data (e.g., Ni-Meister *et al.*, 2005; Dong *et al.*, 2005).

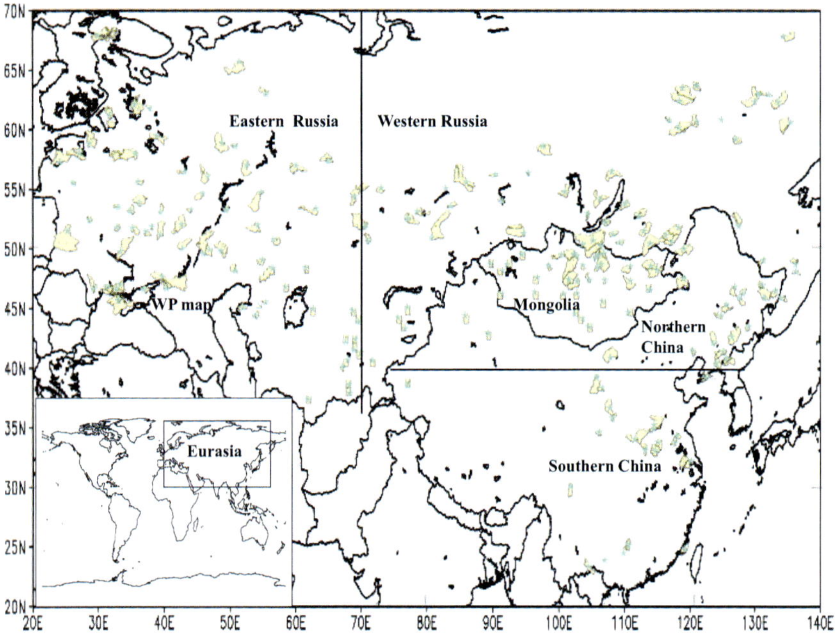

Fig. 12.1. Eurasian catchments and observation stations (shown by stars) used in this study. Only the catchments that contain soil moisture observation station are shown.

12.4.2 *Assimilation of SMMR data into CLSM*

Kalman filter (KF)-based soil moisture data assimilation strategy in the GMAO's catchment model was also used to assimilate SMMR-derived real surface soil moisture data (Owe *et al.*, 2001; Fig. 12.1) into the GMAO/CLSM for the period of 1979–1987 as one of the first studies to demonstrate the assimilation of real satellite data at a large continental scale over a long period. Two types of studies were conducted to assimilate SMMR data into the GMAO's catchment model. One type is conducted by Reichle and Koster (2004, 2005) and the other by Ni-Meister *et al.* (2005, 2006). In the study by Reichle and Koster (2004, 2005): (1) the satellite-based soil moisture contents in their study were scaled to match the model's climatology before assimilation; (2) their model error was generated based on calibrated model error parameters (Reichle *et al.*, 2002b) and not on realistic model error values; and (3) their evaluation was conducted by comparing the correlations between the assimilated and observed data and between the modeled and observed data. The logic behind this type of study is that satellite observed and modeled soil moistures have different physical meanings and the model will not benefit much by direct assimilating satellite observed soil moisture for initializing seasonal climate prediction models.

Ni-Meister *et al.* (2005, 2006) uses reliable model and satellite observation error information from a ground observation comparison. Ni-Meister *et al.* (2005) characterizes the model and SMMR errors, using the historical Eurasian soil moisture observations archived in the Soil Moisture Data Bank (SMDB) (Robock *et al.*, 2000) which is the most extensive soil moisture data set available during the SMMR time period. Moreover, the EnKF soil moisture improvement performance is evaluated using the Eurasian station soil moisture observation network. The following describes more details of the work by Ni-Meister *et al.* (2005, 2006).

12.4.2.1 *SMMR and model errors*

We first assessed the error or the CLSM by comparing the model results with SMDB. The SMDB covers large areas including 43 Chinese, 36 Mongolian, and 130 Russian soil moisture monitoring stations over long time periods. Figure 12.2 shows a map of the Eurasian station measurement network and the corresponding catchments used in the study. Figure 12.3 shows the climatological (time series average) differences between the CLSM and SMDB surface zone and root-zone soil moisture by season. The CLSM is biased dry in the already relatively dry Mongolian and southern Russian climate. Along the relatively wet Chinese east coast and the boundary between Mongolia and Russia, the CLSM shows wet biased surface zone and root-zone soil moisture. In northwestern Russia, the CLSM is biased very dry in the winter and fall and biased wet in the spring and summer. This winter and fall dry bias may be partially related to the CLSM's simple soil freezing treatment and partially due to the fact that the modeled surface zone soil moisture is more tightly coupled with root-zone soil moisture than other models. The spring and summer Chinese and Russian CLSM root-zone soil moisture is biased wet with a dry bias during the winter and fall. Generally, the CLSM surface zone and root-zone soil moisture are biased less than 0.08 v/v dry in dry climate and frozen soil areas, and are biased wet over 0.08 v/v (as high as 0.16 v/v) in wet climate areas.

There are multiple possible sources of CLSM error. First, simplified CLSM physics can lead to systematic prediction error. For instance, the

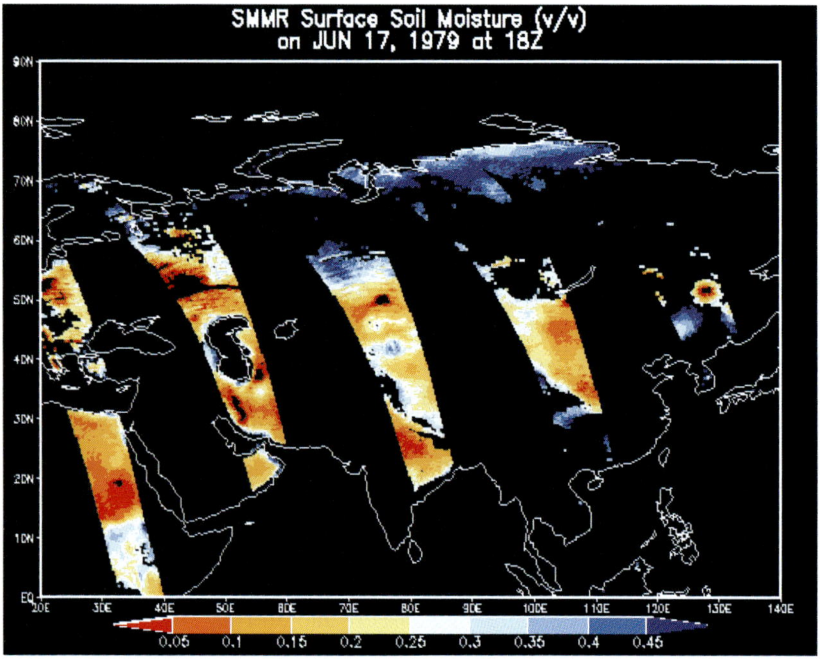

Fig. 12.2. SMMR derived surface soil moisture on June 17, 1979.

simple algorithm used to convert the CLSM's three soil moisture prognostic variables to surface zone soil moisture can lead to a misrepresentation of surface zone soil moisture. Moreover, the simple frozen ground treatment used by CLSM that prevents all soil water movement (including infiltration) when the soil temperature is below 0°C could lead to cold region dry biases. Second, Berg *et al.* (2003) have shown that the CLSM soil moisture is very sensitive to forcing errors. Generally the global observation-corrected forcing improves the CLSM large-scale performance, but some local errors may still exist. Third, prediction errors can arise from incorrect CLSM vegetation and soil parameters.

Although the gravimetric method is considered to be the most accurate way to estimate soil moisture, it can also have serious errors. Sample cores may contain roots, rocks, voids, and unique drainage characteristics that can lead to large differences between adjacent samples. Further, differences in sample extraction, compaction, handling, and processing can lead to error. There may be error related to using a potentially incorrect soil-type-based wilting point for converting plant available soil moisture back to total soil moisture.

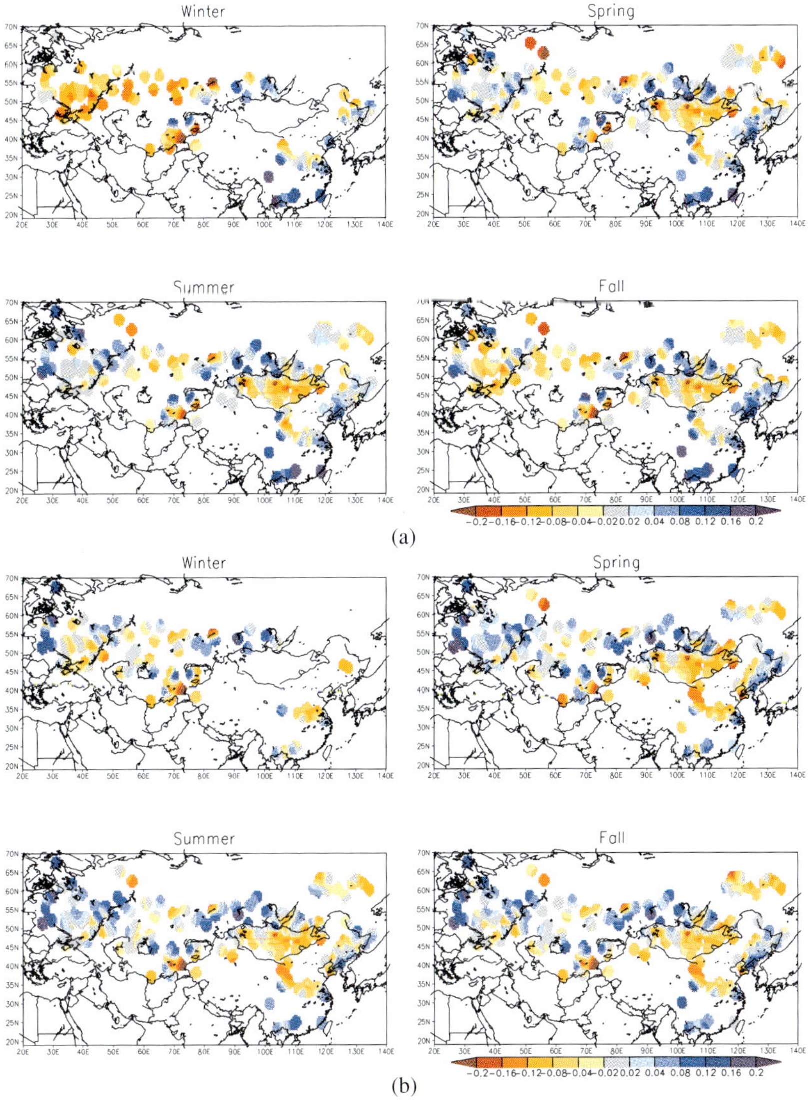

Fig. 12.3. Differences in seasonal climatology between the model and *in situ*: (a) surface and (b) root-zone soil moisture (v/v).

Finally, there are various errors of representation that must be considered. These errors are related to differing SMDB and CLSM averaging scales. The CLSM represents average soil moisture over a 1 to 10,000 km^2 mesoscale catchment, while in most cases only one *in situ* observation from

4-point measurements averaged over a less than 0.2 km^2 area is available. An additional representation error can be attributed to differing soil moisture depths. The CLSM represents surface zone soil moisture in the top 2 cm while the SMDB observations are averaged over the top 5 cm to 10 cm depending on location.

We also assessed the SMMR errors. Figure 12.4 shows the climatological differences between the SMMR, CLSM and SMDB surface zone soil moisture by season. Mongolian and Russian SMMR surface zone soil moisture is drier than SMDB, except in the northeastern Russian wet climate. Conversely, the Chinese SMMR surface zone soil moisture is wetter for all seasons. The east Eurasia SMMR surface zone soil moisture is wetter than the CLSM, and drier in west Eurasia. Generally, SMMR is biased 0.05 v/v dry in dry climate and 0.1 v/v wet in wet climate regions, but with accurate seasonal variations. The largest SMMR bias was found in China and northeastern Russia, where precipitation and vegetation cover is also larger.

Additional factors to consider when studying the discrepancy between SMMR and SMDB surface zone soil moisture include SMDB measurement error (reviewed earlier), and representation errors. The SMMR surface zone soil moisture estimate represents a 1-cm deep average over a 625 km^2 area, whereas the 5 cm to 10 cm deep surface zone SMDB observations represent the field-scale at best. Influences from these factors likely impact the comparison between SMMR and SMDB but are not possible to quantify given the available data.

12.4.2.2 *Assimilation results*

The soil moisture estimate improvement through data assimilation was assessed as the difference between the mean absolute error in model predictions with and without assimilation, for both the surface and root-zone soil moisture for each of the four seasons. The KF updates the system states based on the relative magnitudes of the model and satellite observation errors. In this study, both the predicted model and satellite observation errors are allowed to remain smaller than the actual errors, but their ratio is constrained to the actual ratio. Two types of errors were used. One is the matched error ratio, which agrees more closely with the actual error ratio. This ratio used suggests that the model errors are on average slightly smaller than the SMMR error. The other is the unmatched error ratio, which suggests a much smaller predicted mean model error than SMMR error,

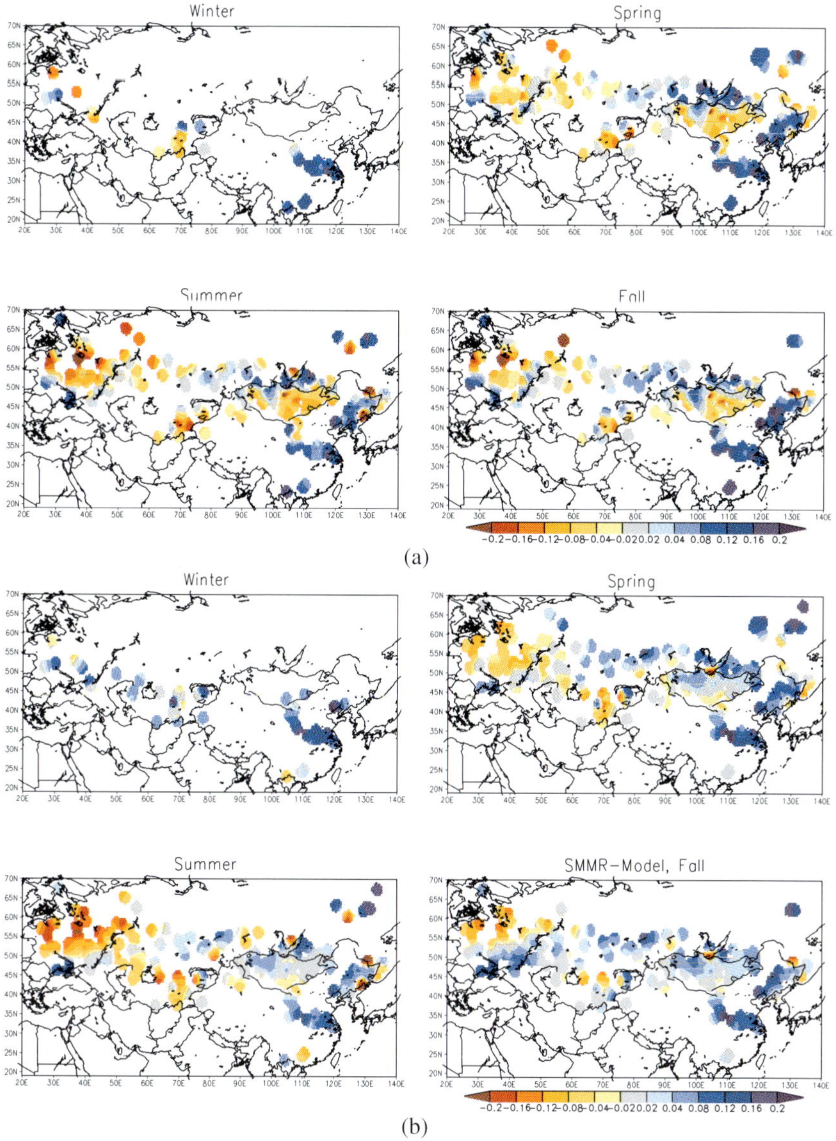

Fig. 12.4. Differences in seasonal climatology between: (a) SMMR and (b) model surface zone soil moisture (v/v) and *in situ* measurements.

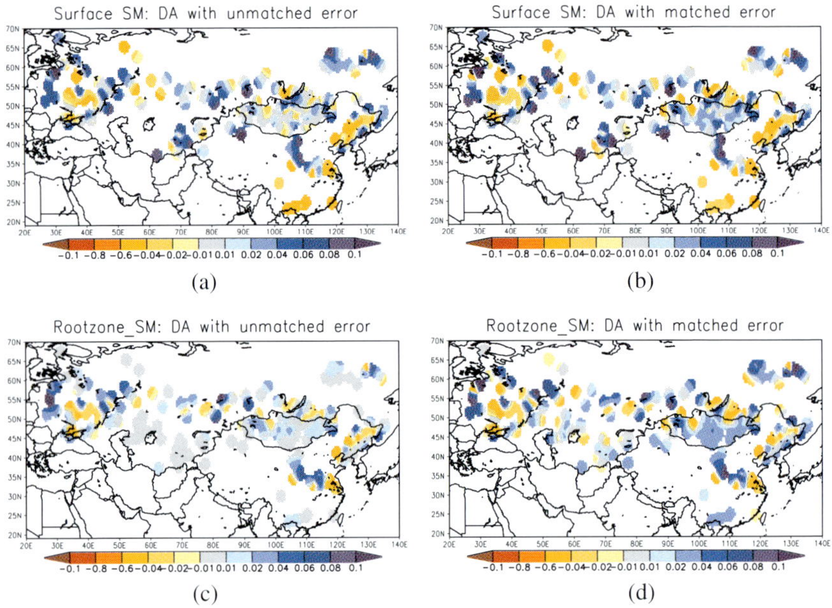

Fig. 12.5. Assimilated summertime surface ((a), (b)) and root-zone ((c), (d)) soil moisture improvement (absolute model error minus absolute assimilation error) with unmatched ((a), (c)) and matched ((b), (d)) error in Eurasia for 1979–1987.

meaning that the data assimilation scheme would trust the model more than the SMMR observation. Figure 12.5 shows the summertime assimilated soil moisture improvement for 1979–1987 for all stations. Blue areas indicate a reduction in soil moisture estimate error through data assimilation while orange areas indicate an increase in soil moisture estimate error due to the data assimilation. Near zero values (within a range of −1 and 1% v/v) in grey indicate a negligible impact from data assimilation on the soil moisture estimate. Figure 12.4 shows more blue than orange areas, particularly for root-zone soil moisture for the assimilation with matched error, but is mixed with a lot of grey areas, indicating data assimilation with the matched error produces better results than those with the unmatched error.

Figure 12.6 is similar to Fig. 12.5, except it only shows results for the stations which have a smaller climatological mean SMMR error compared to the model error. In contrast to the results for all stations, Fig. 12.6 shows soil moisture improvement through data assimilation for most stations in Mongolia and Russia and a few stations in China, indicating when the

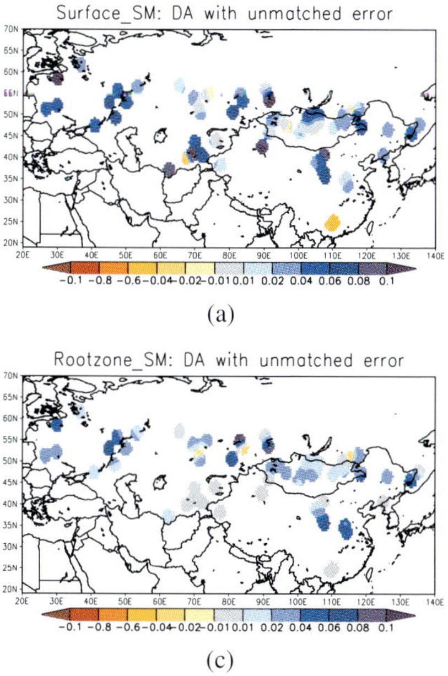

Fig. 12.6. As for Fig. 12.5, but only for cases where the SMMR error is less than the model error.

observation errors are smaller than the model error, the data assimilation improves soil moisture estimation. Many stations in China have wet biased SMMR observations compared to *in situ* soil moisture measurements. Moreover, many Chinese stations show a data assimilation improvement in root-zone soil moisture with corresponding degradation in surface zone soil moisture. When the SMMR mean error is greater or equal to the model error, data assimilation typically leads to an increase in soil moisture bias, particularly in the surface soil moisture estimates. In this case, bias correction is required before the observations can be rigorously used in the EnKF, since one assumption of the KF is that both the model and satellite observation errors are unbiased.

To investigate the temporal response of soil moisture data assimilation, modeled soil moisture with and without assimilation for both matched and unmatched error cases are compared with satellite-derived and station-measured surface zone and root-zone soil moisture for one China station (Zhumadian) (Fig. 12.7). Figure 12.7 shows SMMR-derived surface soil moisture overestimates ground observations, leading to the overestimation

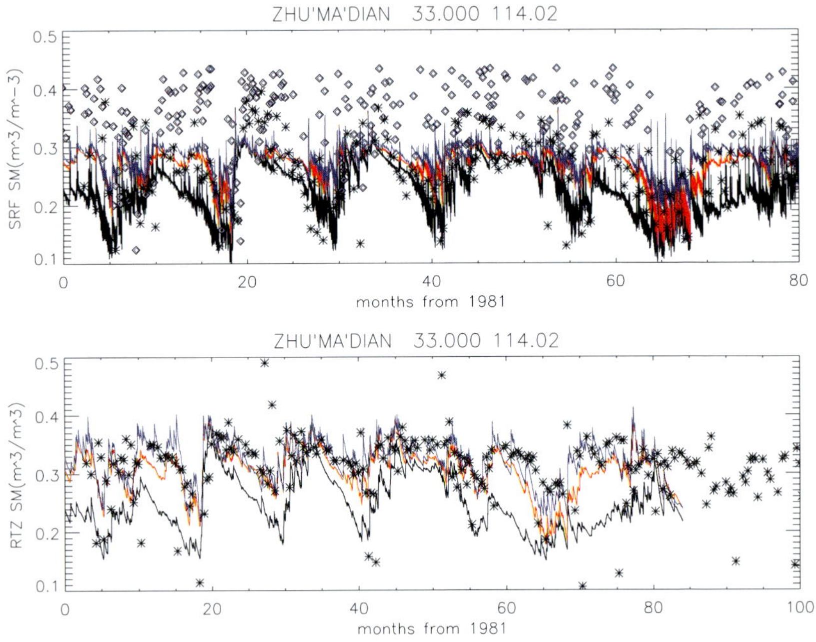

Fig. 12.7. Comparison of assimilated and modeled surface (top panel) and root-zone (bottom panel) soil moisture with station measurements and SMMR observations at Zhumadian, China, for the period of 1981–1987. Black stars: station measurements, black lines: modeled soil moisture, blue diamonds: SMMR observations, red lines: assimilated soil moisture with unmatched error and blue lines: assimilated soil moisture with matched error.

of surface soil moisture after assimilation. However, the model underestimates the root-zone soil moisture, and assimilation helps bring the underestimated soil moisture up to the observed values, resulting in a somewhat better agreement with the station observations.

This study is one of the first to demonstrate the potential using actual remote sensing data together with field observations to provide a better estimate of land surface soil moisture states to improve seasonal climate prediction. This study demonstrates that the KF corrects model-generated soil moisture toward the satellite observation, with the size of the correction dependent upon the relative magnitudes of the satellite observation and model errors. It has been shown that, with the correct model and observation errors, data assimilation provided more accurate root-zone soil moisture estimates. This improvement in root-zone soil moisture estimates has substantial implications for seasonal climate prediction improvement, since better knowledge in root-zone soil moisture is more important than surface zone soil moisture in terms of land surface–atmosphere interactions. Moreover, these results highlight the need for accurate model and observation error estimates. Analysis of the impact of SMMR error on the assimilation results showed that when the climatologic mean SMMR error was less than the model error, the data assimilation improved both surface zone and root-zone soil moisture estimates, with the greatest improvement being in the root zone, which indicates the need of more accurate remotely sensed soil moisture observations to ensure data assimilation works. Additionally, most stations in China resulted in biased surface soil moisture estimates as a result of wet biased SMMR observations in that region. This indicates that any bias must be removed before assimilation to ensure improvement in soil moisture estimates.

12.5 Conclusions and Discussion

New remote sensing technologies and spatially distributed models offer great potential for improving our understanding of hydrologic processes, especially at large scales in the coming decades. The example presented in this study has demonstrated that a KF-based assimilation scheme is able to integrate surface soil moisture observations into LSMs and to provide a more accurate soil moisture estimate particularly at the root zone than modeling alone.

There is, however, much work to be done before soil moisture data assimilation algorithms can be used operationally. The accurate estimation of modeling and observing error parameters remains challenging to successfully implement LDAS globally. Most soil moisture EnKF applications are based on simplistic error models by perturbing meteorological forcings, generated from normally distributed random number perturbations with zero mean and unit variance. However, it is particularly difficult to reflect the "true" uncertainty in the initial or analysis state and model error into an optimal perturbation magnitude and structure for the forcings, model structure, state, and parameters using EnKF (De Lannoy *et al.*, 2009). Houtekamer and Mitchell (1998) applied the maximum-likelihood method to estimate model error variance parameters in an EnKF framework, where initially only the uncertainty in the initial state or analysis was considered. Even the example presented in this review can only characterize the model and measurement errors in Eurasia where extensive *in situ* measurements are available.

Recently Crow and Reichle (2008) developed adaptive filtering approaches that address this issue. It attempts to diagnose the correct covariance of both modeling and observation errors through iteratively whitening EnKF innovations (i.e., the time series of observations minus the model background realized during the analysis cycle). However, such approaches are based on an assumption that observing errors are temporally uncorrelated. To overcome this problem, more recently Crow and Van den Berg (2010) developed a scheme to tune the variance of assumed modeling perturbations based on the error variance of remotely sensed surface soil moisture retrievals until the normalized filtering innovations have a temporal variance of one. However, all those approaches are still limited to theoretical analysis. Much work needs to be accomplished to better understand the errors of model and remote sensing data before these data assimilation system can be used operationally.

Another barrier to the operational use of soil moisture data assimilation systems is the lack of a consistent and good quality global soil moisture data. Currently available global remotely sensed soil moisture datasets themselves show large differences even in their multi-year means and temporal variability and are biased compared to *in situ* soil moisture measurements (Reichle *et al.*, 2007). The study presented above shows that larger errors exist in SMMR data. Even with the newest X-band (10.7 GHz) passive microwave brightness temperature observations by the Advanced Microwave Scanning Radiometer for the Earth Observing System (AMSR-E), the retrieved soil moisture data show different multi-year means and temporal variability from SMMR derived soil moisture data and also different from *in situ* soil moisture observations (Reichle *et al.*, 2007). The differences are likely due to different retrieval algorithms and a different physical meaning of remotely sensed soil moisture as compared to *in situ* measurements, as discussed in Ni-Meister *et al.* (2005). Only when good remote sensing soil moisture data are available, it will be possible to obtain dramatic improvement of soil moisture for large areas and consistent soil moisture estimation through data assimilation (Wagner *et al.*, 2007). Walker and Houser (2004) found that near-surface soil moisture observations must have an accuracy better than $0.05\mathrm{m}^3/\mathrm{m}^3$ to positively impact soil moisture forecasts. The mission of European Space Agency's Soil Moisture and Ocean Salinity (SMOS) will provide global L-band observations from which a better global soil moisture data set can be obtained.

Further research is necessary to advance on current data assimilation methods to make the best use of current remote sensing data. One approach is through bias correction during data assimilation. More work is needed to develop a persistent bias model to ensure the efficiency of bias correction (De Lannoy *et al.*, 2007). Another modification would be directly assimilating microwave brightness temperature into a coupled radiative transfer and land surface process model (Dunne and Entekhabi, 2005, 2006; Huang *et al.*, 2008; Jia *et al.*, 2010). The radiative transfer model would allow to calculate the brightness temperature. This approach would provide a consistent way to estimate soil moisture by assimilating brightness temperature. This approach would also allow to assimilate multi-source remote sensing data to constrain soil moisture estimates. Operational application of a soil moisture data assimilation scheme will also require a sufficiently sophisticated data assimilation algorithm to be able to assimilate multiple data sources that observe different processes acting at different scales.

Acknowledgments

This work was funded by NASA contract number NNG04GL08G.

References

Bennet, A. F. (1992). *Inverse Method in Physical Oceanography*, 346pp. Cambridge, UK: Cambridge University Press.

Berg, A. A., Famiglietti, J. S., Walker, J. P. and Houser, P. R. (2003). Impact of bias correction to reanalysis products on simulations of North American soil moisture and hydrological fluxes. *Journal of Geophysical Research*, **108**(D16), 4490.

Bernard, R., Vauclin, M. and Vidal-Madjar, D. (1981). Possible use of active microwave remote sensing data for prediction of regional evaporation by numerical simulation of soil water movement in the unsaturated zone. *Water Resources Research*, **17**(6), 1603–1610.

Bindlish, R., Jackson, T. J., Wood, E., Gao, H. L., Starks, P., Bosch, D. and Lakshmi, V. (2003). Soil moisture estimates from TRMM microwave imager observations over the southern United States. *Remote Sensing of Environment*, **85**(4), 507–515,

Bolten, J. D., Crow, W. T., Jackson, T. J., Zhan, X. and Reynolds, C. A. (2010). Evaluating the utility of remotely sensed soil moisture retrievals for operational agricultural drought monitoring. *IEEE Journal of Selected Topics in Applied Earth Observations and Remote Sensing*, **3**, 57–66.

Bruckler, L. and Witono, H. (1989). Use of remotely sensed soil moisture content as boundary conditions in soil — atmosphere water transport modeling: 2. Estimating soild water balance. *Water Resources Research*, **25**(12), 2437–2447.

Cosgrove, B. A. *et. al.* (2003). Real-time and retrospective forcing in the North American Land Data Assimilation (NLDAS) project. *Journal of Geophysical Research*, **108**, 8842, doi:10.1029/2002JD003118.

Crow, W. T. and Wood, E. F. (2003). The assimilation of remotely sensed soil brightness temperature imagery into a land surface model using ensemble Kalman filtering: A case study based on ESTAR measurements during SGP97. *Advances in Water Resources*, **26**, 37–149.

Crow, W. T. and Reichle, R. H. (2008). Comparison of adaptive filtering techniques for land surface data assimilation. *Water Resources Research*, **44**, W08423, doi:10.1029/2008WR006883.

Crow, W. T. and Van den Berg, M. J. (2010). An improved approach for estimating observation and model error parameters in soil moisture data assimilation. *Water Resources Research*, **46**, W12519, doi:10.1029/2010WR009402.

Daley, R. (1991). *Atmospheric Data Assimilation*. New York: Cambridge University Press.

De Lannoy, G. J. M., Reichle, R. H., Houser, P. R., Pauwels, V. R. N. and Verhoest, N. E. C. (2007). Correcting for forecast bias in soil moisture assimilation with the ensemble Kalman filter. *Water Resources Research*, **43**, W09410, doi:10.1029/2006WR005449.

De Lannoy, G., Houser, P. R., Verhoest, N. E. C. and Pauwels, V. R. N. (2009). Adaptive soil moisture profile filtering for horizontal information progagation in the independent column-based CLM2.0. *Journal of Hydrometeorology*, **10**, 766–770, DOI: 10.1175/2008JHM1037.1.

Dong, J., Walker, J. P. and Houser, P. R. (2005). Factors affecting remotely sensed snow water equivalent uncertainty. *Remote Sensing of Environment*, **97**(1), 68–82, doi:10.1016/j.rse.2005.04.010.

Dong, J., Walker, J. P., Houser, P. R. and Sun, C. (2007a). Scanning multichannel microwave radiometer snow water equivalent assimilation. *Journal of Geophysical Research*, **112**, D07108, doi:10.1029/2006JD007209.

Dong, J., Ni-Meister, W. and Houser, P. R. (2007b). Impacts of vegetation and cold season processes on soil moisture — Climate relationships over Eurasia. *Journal of Geophysical Research*, **112**, D09106, doi:10.1029/2006JF007774.

Dunne, S. and Entekhabi, D. (2005). An ensemble-based reanalysis approach to land data assimilation. *Water Resources Research*, **41**, W02013, doi:10.1029/2004WR003449.

Dunne, S. and Entekhabi, D. (2006). Land surface state and flux estimation using the ensemble Kalman smoother during the Southern Great Plains 1997 field experiment. *Water Resources Research*, **42**, W01407, doi:10.1029/2005WR004334.

Entekhabi, D. Nakamura, H. and Nioku, E. (1994). Solving the inverse problem for soil moisture and temperature profiles by sequential assimilation of multifrequency remotely sensed observations, IEEE Transactions on Geoscience and Remote Sensing, **32**, 438–448.

Evensen, G. (1994). Sequential data assimilation with a nonlinear quasi-geostrophic model using Monte Carlo methods to forecast error stastics, *Journal of Geophysical Research*, **99**(10), 143– 162.

Gao, H., Wood, E. F., Drusch, M., Jackson, T. and Bindlish, R. (2006). Using TRMM/TMI to retrieve soil moisture over the southern United States from 1998 to 2002. *Journal of Hydrometeorology*, **7**, 23–38.

Gao, H., Wood, E. F., Drusch, M., Crow W. T. and Jakson, T. J. (2004). Using a microwave emission model to estimate soil moisture from ESTAR observation during SGP99. *Journal of Hydrometeorological*, **5**, 49–63.

Gelb, A. (1974). *Applied Optimal Estimation*, p. 373. Cambridge, Massachusetts: The MIT Presss.

Houtekamer, P. L. and Mitchell, H. L. (1998). Data assimilation using an ensemble Kalman filter techniques. *Monthly Weather Review*, **126**, 796–811.

Houser, P. R., Shuttleworth, W. J., Gupta, H. V., Famiglietti, J. S., Syed, K. H. and Goodrich, D. C. (1998). Integration of soil moisture remote sensing and hydrologic modeling using data assimilation. *Water Resources Research*, **34**(12), 3405–3420.

Huang, C. L., Li, X., Lu, L. and Gu, J. (2008). Experiments of one-dimensional soil moisture assimilation system based on ensemble Kalman filter. *Remote Sensing of Environment*, **112**(3), 888–900.

Jackson, T. J. *et al.* (1981). Soil moisture updating and microwave remote sensing for hydrological simulation. *Hydrological Science B*, **26**(3), 305–319.

Jackson, T., Levine, D. L., Swift, C., Schmugge, T. and Schiebe, F. (1995). Large area mapping of soil moisture using the ESTAR passive microwave radiometer in Washita'92. *Remote Sensing of Environment*, **54**(1), 27–37.

Jackson T. (1997). Soil moisture estimation using special satellite microwave/imager (SSM/I) satellite data over a grassland region. *Water Resources Research*, **333**, 1475–1484.

Jackson, T., Levine, D., Hsu, A., Oldak, A., Starks, P., Swift, C., Isham, J. and Haken, M. (1999). Soil moisture mapping at regional scales using microwave radiometry: The Southern Great Plains Hydrology Experiment, *IEEE Transactions on Geoscience and Remote Sensing*, **37**, 2136–2151.

Jia, B. H., Xie, Z. H., Tian, X. J. and Shi, C. X. (2010). Assimilating soil brightness temperature using Kalman filter. *Science in China Series: Earth Science*, **40**(2), 239–251.

Kalman, R. E. (1960). A new approach to linear filtering and prediction problems. *Journal of Basic Engineering*, **82**, 35–45.

Kalnay, E. (2003). *Atmospheric Modeling, Data Assimilation and Predictability.* UK: Cambridge University Press.

Kumar, P. (1999). A multiple scale state — space model for characterizing subgrid scale variability of near-surface soil moisture. *IEEE Transactions on Geoscience and Remote Sensing*, **37**(1), 182–197.

Kumar, S. V., Peters-Lidard, C. D., Tian, Y., Houser, P. R., Geiger, J., Olden, S., Lighty, L., Eastman, J. L., Doty, B., Dirmeyer, P., Adams, J., Mitchell, K., Wood, E. F. and Sheffield, J. (2006). Land information system — An interoperable framework for high resolution land surface modeling. *Environmental Modelling and Software*, **21**, 1402–1415.

Kumar, S. V., Reichle, R. H., Peters-Lidard, C. D., Koster, R. D., Zhan, X., Crow, W. T., Eylander, J. B. and Houser, P. R. (2008). A land surface data assimilation framework using the land information system: Description and applications. *Advances in Water Resources*, **31**, 1419–1432, doi:10.1016/j.advwatres.2008.01.013.

Kumar, S. V., Reichle, R. H., Koster, R. D., Crow, W. T. and Peters-Lidard, C. D. (2009). Role of subsurface physics in the assimilation of surface soil moisture observations. *Journal of Hydrometeorology*, **10**, 1534–1547, doi:10.1175/2009JHM1134.

Koster R. D. and Suarez, M. J. (2000). Soil moisture memory in climate models. *Journal of Hydrometeorology*, **2**, 558–570.

Koster R. D., Suarez, M. J., Ducharne, A., Stieglitz, M. and Kumar, P. (2000). A catchment-based approach to modeling land surface processes in a general circulation model, Part 1: Model structure. *Journal of Geographical Research*, **105**(20), 24809–24822.

Koster, R. D., Suarez, M. J., Liu, P., Jambor, U., Berg, A. A., Kistler, M., Reichle, R. H., Rodell, M. and Famiglietti, J. (2004). Realistic initialization of land surface states: Impacts on subseasonal forecast skill. *Journal of Hydrometeorology*, **5**(6), 1049–1063.

Li, X., Huang, C. L., Che, T., Jing, R., Wang, S. G., Wang, J. M., Gao, F., Zhang, S. W., Qiu, C. J. and Wang, C. H. (2007). Development of a Chinese land data assimilation system (CLDAS): Its progress and prospects. *Progress in Natural Science*, **17**(8), 881–892.

Margulis, S. and Entekhabi, D. (2003). Variational assimilation of radiometric surface temperature and screen-level micrometeorology into a model of the atmospheric boundary layer and land surface. *Monthly Weather Review*, **131**(7), 1272–1288.

Margulis, S. McLaughlin, D., Entekhabi, D. and Dunne, S. (2002). Land data assimilation and soil moisture estimation using measurements from the Southern Great Plains 1997 field experiment. *Water Resources Research*, **38**(12), 1299, doi:10.1029/2001WR001114.

Mitchell, K. *et al.* (1999). GCIP Land Data Assimilation System (LDAS) project now underway. *GEWEX News*, **9**(4), 3–6.

Nichols, N. K. (2002). Data assimilation: Aims and basic concepts. In *Data Assimilation for the Earth System*, Swinbank, R., Shutyaev, V. and Lahoz, W. A. (eds.), NATO Science Series IV: Earth and Environmental Sciences, Kluwer Academic Publishers.

Ni-Meister, W., Walker, J. P. and Houser, P. R. (2005). Soil moisture initialization for climate prediction: Characterization of model error and measurement error. *Journal of Geophysical Research*, **110**, D13111, doi:10.1029/2004JD005745.

Ni-Meister, W., Houser, P. and Walker, J. (2006). Soil moisture initialization for climate prediction: Assimilation of SMMR soil moisture data into a land surface model. *Journal of Geophysical Research*, **111**, D20102, doi:10.1029/2006JD007190.

Owe, M., De Jeu, R. and Walker, J. P. (2001). A methodology for surface soil moisture and vegetation optical depth retrieval using microwave polarization difference index. *IEEE Transactions on Geoscience and Remote Sensing*, **39**, 1643–1654,.

Pan, M., Wood, E., McLaughlin, D., Entekhabi, D. and Luo, L. (2009). A multiscale ensemble filtering system for hydrologic data assimilation, Part I: Implementation and synthetic experiment. *Journal of Hydrometeorology*, **10**(3), 794–806, doi: 10.1175/2009JHM1088.1.

Parada, L. M. and Liang, X. (2004). Optimal multiscale Kalman filter assimilation for near-surface soil moisture into land surface models. *Journal of Geophysical Research*, **109**, D24109, doi:10.1029/2004JD004745.

Peters-Lidard, C. D., Houser, P. R., Tian, Y., Kumar, S. V., Geiger, J., Olden, S., Lighty, L., Doty, B., Dirmeyer, P., Adams, J., Mitchell, K., Wood, E. F. and Sheffield, J. (2007). High-performance earth system modeling with NASA/GSFC's land information system. *Innovations in Systems and Software Engineering*, **3**(3), 157–165.

Prevot, L. *et al.* (1984). Evaporation from a bare soil evaluated using a soil water transfer model and remotely sensed surface soil moisture data. *Water Resources Research*, **20**(2), 311–316.

Reichle, R., McLaughlin, D. B. and Entekhabi, D. (2001a). Variational data assimilation of microwave radio brightness observations for land surface hydrologic applications. *IEEE Transactions on Geoscience and Remote Sensing*, **39**(8), 1708–1718.

Reichle, R. D. and McLaughlin, D. B. (2001b). Downscaling of radio brightness measurements for soil moisture estimation: A four-dimensional variational data assimilation approach. *Water Resources Research*, **37**(9), 2353–2364.

Reichle R. H., McLaughlin, D. B. and Entekhabi, D. (2002a). Hydrologic data assimilation with the ensemble Kalman filter. *Monthly Weather Review*, **130**(1), 103–114.

Reichle R. H., Walker, J. P., Koster, R. D. and Houser, P. R. (2002b). Extended versus ensemble Kalman filtering for land data assimilation. *Journal of Hydrometeorology*, **3**(6), 728–740.

Reichle, R. H. and Koster, R. D. (2005). Global assimilation of satellite surface soil moisture retrievals into the NASA catchment land surface model. *Geophysical Research Letters*, **32**(2), L02404, doi:10.1029/2004GL021700.

Reichle, R. H. and Koster, R. D. (2007). Comparison and assimilation of global soil moisture retrievals from the advanced microwave scanning radiometer for the earth observing system (AMSR-E) and the scanning multichannel microwave radiometer (SMMR). *Journal of Geophysical Research*, **112**, D09108, doi: 10.1029/2006JD008033.

Reichle, R. H. and Koster, R. D. (2004). Bias reduction in short records of satellite soil moisture, *Geophysical Research Letters*, **31**, L19591.

Robock A., Vinnikov, K. Y., Srinivasan, G., Entin, J. K., Hollinger, S. E., Speranskaya, N. A., Liu, S. and Namkhai, A. (2000). The global soil moisture data bank. *Bulletin of the American Meteorological Society*, **81**, 81,1281–81,1299.

Rodell, M., Houser, P. R., Jambor, U., Gottschalck, J., Mitchell, K., Meng, C.-J., Arsenault, K., Cosgrove, B., Radakovich, J., Bosilovich, M., Entin, J. K., Walker, J. P., Lohmann, D. and Toll, D. (2004). The global land data assimilation system. *Bulletin of the American Meteorological Society*, **85**(3), 381–394.

Rodell, M. and Houser, P. R. (2004). Updating a land surface model with MODIS-derived snow cover. *Journal of Hydrometeorology*, **5**, 1064–1075.

Shi, C. X., Xie, Z. H., Qian, H., Liang, M. L. and Yang, X. C. (2011). Chinese land data assimilation system, *Science in China Series: Earth Science*, **41**(3), 375–385.

Sun, C., Walker, J. P. and Houser, P. R. (2004). A methodology for snow data assimilation in a land surface model. *Journal of Geophysical Research–Atmospheres*, **109**, D08108, doi:10.1029/2003JD003765.

Van den Hurk, B. (2002). European LDAS established. *GEWEX News*, **12**(2), p. 9. Available at: http://www.gewex.org/.

Wagner, W., Bloschl, G., Pampaloni, P., Calvet, J. C., Bizzarri, B., Wlgneron,J. P. and Kerr, Y. (2007). Operational readiness of microwave remote sensing of soil moisture for hydrologic applications. *Nordic Hydrology*, **38**(1), 1–20.

Walker J. P. and Houser, P. R. (2001). A methodology for initializing soil moisture in a global climate model: Assimilation of near-surface soil moisture observations. *Journal of Geophysical Research*, **106**(D11), 11761–11774.

Walker, J. P., Willgoose, G. R. and Kalma, J. D. (2001). One-dimensional soil moisture profile retrieval by assimilation of near-surface observations: A comparison of retrieval algorithms. *Advances in Water Resources*, **24**(6), 631–650.

Walker, J. P., Willgoose, G. R. and Kalma, J. D. (2002). Three-dimensional soil moisture profile retrieval by assimilation of near-surface measurements: Simplified Kalman filter covariance forecasting and field application. *Water Resources Research*, **38**(12), 1301, doi:10.1029/2002WR001545.

Walker, J. P. and Houser, P. R. (2004). Requirements of a global near-surface soil moisture satellite mission: Accuracy, repeat time, and spatial resolution. *Advances in Water Resources*, **27**, 785–801, doi:10.1016/j.adv watres.2004.05.006.

Zhang, S. W., Li, H. R., Zhang, W. D., Qiu, C. J. and Li, X. (2005). Estimating the soil moisture profile by assimilating near-surface observations with the ensemble Kalman filter (EnKF). *Advances in Atmospheric Sciences*, **22**(6), 936–945.

Chapter 13

Assimilation of Remote Sensing Data and Crop Simulation Models for Agricultural Study: Recent Advances and Future Directions

Hongliang Fang*, Shunlin Liang† and Gerrit Hoogenboom‡

*LREIS, Institute of Geographic Sciences and Natural Resources Research, Chinese Academy of Sciences, Beijing, 100101, China

†Department of Geographical Sciences, University of Maryland, College Park, MD 20742, USA

‡Department of Biological Systems Engineering, Washington State University, Prosser, WA 99350, USA

Crop simulation models (CSMs) estimate crop production, water and nitrogen balances, and carbon dynamics through a deterministic scheme with input data such as field and climate conditions, crop characteristics, and management practice. Many parameters and input variables, required by crop models but usually poorly known, can be supplied through the remote sensing technique. In the past decades, integration of remote sensing data and CSMs has gained increasing attention in both scientific communities and agricultural practice. This chapter starts with a general introduction of the procedure to assimilate remote sensing data into CSMs, followed by a brief description of the Decision Support System for Agrotechnology Transfer (DSSAT) model (Hoogenboom *et al.*, 1999, 2004). Several assimilation methods of various degrees of complexity and integration are described, including the direct input, sequential, and variational assimilation techniques. Remote sensing data in solar, microwave, and thermal domains are presented for their capabilities in assimilating with crop growth models. Regional application examples are presented to illustrate the procedure of sensitivity study, cost function construction, crop yield estimation, and the hydrological simulation process. Challenges facing both remote sensing data and CSMs are discussed and some recommendations about the future research priorities are proposed at the end.

S. Liang, X. Li & X. Xie (eds.), *Land Surface Observation, Modeling and Data Assimilation*, 403–438.

13.1 Introduction

Reliable and timely prediction of crop yield is important for agricultural land management and policy-making. Many studies have demonstrated the utilization of satellite data in crop yield estimation. Country-level crop yield estimation using remote sensing data has been carried out operationally for more than three decades (MacDonald and Hall, 1980). Vegetation indices calculated from Landsat Thematic Mapper (TM) data have improved the accuracy of ground wheat yield surveys through stratification of primary sampling units (Singh *et al.*, 1992). Promising results have been obtained through the statistical relationship between satellite reflectance values, vegetation indices, or leaf area index (LAI) and crop yield (Baez-Gonzalez *et al.*, 2005; Jiang *et al.*, 2003; Kogan *et al.*, 2005; Thenkabail *et al.*, 1992; Weissteiner and Kühbauch, 2005). However, these methods are mainly empirical and they may work only for specific crop cultivars, a particular crop growth stage, or environmental conditions at specific times.

Crop simulation models (CSMs) describe the fundamental processes in crop growth such as photosynthesis, respiration, biomass partitioning, water and nitrogen transfer. With a deterministic scheme, they directly estimate variables of interest, such as the biomass productions, water and nitrogen balance, as well as crop yields, using input data such as climate variables and soil characteristics. Crop models are also able to describe the dynamics of vegetation structure, such as LAI, and may be adapted to describe other optical properties. They can thus be coupled with vegetation radiative transfer models to simulate the radiometric response.

The use of crop models is often limited by uncertainties in their input parameters such as soil conditions, sowing date, planting density, and initial field conditions. Except in some controlled experimental fields, many of these parameters are poorly known. Remote sensing can play a critical role in helping identify the field and crop status from estimated biophysical parameters (Clevers and Van Leeuwen, 1996). Remote sensing data, therefore, can be assimilated with crop growth models to improve their overall performance (Baret *et al.*, 2000; Delécolle *et al.*, 1992b; Doraiswamy *et al.*, 2004; Fang *et al.*, 2008, 2011; Maas, 1988a; Moulin *et al.*, 1998; Plummer, 2000). Integration of remote sensing and crop growth simulation models has become recognized increasingly as a potential tool for crop growth monitoring and yield estimation (Abou-Ismail *et al.*, 2004; Dente *et al.*, 2008; Jongschaap, 2006; Liang and Qin, 2008; Mo *et al.*, 2005).

Several assimilation schemes with different degrees of complexity and integration have been developed (Baret *et al.*, 2000; Delécolle *et al.*, 1992a; Maas, 1988a; Moulin *et al.*, 1998; Plummer, 2000). Maas (1988a, b) initially reviewed different methods for combining a CSM with radiometric observations (ground measurements or satellite data). Delécolle *et al.* (1992a) generalized them into four methods: (1) directly using remotely sensed variables in the model; (2) updating state variables in the model; (3) reinitialization of the model; and (4) reparameterization of the model. Reparameterization is similar to reinitialization except that the model parameters are adjusted (Plummer, 2000). Putting satellite products into crop growth models in the reparameterization or reinitialization procedure corresponds to the highest degree of integration (Baret *et al.*, 2000). These assimilation schemes can also be applied by the generic Soil–Vegetation–Atmosphere Transfer (SVAT) models (Olioso *et al.*, 2005) (see other chapters of this book).

In this chapter, we aim at introducing the recent advances in the assimilation of CSMs and remotely sensed data with emphasis put on the crop yield estimation. The chapter is organized into five main sections. Among the many crop growth models that could be used in data assimilation, we will focus on the Decision Support System for Agrotechnology Transfer (DSSAT) (Hoogenboom *et al.*, 1999, 2004) as an example in Sec. 13.2. Sections 13.3 and 13.4 describe various data assimilation approaches and remote sensing data in different domains that could be assimilated with crop models, respectively. A data assimilation example is presented in Sec. 13.5 for illustration of regional corn yield and water balance information estimation. The prospects for better remote sensing data and crop growth model are discussed in Sec. 13.6.

13.2 Crop Growth Modeling

Crop models describe the main processes that occur during plant growth and crop production: phenology, photosynthesis, dry matter production, water processes, nitrogen processes, biomass partitioning, organ building, and solar radiation absorption. Numerous crop models have been developed with different complexity levels (see reviews by Hammer, 1998; Sinclair and Seligman, 1996). Specific parameterization schemes may be required for each crop type, even for models that aim to be generic (Olioso *et al.*, 2005). The DSSAT-Cropping System Model (CSM) simulates growth and development of a crop over time, as well as the soil water, carbon

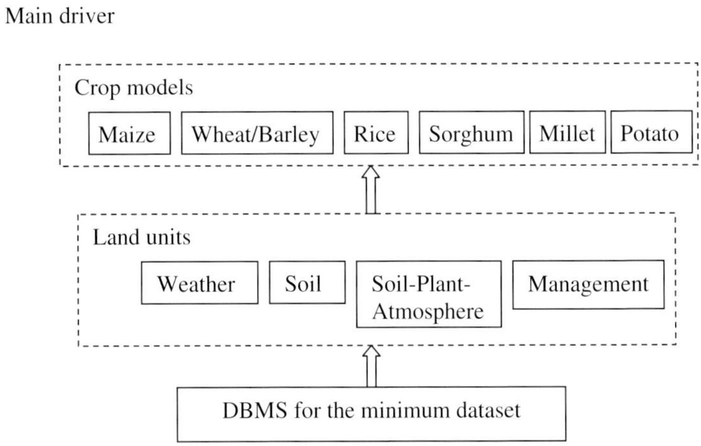

Fig. 13.1. Main components in the DSSAT-CSM.

and nitrogen processes and management practices (Hoogenboom *et al.*, 1999, 2004). The model consists of three components (Fig. 13.1): (1) a Data Base Management System (DBMS) to enter, store, and retrieve the "minimum data set" needed to calibrate, evaluate, and use the crop models for solving problems; (2) a land unit module, which simulates the various processes that affect the land unit, including weather, plant growth, soil processes, soil–plant–atmosphere interface and management practices, and (3) a set of crop models for simulating processes and outcomes of by environment interactions. Collectively, these components simulate the changes in the soil and plants with a main driver program, which controls timing for each simulation. The DSSAT-CSM incorporates models of all crops within a single set of code. The outcomes of long-term simulated agronomic experiments can be analyzed and displayed. This design feature greatly simplifies the simulation of crop rotations since soil processes operate continuously, and different crops are planted, managed, and harvested according to the cropping system information for the model (Hoogenboom *et al.*, 2003).

The DSSAT-CSM model enables users to input the biological and physical parameters and obtain user-specified objectives (Tsuji *et al.*, 1994, 1998). It has been used to predict grain yield and kernel numbers (Asadi and Clemente, 2001; Kiniry *et al.*, 1997; Lizaso *et al.*, 2001; Tojo Soler *et al.*, 2007), the effects of ground water and precipitation (O'Neal *et al.*, 2002; Xie *et al.*, 2001), and the effect of increasing atmospheric carbon dioxide

(Tubiello *et al.*, 1999). The CSM-CERES (Crop-Environment Resource Synthesis) is one set of models under the DSSAT shell that has different modules for different kinds of crops (Jones *et al.*, 2003). Timsina and Humphreys (2006) reviewed the performance of CERES-rice and CERES-wheat models in rice–wheat systems and indicated that both models generally perform satisfactorily under water and nitrogen non-limiting conditions. The model has performed very well for predicting corn yield. For example, the relative errors are as low as 6.2% and 5% as reported by Asadi and Clemente (2001) and Kiniry *et al.* (1997), respectively.

The CSM-CERES-maize model simulates the growth, development, and yield of corn, taking into account the effects of genetics, weather, and soil conditions and crop management. The model offers the ability to evaluate options for increasing yield and water- and nitrogen-use efficiency. The model was modified in an inversion looping structure to dynamically assimilate with remotely sensed LAI and vegetation indices data (see Sec. 13.5). Besides the LAI dynamics, the CSM-ERES-maize model also simulates the growth and development for a maize canopy as well as the water and nitrogen balances of the soil–vegetation–atmosphere system at a daily step.

13.3 Data Assimilation Methods

Various techniques have been proposed and used for assimilating remote sensing data in dynamic crop growth models (Delécolle *et al.*, 1992b; François *et al.*, 2001; Olioso *et al.*, 1999). In the simple *direct input* approach, the crop model makes use of variables estimated by remote sensing as initial conditions or input data, such as crop characteristics, soil condition, management practice, and weather information (Fig. 13.2(a)). The *sequential assimilation* approach estimates model "state variables" from remote sensing data and directly introduces them into the model at occurrences compatible with the time step of the model or through model interpolation (Fig. 13.2(b)). The third *variational assimilation* approach minimizes the difference between values estimated from remote sensing and crop models, respectively, by adjusting the "free variables" — initial conditions or model parameters (Fig. 13.2(c)). In variational assimilation techniques, all the available observations for a certain period of time are processed together. Liang and Qin (2008) provided a very good review of data assimilation methods for land surface variable estimation. In the following sections, we will introduce various data assimilation methods and present examples of assimilation studies.

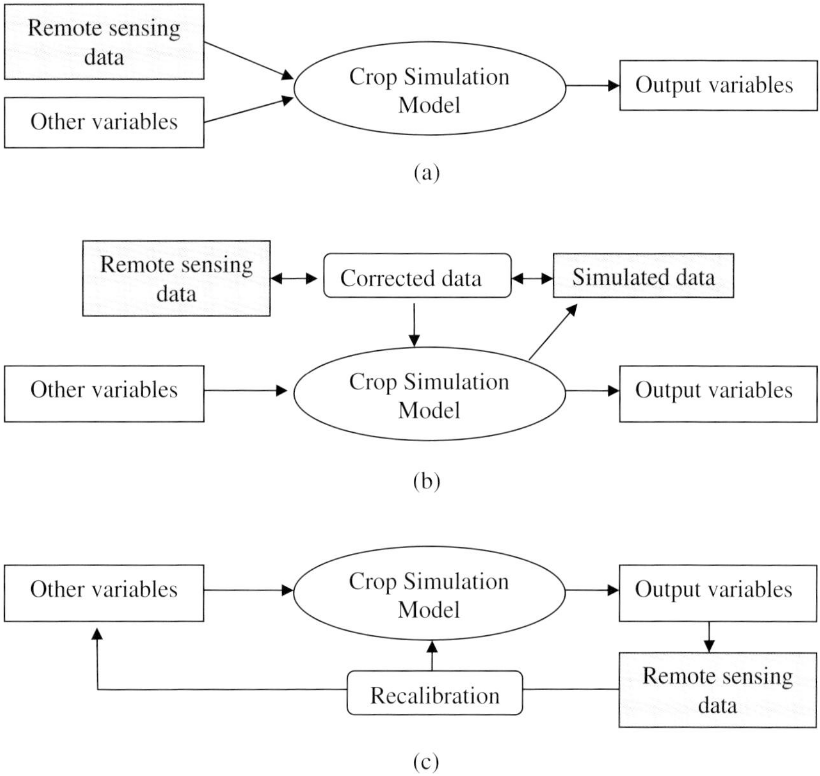

Fig. 13.2. Various schemes to assimilate remote sensing data and CSMs.

13.3.1 *Direct input approach*

Dynamic crop models need detailed information on environmental conditions and have to be calibrated under local conditions to generate reliable results. Remote sensing plays a key role on obtaining the initial conditions and crop growth information for crop models. The direct input approach is to drive the crop model using remote sensing products as the values of some variables (Fig. 13.2(a)). Doraiswamy *et al.* (2005) evaluated the Moderate Resolution Imaging Spectroradiometer (MODIS) 250 m resolution data for retrieval of crop growth variables that could be integrated in crop yield simulation models. Crop planting and emergence dates were assessed from MODIS derived LAI change during the vegetative phase of crop development (Doraiswamy *et al.*, 2005). Jongschaap and Schouten (2005) applied a point-based model to estimate a 45 km × 50 km regional wheat production in the Southeast of France. They used the SPOT HRV (High Resolution Visible) sensor to estimate wheat growth area and microwave data (ERS SAR-C) to estimate regional wheat flowering dates to calibrate a wheat growth model. The results demonstrate that point-based simulation models can be applied at spatially higher levels with the aid of remote sensing data. Olioso *et al.* (2005) estimated LAI from field normalized difference vegetation index (NDVI) and then forced LAI information into two crop models: ISBA (Noilhan and Mahfouf, 1996) and ISBA-Ags SVAT (Calvet *et al.*, 1998) in Avignon, France. Simulated evapotranspiration (ET) was in good agreement with the measurements despite a significant underestimation due to low LAI values at the beginning and end of the season (Olioso *et al.*, 2005). In Europe, global solar radiation and relative ET information derived from meteorological satellite Meteosat were used in a crop growth model for crop yield prediction at national level (Roebeling *et al.*, 2004). Mo *et al.* (2005) developed a process-based crop growth model to predict regional crop yield, water consumption, and water-use efficiency (WUE) in the North China Plain (NCP). The LAI data required by the crop model were retrieved from the 10-day maximum value composited NOAA AVHRR NDVI data (Mo *et al.*, 2005).

13.3.2 *Sequential assimilation approach*

The sequential assimilation approach consists in correcting the course of model dynamic variables by comparing them with remote sensing measurements (Fig. 13.2(b)). Comparison between simulated and measured radiometric data can be applied to force the model to run in a more realistic way. Whenever the new observations are available, the model state variable is updated with the new observation. The underlying hypothesis of the sequential assimilation approach is that the remotely sensed crop physiological information has greater accuracy than the simulated ones (Jongschaap, 2006). Depending on how the stable variable is adjusted, there are two forms of sequential assimilation methods: the successive correction and the interpolation methods. The successive correction method resets the model value *on remote sensing observation dates* whenever the new observations are available. A deviation threshold (e.g., 10%) between the model simulation and the remote sensing estimate can be set *a priori* for the adjustment. For the interpolation method, simulation proceeds until the first remote sensing observation date, after which the relevant variables are derived from *interpolation between observations*, and so forth.

After the last remote sensing observation, simulation is resumed until crop harvest.

Jongschaap (2006) tested various run-time calibration scenarios for replacing simulated values by remotely observed values. The simulated LAI and canopy nitrogen contents were replaced with values estimated from remote sensing. LAI and canopy nitrogen were related with various vegetation indices. The results show that run-time calibration of mechanistic simulation models may enhance crop simulation accuracy. In a more recent study, Oppelt (2010) assimilated the CHRIS/PROBA data with the PROMET (PROcess of Radiation Mass and Energy Transfer) model (Mauser and Bach, 2009) on a field scale to simulate crop growth and yield. During the model runs, information on the absorptive capacity of the leaves for two canopy layers (sunlit and shaded layer) was updated using remotely sensed chlorophyll measurements from CHRIS data. The performance of successive correction and interpolation methods were compared by Clevers *et al.* (2002) who used LAI estimated from the SPOT satellite to calibrate the crop growth model ROTASK to simulate wheat development and the partitioning of dry matter over the various organs in Southern France. In the successive correction method, the simulated LAI was reset to the actual value estimated from the SPOT images at the SPOT observation dates. In the interpolation method, the crop growth model simulated growth and development until the first SPOT observation date (and eventually after the last one). After that the LAI was estimated from day to day by interpolating the estimated LAI values between the SPOT observation dates. A better fit to the observations was obtained with the interpolation method.

Selecting suitable dates for remote sensing images was an important factor which could influence estimation accuracy. High frequencies of remote sensing observations and interpolation between them, allow reconstructing the evolution of model state variables to be integrated in the simulation, thereby increasing the simulation accuracy (Jongschaap, 2006). Abou-Ismail *et al.* (2004) experimented the effect of different number of remote sensing images on rice yield estimation with the rice growth simulation model ORYZA1 and suggested four remote sensing images for early and late rice, and five images for the single season rice. Estimating rice yield using two or three images was possible if images were obtained during the panicle initiation and heading stages (Abou-Ismail *et al.*, 2004). The deviation between simulation and observation usually increases due to an increase in leaf senescence during the later stages of crop development (Fang *et al.*, 2011; Jongschaap, 2006).

13.3.3 *Variational assimilation approach*

In variational assimilation techniques, all the available observations for a certain period of time are processed together (Fig. 13.2(c)). They are usually used for retrieving initial values or model parameters using optimization algorithms, based either on traditional optimization algorithms (such as quasi-Newtonian and conjugate gradient) and heuristic optimization algorithms (such as genetic method and simulated annealing) (Demarty *et al.*, 2004). The variational assimilation method allows correction of crop model variables and parameters identified in a previous sensitivity analysis. A common inversion procedure of the data assimilation is similar to the 4D variational analysis in Liang (2004). A general cost function J is constructed as:

$$J(X(t_0)) = [X(t_0) - X_b]^T B^{-1}[X(t_0) - X_b] + \sum_{i=1}^{n}[H_i(X(t_i)) - Y_i)]^T R_i^{-1}[H_i(X(t_i)) - Y_i] \qquad (1)$$

where X and Y are model state and observations, respectively. X_b is the background value (or first guess), $X(t_0)$ is the observed value at time t_0, H is the observation operator, and R and B are the observation and background error covariance matrices, respectively. The first term in Eq. (1) is to force the optimal parameters as close as possible to model states, and the second term is to adjust variables so that model outputs will be as close to the sequential observations as possible.

The objective now is to minimize the merit function J. Many different types of methodology and mathematical tools have been developed for implementing the optimization procedures (Press *et al.*, 1992). Rodriguez *et al.* (2004) used the Simplex algorithm (Nelder and Mead, 1965) included in the STICS (Simulateur mulTIdisciplinaire pour les Cultures Standard) model (Brisson *et al.*, 1998, 2002, 2003) for optimization of free parameters in the wheat yield simulation. A conjugate direction method (e.g., the POWELL algorithm from Press *et al.*, 1992) was applied to minimize the differences between simulated and remotely sensed LAIs and vegetation indices (Fang *et al.*, 2008, 2011). Ma *et al.* (2008) linked the WOFOST model and the SAIL-PROSPECT model through LAI in order to simulate and optimize the soil adjusted vegetation index (SAVI). The FSEOPT optimization software (Stol *et al.*, 1992), based on the downhill-simplex algorithm, was used to search for the best initial combinations of emergence date and regional biomass at green-up stage (Ma *et al.*, 2008).

Other types of methodology and mathematical tools, such as the Kalman filter (KF) (Reichle *et al.*, 2003), have been developed for implementing assimilation procedures. Their use for assimilating data in crop models has shown great potential. Pauwels *et al.* (2007) coupled a fully process-based water and energy balance model with the WOFOST crop growth model through the assimilation of observed LAI and soil moisture values using the Ensemble Kalman Filtering (EnKF). Vazifedoust *et al.* (2009) assimilated LAI and relative ET estimated from MODIS to estimate wheat yield with a constant gain KF data assimilation algorithm. The yield predictions for one month in advance at a regional scale were very promising with respect to the statistical data (bias ±10%). Evolutionary algorithms are very powerful methods that can find the optimum of complex real-world questions (Jarlan *et al.*, 2005). However, their computation efficiency is still a concern for practical applications. Using the stochastic method rather than the systematic exploration of the whole parameter space could improve the computation efficiency (Jarlan *et al.*, 2005). The Shuffled Complex Evolution-University of Arizona (SCE-UA) method, a variation of the evolutionary algorithms developed by Duan *et al.* (1992, 1993) has been used in crop data assimilation studies to determine the optimal set of input variables (Jin *et al.*, 2010). Compensations do occur among the initial parameters during implementing variational assimilation procedures (Demarty *et al.*, 2004, 2005; Wigneron *et al.*, 2002). For example, a lower nitrogen application could be counterbalanced by an earlier sowing date, leading to a similar LAI profile (Fang *et al.*, 2008). Further assimilation procedures should be explored considering the compensation issue and the computation efficiency and accuracy.

13.4 Remote Sensing Data and Preprocessing

Remote sensing data in various wavelength domains (solar, microwave, and thermal) may be used for constraining or driving CSMs in order to provide continuous simulation of canopy variables, crop production, and field conditions. Studies in using solar, microwave (active and passive), and thermal remote sensing data are presented to illustrate their capabilities in assimilating with crop growth models.

13.4.1 *Visible and near-infrared (NIR) information*

Most of the studies devoted to assimilation of remote sensing data into crop models focused on the use of visible and near-infrared reflectances since these measurements give information on crop structure and biomass seasonal dynamics (Olioso *et al.*, 2005). Different researchers have used LAI, spectral reflectance, or vegetation indices as a control variable to adjust or determine the optimal set of input parameters (e.g., Bach *et al.*, 2001, 2003; Doraiswamy *et al.*, 2003; Fang *et al.*, 2008, 2011). Maas (1988a) and Dente *et al.* (2008) used remotely sensed LAI in the adjustment of model initial conditions to minimize the difference between model-predicted and remotely estimated LAI. Moulin *et al.* (1998) coupled a crop production model and a radiative transfer model, compared a simulated reflectance profile with a satellite measured reflectance, and successfully retuned crop model parameters and initial conditions. de Wit (1999) integrated the WOFOST model (Boogaard *et al.*, 1998) and satellite vegetation index data to retrieve wheat and sunflower conditions. Inoue *et al.* (1998) recalibrated a simple model with the remotely sensed fraction of photosynthetically actives radiation (FPAR) absorbed by vegetation based on a Simplex algorithm for predicting rice growth and yield. Guérif and Duke (2000) also reported the usefulness of vegetation indices, especially for estimating the planting date and emergence parameters. Weiss *et al.* (2001) successfully assimilated the radiative transfer model SAIL and a crop growth model (STICS); simulated and measured reflectances agreed very well. Schneider (2003) used Landsat Thematic Mapper (TM) images to estimate management related parameters and obtained improved modeling results from the PROcess Oriented Modular Environment and Vegetation (PROMET-V) model. El Hajj *et al.* (2009) combined the SPOT-5 NDVI time series, crop growth modeling and expert knowledge to monitoring sugar cane harvest date with a fuzzy inference system. Todoroff *et al.* (2010) integrated SPOT-NDVI with a crop model to simulate the total available soil water content (TAWC) for sugar cane.

Some work has been done to compare the performances by assimilating different satellite products, such as LAI and vegetation index. Guérif and Duke (2000) compared the effect of spectral reflectance and vegetation index and discovered that using a vegetation index was a better choice than spectral reflectance because the assimilation was very sensitive to variables such as soil reflectance and leaf optical properties in the canopy radiative transfer models. Fang *et al.* (2008, 2011) assimilated MODIS LAI and vegetation indices into CERES-maize crop growth model. Fang *et al.*

(2011) further compared the effect of assimilating LAI separately and LAI and vegetation index synergistically and found that LAI alone was sufficient for a regional assimilation study.

13.4.1.1 *Remote sensing data preprocessing*

The quantity and timing of remote sensing data with regard to crop development will influence the assimilation performance (Guérif and Duke, 2000; Moulin *et al.*, 1998). Due to cloud and snow contamination, spatially and temporally complete remote sensing dataset are usually not available at the current stage. Therefore, the original remote sensing data need to be preprocessed before they can be applied in the optimization process. This is illustrated by a simple method to process the MODIS LAI products (Fang *et al.*, 2011).

The standard MODIS LAI products are provided globally every 8 days at a spatial resolution of 1 km (Justice *et al.*, 1998; Myneni *et al.*, 2002). The essence of this LAI adjustment procedure is based on the common norm that crop LAI before the greenup onset should be nil. The LAI annual profile was first smoothed with the Savitzky–Golay (SG) filter (Savitzky and Golay, 1964) to remove the temporal gaps and low-quality pixels. The SG filter has been shown to be very helpful for LAI smoothing. The hollow squares in Fig. 13.3(b) are the LAI points after the SG smoothing. The MODIS enhanced vegetation index (EVI) products were used to model the vegetation phenological profile and determine the transition dates based on an approach developed by Zhang *et al.* (2003). This method uses a series of piecewise logistic functions fit to EVI to represent the intra-annual vegetation dynamics and identifies the phenological transition dates by the rate of change in curvature of the fitted logistic models. An example of the greenup and maturity onset dates generated by the piecewise logistic method is shown in Fig. 13.3(a) for a corn pixel in Indiana.

Since the LAI values before the greenup onset are zero, the SG-filtered LAI was adjusted using a piecewise logistic function:

$$\mathrm{LAI}(t) = \frac{c}{1 + e^{a+bt}} \tag{2}$$

where t is the time in day of year, $\mathrm{LAI}(t)$ is the LAI value at time t, a and b are fitting parameters, and c is the maximum LAI value. Parameters a and b are calculated based on the LAI values at the greenup $(=0)$ and maturity onset dates. The solid squares in Fig. 13.3(b) are the adjusted LAI points for one corn pixel. In practice, we mainly focused on adjusting the LAI values between the greenup and maturity onset dates.

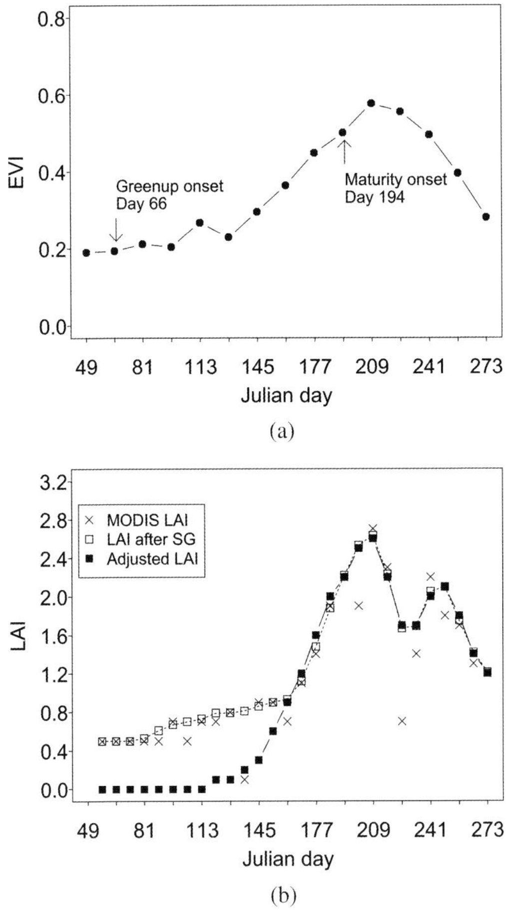

Fig. 13.3. (a) A sample time series of MODIS EVI data and the estimated phenological transition dates for a corn pixel in Indiana, 2000. (b) The original MODIS LAI (cross), the LAI after the SG filter (hollow square), and the adjusted LAI (solid square) based on the phenological dates from (a) and a piecewise logistic function in Eq. (4) (from Fang *et al.*, 2011. Reproduced by permission of Taylor & Francis, Ltd.).

13.4.2 *Microwave information*

Microwave radar data overcome some of the limitations of the remotely sensed reflectance and LAI data, such as gaps during the growing season caused by cloudy and rainy conditions (Shen *et al.*, 2009). Chauki *et al.* (2000) assimilated the European Remote Sensing Satellite (ERS) Synthetic Aperture Radar (SAR) data and RadarSAT SAR with the STICS crop model (Brisson *et al.*, 1998) to estimate wheat canopy biomass and LAI at

These calibrations significantly improved the simulation of the LAI and LE (RMSE decreased by 21% for LE) (Cayrol *et al.*, 2000). Olioso *et al.* (2005) demonstrated the use of reflectances and TIR data to monitor evapotranspiration and irrigation events over wheat crops within a small agricultural area. Demarty *et al.* (2005) performed a combined assimilation of both passive microwaves and thermal infrared remotely sensed observations. A multi-objective calibration iterative procedure (MCIP) is applied on the Simple Soil Plant Atmosphere Transfer-Remote Sensing (SiSPAT-RS) model. The soil water content of the upper 5 cm (related to L-band microwave brightness temperature) and the thermal infrared brightness temperature were used to calibrate the SiSPAT-RS model. The MCIP approach performs successive contractions of the feasible parameter space with the multi-objective generalized sensitivity analysis algorithm (Demarty *et al.*, 2004).

13.5 Corn Yield Estimation at a Regional Level

Assimilation of remote sensing data into CSMs give priceless information for crop growth and production, water balance assessment, agricultural management, and more recently for carbon cycle studies in relation with climate research. This section introduces the Integrated Crop Simulation System (ICSS) developed by Fang *et al.* (2008, 2011) to illustrate the assimilation method for corn yield estimation and water balance study at a regional level.

The general methodology of the data assimilation example is illustrated in Fig. 13.4. Three steps were involved in this procedure:

(1) Input data, such as crop characteristics, soil condition, management practice, and weather information, were prepared in advance for the CSM-CERES-maize model. Daily weather data, soil properties, and crop ecology and management information are required to run the CSM. The planting date, planting population, and row spacing were free variables and were estimated on a pixel basis. The US Department of Agriculture (USDA) National Agricultural Statistical Service (NASS) reports provide the average planting date for each crop at the state level. The nutrient supply is represented by the application of nitrogen, which was set as a free variable. The date of nitrogen application was set to the same date as the planting date. An initial parameterization was conducted by adjusting the CSM-CERES-maize model to the USDA NASS corn yield at the state level. With these input data, crop

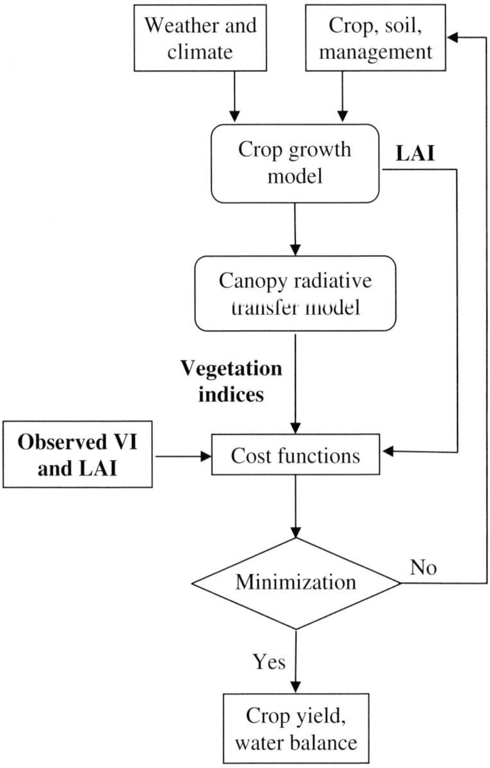

Fig. 13.4. General methodology of the data assimilation approach that integrates remote sensing data with crop growth and canopy reflectance models for crop yield and water balance studies (revised from Fang *et al.*, 2011). Reproduced by permission of Taylor & Francis, Ltd.

biophysical information (e.g., LAI) was generated by the crop growth model.

(2) The parameters simulated by the crop model (e.g., LAI) and other ancillary information (e.g., soil reflectance) were used in the canopy reflectance model. Vegetation indices were calculated from the simulated reflectances. The Markov Chain Reflectance Model (MCRM) (Kuusk, 1998) was used to simulate the reflectance at the top of the canopy. Reflectances simulated by the Markov model for the top of canopy were used to compute the vegetation indices.

(3) The simulated vegetation indices and LAI were compared with the corresponding MODIS products, and residuals between the simulated and MODIS LAI or vegetation index were minimized by adjusting the

input parameters in Step 1. With the optimized set of input parameters, the model was executed to update the crop yield prediction.

13.5.1 *Sensitivity study*

The sensitivity study tests the sensitivities of model outputs to variations of each input parameter. Such an analysis helps understand individual parameters and how processes are modeled, identifies the parameters that need to be accurately determined, and provides insight into the limitations of the model. Fang *et al.* (2008) conducted sensitivity studies for the CERES-maize model by adjusting the values of input variables (Table 13.1). The initial planting date was set on May 7 (DOY 128). The plant density and row spacing were 6.5 plants m^{-2} and 80 cm, respectively. The simulated yield (9723 kg ha^{-1}) was slightly higher than the USDA NASS data (9172 kg ha^{-1}). Various input values were changed to determine the impact on the simulated yield from the assimilated model. Changes in planting date, planting density, and MODIS LAI will all impact the final simulated yield. With the same parameter setting, moving the planting date 5 days earlier caused a yield decrease of up to 57.1%. By running the coupled inversion model, the range and uncertainties of the sowing date and density can be evaluated by choosing different input options. On the other hand, yield was relatively stable for different row spacings. Alternative management scenarios, such as cultivar selections and crop rotation, were also tested. Management measures are important for crop growth, but they were kept constant and were not adjusted in the data assimilation process.

Through sensitivity analysis, users can determine the impact of not only crop types and management practice but also soil quality and other environmental factors (Heinemann *et al.*, 2002; O'Neal *et al.*, 2002). Other researchers have found various sensitive stable variables. Maas (1988a)

Table 13.1. Impact of input variables on the simulated yield with the CERES-maize model.

	Simulated values	Planting date (DOY 128)		Plant density (6.5 plants/m^2)		Row spacing (80 cm)		MODIS LAI	
Cases		−5	+5	−0.5	+0.5	−10	+10	−10%	+10%
Yield (kg/ha)	9723	4173	11,926	10,046	9250	9745	9717	8301	10,495
Diff. (%)		−57.1	22.7	3.3	−4.9	0.2	−0.1	−14.6	7.9

Note: The first row lists parameters tested and their initial values. The last row shows the difference with the simulated yield.

found the "initial LAI" important for crop yield estimation. Bouman (1992) conducted a sensitivity analysis to select the crop species parameters and initial conditions that have a large effect on both the remote sensing signal and the canopy biomass and the LAI. Bouman (1992) and Clevers *et al.* (1994) identified four sensitive parameters, i.e., sowing date, relative growth rate, light use efficiency, and maximum leaf area. Brown and Rosenberg (1997) tested the impact of temperature, precipitation, solar radiation, humidity, and atmospheric carbon dioxide on yields and water use in corn, soybean, winter wheat, and sorghum. Chauki *et al.* (2000) used the sowing date, the soil dry bulk density, the initial soil moisture and the initial amount of nitrogen in soil as the free parameters. O'Neal *et al.* (2002) simulated the impact of precipitation on yield with CERES in an Indiana field. Kravchenko *et al.* (2000) used a multi-fractal analysis to study the relationship between topography and yield. Further, Kravchenko and Bullock (2000) analyzed the yield-topography-soil relationships and found that soil properties and topography explained about 40% of yield variability. Prévot *et al.* (2003) identified five free parameters in their sensitivity analysis: the sowing date, the compensation between stem number and stem density, the duration of the period between the maximum of LAI and the beginning of net senescence, the field capacity of the first horizon of soil, and the parameter defining the end of the maximum evaporation stage of the soil. Rodriguez *et al.* (2004) selected four parameters, the maximum rate of LAI increase, the LAI value from which starts the interplant competition, the stem density, and the maximal lifespan of an adult leaves, based on the results and sensitivity analysis in Brisson *et al.* (2002) and Ruget *et al.* (2002).

13.5.2 *Cost function construction*

The crop growth and canopy reflectance models were coupled such that the crop biophysical parameters from the CSM-CERES-maize model were used to drive the Markov model (Kuusk, 1998). The cost function was based on the differences between the simulated and measured LAI and/or vegetation indices using the conjugate direction optimization algorithm POWELL provided by Press *et al.* (1992). A simplified minimization was proposed to calculate the cost functions (Fang *et al.*, 2011):

$$J_1 = \sum_{i=1}^{m} \text{abs}(\text{LAI}_S(t_i) - \text{LAI}_M(t_i))/\text{LAI}_M(t_i) \quad (3)$$

$$J_2 = \sum_{j=1}^{n} \text{abs}(\text{VI}_S(t_j) - \text{VI}_M(t_j))/\text{VI}_M(t_j) \quad (4)$$

$$J_3 = \frac{1}{m}\sum_{i=1}^{m} \text{abs}(\text{LAI}_S(t_i) - \text{LAI}_M(t_i))/\text{LAI}_M(t_i) + \frac{1}{n}\sum_{j=1}^{n} \text{abs}(\text{VI}_S(t_j) - \text{VI}_M(t_j))/\text{VI}_M(t_j) \quad (5)$$

where $\text{LAI}_S(t_i)$, $\text{LAI}_M(t_i)$ are the simulated and measured LAI at time t_i, respectively; VI_S, VI_M are the simulated and measured VI, respectively; and m and n are the number of LAI and vegetation index values. Many factors influence the assimilation performance, such as the number and timing of remote sensing data with regard to the crop development (Guérif and Duke, 2000; Moulin *et al.*, 1998). Some researchers indicate only very few data was enough for monitoring crop processes — four images acquired over the wheat crop cycle with the SPOT-HRV instrument is feasible for the calibration and correction of a crop model (Clevers *et al.*, 2002; Moulin *et al.*, 2002). Others have found that the best situation was to use the LAI and vegetation indices data that covered the whole growth period (Fang *et al.*, 2008, 2011; Guérif and Duke, 2000; Rodriguez *et al.*, 2004).

The optimization process starts from an initial parameterization and adjusts the free parameters in the coupled crop yield and radiative transfer model in order that the model gives LAI and/or vegetation index simulation in agreement with the MODIS observations. Six model variables, four for the CSM-CERES-maize (planting date, plant population, row spacing, and nitrogen fertilizer amount) and two for the Markov model (red band soil reflectance and the effective number of elementary layers in a leaf) were adjusted and optimized iteratively based on the cost functions. The simulated LAI or vegetation index values depend on the values of the free variables (e.g., planting date, see Table 13.1) that are estimated by minimizing the cost function J. To compare the effect of using different control variables, five schemes were designed based on the cost function varieties: S1 (LAI), S2 (EVI), S3 (NDVI), S4 (EVI + LAI) and S5 (NDVI + LAI). The cost function Eqs. (4), (2), and (3) were applied for schemes S1, S2 and S3, and S4 and S5, respectively.

13.5.3 *Corn yield estimation*

The estimated corn yield was compared with the USDA NASS statistical data for selected counties in Indiana, 2000 (Fig. 13.5). A total of 43 counties contained valid yield data by having at least $5\,\text{km}^2$ planted in corn

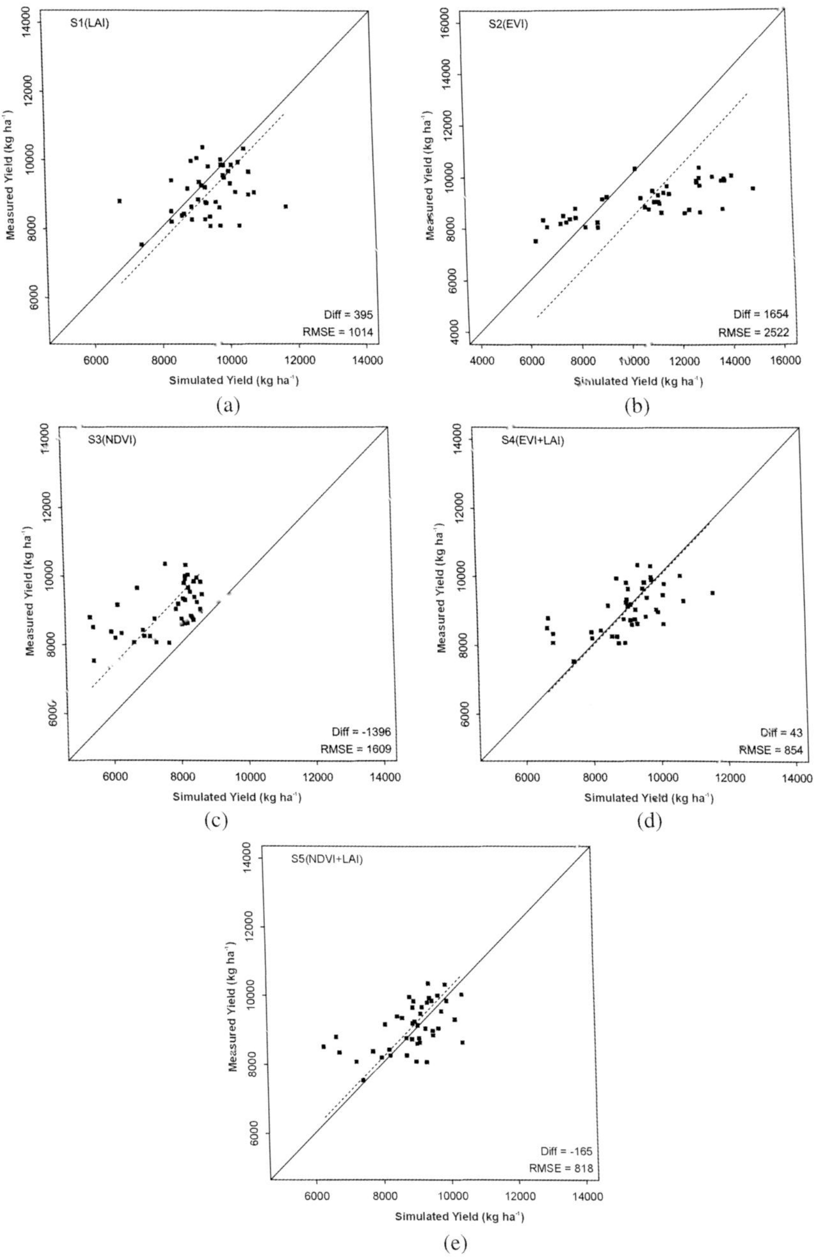

Fig. 13.5. Comparison of the simulated corn yield and USDA NASS yield data for selected counties (43) in Indiana, 2000. (a) S1 (LAI); (b) S2 (EVI); (c) S3 (NDVI); (d) S4 (EVI + LAI); and (e) S5 (NDVI + LAI). The abscissa and ordinate are the simulated and measured corn yield (kg ha^{-1}), respectively. The dashed line shows the mean offset indicated by "Diff". RMSE: root-mean-squared error (from Fang *et al.*, 2011. Reproduced by permission of Taylor & Francis, Ltd.).

(≥5 pixels) within the county. The results from the five schemes were compared. The best results were based on schemes with both vegetation index and LAI (S4 and S5), which had the lowest root-mean-squared errors (RMSEs) and biases. When only vegetation index was used (S2 and S3), the results were unacceptable. Using only LAI (S1) generated moderately good results. Comparing the distribution of the estimated and NASS yields for all counties in 2000, the NASS county yields were mostly clustered between 8000 kg ha^{-1} and 10,500 kg ha^{-1} while the simulated yield scatters between 5000 kg ha^{-1} and 15,000 kg ha^{-1}. The yield distribution of schemes S1, S4, and S5 were comparable to that of the NASS data, while the mean values of S2 and S3 were higher and lower, respectively, than the NASS data. It is noted that the data assimilation results are based on pixels, while statistical yields are based on administrative regions.

Statistical comparison of the estimated and NASS yields showed the corn yield deviation from scheme S5 was only −0.2%. Scheme S4 (EVI + LAI) gave very good results, and the relative difference is 3.5%. In contrast, the results from S2 and S3 differed greatly from the measurements. The standard deviations from the estimated results were all higher than those of the NASS data. Using only NDVI significantly underestimated the corn yield while using only EVI overepredicted the yield. Scheme S2 (EVI) and S3 (NDVI) showed the largest overestimation (20.8%) underestimation (−13.2%), respectively. This proved the importance of LAI in the data assimilation process and that the incorporation of both LAI and vegetation index can improve crop yield prediction. The inefficiency of using a vegetation index might be attributed to uncertainties in the radiative transfer model and the spatial variation of the MODIS products.

13.5.4 *Water balance studies*

Many crop models were developed to study the interactions between water processes and crop production; thus, they may be useful in irrigation management (Ambast *et al.*, 2002; Olioso *et al.*, 2005). Remote sensing provides information related to the spatial variability and the time variation of crop water status (Droogers and Bastiaanssen, 2002; Moran *et al.*, 1996). Other management related parameters, such as the species information, planting density, irrigation, and harvest date, are hard to be modeled accurately and must rely on observations (Schneider, 2003).

Regional hydrological information can be predicted by the ICSS (Fang *et al.*, 2011). Potential evapotranspiration (PET) was calculated with the Priestley–Taylor formulation (Priestley and Taylor, 1972) using

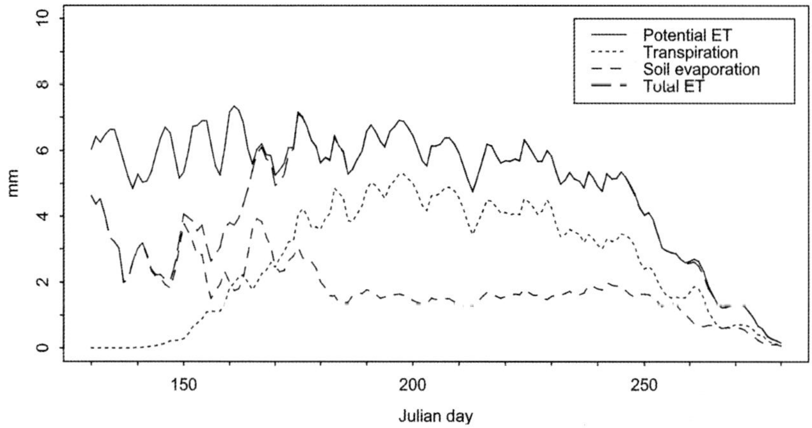

Fig. 13.6. The ICSS simulated PET, transpiration, soil evaporation, and total ET for Indiana, 2000.

monthly air temperature and amount of net radiation estimated from monthly cloudiness and temperature data. The soil potential evaporation was calculated from the total PET and LAI. Total ET was the sum of the vegetation transpiration and soil evaporation. Figure 13.6 shows the simulated daily PET, transpiration, soil evaporation, and total ET for the growing season with scheme S4 (EVI + LAI). The mean PET remains at a relatively stable level with a mean value at 5.88 mm and a standard deviation of 0.55 mm before September 6 (DOY 250). Soil evaporation fluctuates between 1.5 mm and 4.6 mm before day 182, while it stays at a relatively stable level (about 1.60 mm) for the remainder of the season. Transpiration shows a clear seasonal feature, increasing from zero at pre-emergence to the highest value after day 195. Before day 165, ET is less than PET, but after that day ET and PET are nearly the same. After day 250, the hydrological factors decrease gradually when the crop reaches senescence.

One example of the simulated potential ET, transpiration, soil evaporation, and total ET from S4 on July 11, 2000 (day 193) is illustrated in Fig. 13.7. The potential ET shows a vivid increasing trend from north to south in Indiana, due to the temperature gradient (Fig. 13.7(a)). For corn transpiration, about 81% pixels have values less than 5.0 mm. Other higher transpiration pixels are mainly located in the south (Fig. 13.7(b)). The soil evaporation increases northbound (Fig. 13.7(c)). Except for a few pixels in the southwest corner, the total ET also increases toward the north (Fig. 13.7(d)).

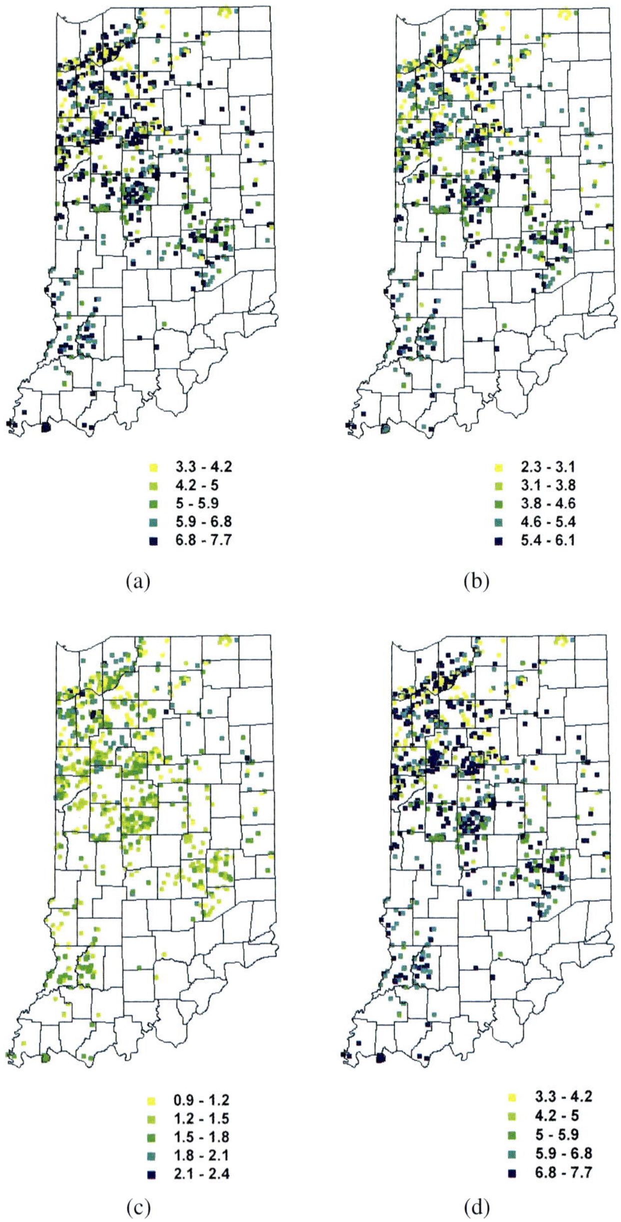

Fig. 13.7. Distribution of (a) the simulated potential evapotranspiration (PET), (b) transpiration, (c) soil evaporation, and (d) total evapotranspiration (ET) on day 193, 2000. Unit: mm.

13.6 Challenges and Future Studies

We outline in this section some further improvements that can be made for the remote sensing data and crop growth models.

(1) ***Remote sensing data quality.*** The effect of remote sensing data quality on crop yield estimation and assimilation results is obvious. Fang *et al.* (2008) found that a −10% LAI bias will lead to as large as 30% nitrogen estimation error. Better results may be obtainable with the use of more accurate parameter estimation from remote sensing. However, the difficulty of obtaining an accurate LAI estimation still limits the application of the data assimilation method (Doraiswamy *et al.*, 2004). Some improvements of the remote sensing derived LAI and vegetation index products are necessary, especially during the beginning of the growing season (Fang *et al.*, 2011). Conversely, crop growth models can estimate LAI when no remote sensing products are available; therefore, it is feasible to monitor LAI and crop conditions all along the phenological cycle and facilitate studies that require LAI with a higher temporal resolution (Kang *et al.*, 2003).

(2) ***Tackling mixed pixels.*** High spatial resolution data (e.g., IKONOS, SPOT, and Landsat) are optimal for deriving inputs for crop growth models, but impractical for large-scale application. Non-pure pixels were inevitable and tolerated in regional applications. Many crop models (e.g., the DSSAT family) simulates different species of crop types separately, making it a challenging task to separate different crops within a mixed pixel. One way to tackle this issue is by working on "pure" pixels only. Guissard *et al.* (2005) noted that a minimum purity of 98% of the pixel surface was needed for winter wheat and of 90% for sugar beet in order to compute the LAI distributions from ENVISAT MERIS images. Fang *et al.* (2008, 2011) only processed pixels having more than 80% of corn. Another way is by increasing the data temporal frequency. Hadria *et al.* (2006) reported the daily satellite frequency always provides the best results on LAI estimation, with no strong differences between the data assimilation approaches. A good compromise for the satellite revisit capacity appears to be the 10-day frequency (Hadria *et al.*, 2006). With sufficient time series of medium-resolution imagery, it is possible to extract regional scale, crop-specific information (Guissard *et al.*, 2004; Wardlow and Egbert, 2008, 2010).

(3) ***Crop model parameterization.*** The assimilation model would certainly benefit from better parameterization of the crop growth models, which usually have a large number of parameters. The analysis of model sensitivity is a key issue for identifying the main influential parameters in data assimilation. Nevertheless, various models differ on the most sensitive parameters (e.g., Sec. 13.5.1). The parameters choice would have a large influence on the remote sensing response and the canopy variables such as biomass and LAI (Dente *et al.*, 2004; Prévot *et al.*, 2003). For these reasons, the sensitivity analysis has to be conducted for each assimilation study to determine a relatively small subset of model inputs for the assimilation. It is acknowledged that these variables are inter-related, and compensations occur between various parameters (Demarty *et al.*, 2004, 2005; Wigneron *et al.*, 2002).

(4) ***Regional model development.*** The application of crop models over large areas is often limited by the required input information. Crop models require continuous inputs of meteorological data and may be highly sensitive to the spatial variation of such data. Crop models require plant physiological information and soil information, which are not easy to obtain over large areas. We argue for the need to develop a new generation of crop models at regional scales which would facilitate both easily obtainable input variables and a seamless integration with remotely sensed data. With a proper model setting, the assimilation method described in Fang *et al.* (2008, 2011) is very promising for forecasting the crop yield and hydrological conditions in a larger region.

13.7 Conclusions

Remote sensing data have become increasingly used to be assimilated into crop growth models to monitor plant growth, field condition, water status, and management practice. After introducing the DSSAT cropping system models, we presented various types of data assimilation methods to integrate remote sensing data and crop models. Although there is a huge literature on various assimilation methods, the search for rapid and accurate method is still necessary. Recent studies using remote sensing data in solar, microwave and thermal wavelength domains were presented to illustrate their capabilities in assimilating with crop growth models.

This chapter introduced the Integrated Crop Simulation System (ICSS) as a promising framework to estimate crop yield at the regional scale. The integrated model assimilates both LAI and vegetation index products into a coupled model consisting of the CSM-CERES-maize crop growth model and the Markov canopy reflectance model. Corn yield and field water conditions were estimated with the example illustrated above. Further studies should focus on improvements of both remote sensing data and crop models. The quality of remote sensing data is critical for data assimilation studies. As discussed earlier, the issues of data uncertainties and mixed pixels need to be solved by the remote sensing community. On the other hand, more research is necessary for better crop model parameterization and regional crop model development.

References

Abou-Ismail, O., Huang, J. and Wang, R. (2004). Rice yield estimation by integrating remote sensing with rice growth simulation model. *Pedosphere*, **14**, 519–526.

Ambast, S. K., Keshari, A. K. and Gosain, A. K. (2002). Satellite remote sensing to support management of irrigation systems: Concepts and approaches. *Irrigation and Drainage*, **51**, 25–39.

Asadi, M. E. and Clemente, R. S. (2001). Simulation of maize yield and N uptake under tropical conditions with the CERES-maize model. *Tropical Agriculture*, **78**, 211–217.

Attema, E. P. W. and Ulaby, F. T. (1978). Vegetation modelled as a water cloud. *Radio Science*, **13**, 357–364.

Bach, H., Mauser, W. and Schneider, K. (2003). The use of radiative transfer models for remote sensing data assimilation in crop growth models. In *Precision Agriculture: Papers from the 4th European Conference on Precision Agriculture*, Berlin, Germany, June 15–19, 2003, Stafford, J. and Werner, A. (eds.), pp. 35–40. Berlin, Germany: Wageningen Academic Publishers.

Bach, H., Schneider, K., Verhoef, W., Stolz, R., Mauser, W., Leeuwen, H., Schouten, L. and Borgeaud, M. (2001). Retrieval of geo- and bio-physical information from remote sensing through advanced combination of a land surface process model with inversion techniques in the optical and microwave spectral range. In *8th International Symposium Physical Measurements & Signatures in Remote Sensing*. Centre Paul Langevin, Aussois, France, pp. 639–647.

Baez-Gonzalez, A. D., Kiniry, J. R., Maas, S. J., Tiscareno, M. L., Macias, C. J., Mendoza, J. L., Richardson, C. W., Salinas, G. J. and Manjarrez, J. R. (2005). Large-area maize yield forecasting using leaf area index based yield model. *Agronomy Journal*, **97**, 418–425.

Baret, F., Weiss, M., Troufleau, D., Prevot, L. and Combal, B. (2000). Maximum information exploitation for canopy characterisation by remote sensing. *Aspects of Applied Biology*, **60**, 71–82.

Boogaard, H. L., Diepen, C. A. V., Rötter, R. P., Cabrera, J. M. C. A. and Laar, H. H. V. (1998). WOFOST 7.1 User's guide for the WOFOST 7.1

crop growth simulation model and MOFOST control center 1.5. Wageningen: DLO Winand Staring Centre, Technial Document 53.

Bouman, B. A. M. (1992). Linking physical remote sensing models with crop growth simulation models, applied for sugar beet. *International Journal of Remote Sensing*, **13**, 2565–2581.

Bouman, B. A. M., Kropff, M. J., Tuong, T. P., Wopereis, M. C. S., Berge, H. F. M. T. and Laar, H. H. V. (2001). *ORYZA2000: Modeling Lowland Rice*. Wageningen, The Netherlands: International Rice Research Institute, Los Baños, Philippines and Wageningen University and Research Centre.

Brisson, N., Mary, B., Ripoche, D. *et al.* (1998). STICS: A generic model for the simulation of crops and their water and nitrogen balances, Part I: Theory and parameterization applied to wheat and corn. *Agronomie*, **18**, 311–346.

Brisson, N., Ruget, F., Gate, F., Lorgeou, J., Nicoullard, B., Tayot, X., Plenet, D., Jeuffroy, M.-H., Brouthier, A., Ripoche, D., Mary, B. and Justes, E. (2002). STICS: A generic model for the simulation of crops and their water and nitrogen balances, Part II: Model validation for wheat and maize. *Agronomie*, **22**, 69–92.

Brisson, N., Gary, C., Justes, E. *et al.* (2003). An overview of the crop model STICS. *European Journal of Agronomy*, **18**, 309–332.

Brown, R. A. and Rosenberg, N. J. (1997). Sensitivity of crop yield and water use to change in a range of climatic factors and CO_2 concentrations: A simulation study applying EPIC to the central USA. *Agricultural and Forest Meteorology*, **89**, 171–203.

Calvet, J.-C., Noilhan, J., Roujean, J.-L., Bessemoulin, P., Cabelguenne, M., Olioso, A. and Wigneron, J.-P. (1998). An interactive vegetation SVAT model tested against data from six contrasting sites. *Agricultural and Forest Meteorology*, **92**, 73–95.

Cayrol, P., Kergoat, L., Moulin, S., Dedieu, G. and Chehbouni, A. (2000). Calibrating a coupled SVAT/vegetation growth model with remotely sensed reflectance and surface temperature: A case study for the HAPEX-Sahel grassland sites. *Journal of Applied Meteorology*, **39**, 2452–2472.

Chauki, H., Prévot, L., Troufleau, D., Brisson, N., Olioso, A., Baret, F. and Guérif, M. (2000). Assimilation of ERS and RadarSat SAR observations within STICS crop growth model. *Proceedings of the CEOS SAR Workshop*, Toulouse, pp. 485–490.

Clevers, J. G. P. W., Buker, C., Leeuwen, H. J. C. V. and Bouman, B. A. M. (1994). A framework for monitoring crop growth by combining directional and spectral remote sensing information. *Remote Sensing of Environment*, **50**, 161–170.

Clevers, J. P. G. W. and Van Leeuwen, H. J. C. (1996). Combined use of optical and microwave remote sensing data for crop growth monitoring. *Remote Sensing of Environment*, **56**, 42–51.

Clevers, J. G. P. W., King, C., Prvot, L., Bruguier, N., Vonder, O. W., Jongschaap, R. E. E. and Desprats, J. F. (2002). Using SPOT data for calibrating a wheat growth model under Mediterranean conditions. *Agronomie*, **22**, 687–694.

de Wit, A. J. W. (1999). The application of a genetic algorithm for crop model steering using NOAA-AVHRR data. In *Proceedings of SPIE*, Cecchi, G., Engman, E. T. and Zilioli, E. (eds.), pp. 167–181. Florence, Italy.

Delécolle, R., Maas, S. J., Guérif, M. and Baret, F. (1992a). Remote sensing and crop production models: Present trends. *ISPRS Journal of Photogrammetry and Remote Sensing*, **47**, 145–161.

Delécolle, R., Mass, S. J., Guérif, M. and Baret, F. (1992b). Remote sensing and crop production models: Present trends. *ISPRS Journal of Photogrammetry and Remote Sensing*, **47**, 145–161.

Demarty, J., Ottlé, C., Braud, I., Olioso, A., Frangi, J. P., Bastidas, L. A. and Gupta, H. V. (2004). Using a multiobjective approach to retrieve information on surface properties used in a SVAT model. *Journal of Hydrology*, **287**, 214–236.

Demarty, J., Ottlé, C., Braud, I., Olioso, A., Frangi, J. P., Gupta, H. V. and Bastidas, L. A. (2005). Constraining a physically based soil-vegetation-atmosphere transfer model with surface water content and thermal infrared brightness temperature measurements using a multiobjective approach. *Water Resources Research*, **41**, W01011.

Dente, L., Rinaldi, M., Mattia, F. and Satalino, G. (2004). Assimilation of polarimetric C-band radar data into CERES-Wheat model. In *IGARSS, 2004*. Anchorage, Alaska, pp. 1284–1287.

Dente, L., Satalino, G., Mattia, F. and Rinaldi, M. (2008). Assimilation of leaf area index derived from ASAR and MERIS data into CERES-Wheat model to map wheat yield. *Remote Sensing of Environment*, **112**, 1395–1407.

Doraiswamy, P. C., Moulin, S., Cook, P. W. and Stern, A. (2003). Crop yield assessment from remote sensing. *Photogrammetric Engineering and Remote Sensing*, **69**, 665–674.

Doraiswamy, P. C., Hatfield, J. L., Jackson, T. J., Akhmedov, B., Prueger, J. and Stern, A. (2004). Crop condition and yield simulation using landsat and MODIS. *Remote Sensing of Environment*, **92**, 548–559.

Doraiswamy, P. C., Sinclair, T. R., Hollinger, S., Akhmedov, B., Stern, A. and Prueger, J. (2005). Application of MODIS derived parameters for regional crop yield assessment. *Remote Sensing of Environment*, **97**, 192–202.

Droogers, P. and Bastiaanssen, W. (2002). Irrigation performance using hydrological and remote sensing modeling. *Journal of Irrigation and Drainage Engineering*, **128**, 11–18.

Duan, Q., Sorooshian, S. and Gupta, V. (1992). Effective and efficient global optimization for conceptual rainfall-runoff models. *Water Resources Research*, **28**, 1015–1031.

Duan, Q. Y., Gupta, V. K. and Sorooshian, S. (1993). Shuffled complex evolution approach for effective and efficient global minimization. *Journal of Optimization Theory and Applications*, **76**, 501–521.

El Hajj, M., Bégué, A., Guillaume, S. and Martiné, J.-F. (2009). Integrating SPOT-5 time series, crop growth modeling and expert knowledge for monitoring agricultural practices — The case of sugarcane harvest on Reunion Island. *Remote Sensing of Environment*, **113**, 2052–2061.

Fang, H., Liang, S., Hoogenboom, G., Teasdale, J. and Cavigelli, M. (2008). Corn yield estimation through assimilation of remote sensed data into the CSM-CERES-Maize model. *International Journal of Remote Sensing*, **29**, 3011–3032.

Fang, H., Liang, S. and Hoogenboom, G. (2011). Integration of MODIS LAI and vegetation index products with the CSM-CERES-maize model for corn yield estimation. *International Journal of Remote Sensing*, **32**, 1039–1065.

François, C., Cayrol, P., Kergoat, L. and Moulin, S. (2001). Assimilation techniques of remote sensing measurements into vegetation models: Overview, limits and promises. In *8th International Symposium on Physical Measurements and Signatures in Remote Sensing*, Aussois, France, pp. 649–658.

Goudriaan, J. (1977). *Crop Micrometeorology: A Simulation Study (Simulation Monographs)*. Wageningen: Pudoc, Center for Agricultural Publishing and Documentation.

Guérif, M. and Duke, C. L. (2000). Adjustment procedures of a crop model to the site specific characteristics of soil and crop using remote sensing data assimilation. *Agriculture, Ecosystems and Environment*, **81**, 57–69.

Guissard, V., Defourny, P. and Ledent, J. F. (2004). Crop specific information extraction based on coarse resolution pixel sampling. In *VEGETATION 2004 Proceedings*, Antwerpen.

Guissard, V., Lucau-Danila, C. and Defourny, P. (2005). Crop specific LAI retrieval using optical and radar satellite data for regional crop growth monitoring and modeling. In Owe, M. and D'Urso, G. (eds.), *Proceedings of the SPIE*, pp. 198–209.

Hadria, R., Duchemin, B., Lahrouni, A., Khabba, S., Er-raki, S., Dedieu, G., Chehbouni, A. G. and Olioso, A. (2006). Monitoring of irrigated wheat in a semi-arid climate using crop modelling and remote sensing data: Impact of satellite revisit time frequency. *International Journal of Remote Sensing*, **27**, 1093–1117.

Hammer, G. L. (1998). Crop modelling: Current status and opportunities to advance. *ISHS Acta Horticulturae*, **456**, 27–36.

Heinemann, A. B., Hoogenboom, G. and Chojnicki, B. (2002). The impact of potential errors in rainfall observation on the simulation of crop growth, development, and yield. *Ecological Modeling*, **157**, 1–21.

Hoogenboom, G., Wilkens, P. W. and Tsuji, G. Y. (eds.) (1999). *Decision Support System for Agrotechnology Transfer*, Version 3, Volume 4. Honolulu, Hawaii: University of Hawaii.

Hoogenboom, G., Jones, J. W., Porter, C. H., Wilkens, P. W., Boote, K. J., Batchelor, W. D., Hunt, L. A. and Tsuji, G. Y. (2003). *Decision Support System for Agrotechnology Transfer*, Version 4.0, Vol. 1: Overview. Honolulu, Hawaii: University of Hawaii.

Hoogenboom, G., Jones, J. W., Wilkens, P. W. *et al.* (2004). *Decision Support System for Agrotechnology Transfer*, Version 4.0 [CD-ROM]. Honolulu, Hawaii: University of Hawaii.

Inoue, Y., Moran, M. S. and Horie, T. (1998). Analysis of spectral measurements in paddy field for predicting rice growth and yield based on a simple crop simulation model. *Plant Production Science*, **1**, 269–279.

Jarlan, L., Mougin, E., Mazzega, P., Schoenauer, M., Tracol, Y. and Hiernaux, P. (2005). Using coarse remote sensing radar observations to control the trajectory of a simple Sahelian land surface model. *Remote Sensing of Environment*, **94**, 269–285.

Jiang, D., Wang, N. B., Yang, X. H. and Wang, J. H. (2003). Study on the interaction between NDVI profile and the growing status of crops. *Chinese Geographical Science*, **13**, 62–65.

Jin, H., Wang, J., Bo, Y., Chen, G. and Xue, H. (2010). Data assimilation of MODIS and TM observations into CERES-Maize model to estimate regional maize yield. In *Proceedings of the SPIE*, Gao, W., Jackson, T. J. and Wang, J. (eds.), pp. 780908–780910.

Jones, J. W., Hoogenboom, G., Porter, C. H., Boote, K. J., Batchelor, W. D., Hunt, L. A., Wilkens, P. W., Singh, U., Gijsman, A. J. and Ritchie, J. T. (2003). The DSSAT cropping system model. *European Journal of Agronomy*, **18**, 235–265.

Jongschaap, R. E. E. and Schouten, L. S. M. (2005). Predicting wheat production at regional scale by integration of remote sensing data with a simulation model. *Agronomy for Sustainable Development*, **25**, 481–489.

Jongschaap, R. E. E. (2006). Run-time calibration of simulation models by integrating remote sensing estimates of leaf area index and canopy nitrogen. *European Journal of Agronomy*, **24**, 316–324.

Justice, C. O., Vermote, E., Townshend, J. R. G. *et al.* (1998). The moderate resolution imaging spectroradiometer (MODIS): Land remote sensing for global change research. *IEEE Transactions on Geoscience and Remote Sensing*, **36**, 1228–1249.

Kang, S., Running, S. W., Lim, J.-H., Zhao, M., Park, C.-R. and Loehman, R. (2003). A regional phenology model for detecting onset of greenness in temperate mixed forests, Korea: An application of MODIS leaf area index. *Remote Sensing of Environment*, **86**, 232–242.

Kiniry, J. R., Williams, J. R., Vanderlip, R. L., Atwood, J. D., Reicosky, D. C., Mulliken, J., Cox, W. J., Mascagni, H. J. J., Hollinger, S. E. and Wiebold, W. J. (1997). Evaluation of two maize models for nine US locations. *Agronomy Journal*, **89**, 421–426.

Kogan, F., Yang, B., Guo, W., Pei, Z. and Jiao, X. (2005). Modelling corn production in China using AVHRR-based vegetation health indices. *International Journal of Remote Sensing*, **26**, 2325–2336.

Kravchenko, A. N. and Bullock, D. G. (2000). Correlation of corn and soybean grain yield with topography and soil properties. *Agronomy Journal*, **92**, 75–83.

Kravchenko, A. N., Bullock, D. G. and Boast, C. W. (2000). Joint multifractal analysis of crop yield and terrain slope. *Agronomy Journal*, **92**, 1279–1290.

Kuusk, A. (1998). Monitoring of vegetation parameters on large areas by the inversion of a canopy reflectance model. *International Journal of Remote Sensing*, **19**, 2893–2905.

Liang, S. (2004). *Quantitative Remote Sensing of Land Surfaces*. New York: John Wiley and Sons, Inc.

Liang, S. and Qin, J. (2008). Data assimilation methods for land surface variable estimation. In *Advances in Land Remote Sensing: System, Modeling, Inversion and Application*, Liang, S. (ed.), pp. 319–339. New York: Springer.

Lizaso, J. I., Batchelor, W. D. and Adams, S. S. (2001). Alternate approach to improve kernel number calculation in CERES-maize. *Transactions of the ASAE*, **44**, 1011–1018.

Ma, Y., Wang, S., Zhang, L., Hou, Y., Zhuang, L., He, Y. and Wang, F. (2008). Monitoring winter wheat growth in North China by combining a crop model and remote sensing data. *International Journal of Applied Earth Observation and Geoinformation*, **10**, 426–437.

Maas, S. J. (1988a). Using satellite data to improve model estimates of crop yield. *Agronomy Journal*, **80**, 662–665.

Maas, S. J. (1988b). Use of remotely-sensed information in agricultural crop growth models. *Ecological Modeling*, **41**, 247–268.

MacDonald, R. and Hall, F. (1980). Global crop forecasting. *Science*, **208**, 670–679.

Mauser, W. and Bach, H. (2009). PROMET — Large scale distributed hydrological modelling to study the impact of climate change on the water flows of mountain watersheds *Journal of Hydrology*, **376**, 362–377.

Mo, X., Xiang, Y., McVicar, T. R., Liu, S., Lin, Z. and Xu, Y. (2005). Prediction of crop yield, water consumption and water use efficiency with a SVAT-crop growth model using remotely sensed data on the North China Plain. *Ecological Modeling*, **183**, 301–322.

Moran, M. S., Maas, S. J., Clarke, T. R., Pinter, P. J. J., Qi, J., Mitchell, T. A., Kimball, B. A. and Neale, C. M. U. (1996). Modeling/remote sensing approach for irrigation scheduling. In *Evapotranspiration and Irrigation Scheduling*, Camp, C., Sadler, E. and Yoder, R. (eds.), pp. 231–238. San Antonio, Texas: ASAE-IA-ICID, American Society of Agricultural Engineers.

Moulin, S., Bondeau, A. and Delécolle, R. (1998). Combining agricultural crop models and satellite observations: From field to regional scale. *International Journal of Remote Sensing*, **19**, 1021–1036.

Moulin, S., Kergoat, L., Cayrol, P., Dedieu, G. and Prévot, L. (2002). Calibration of a coupled canopy functioning and SVAT model in the ReSeDA experiment. Towards the assimilation of SPOT/HRV observations into the model. *Agronomie*, **22**, 681–686.

Myneni, R. B., Hoffman, S., Knyazikhin, Y. *et al.* (2002). Global products of vegetation leaf area and fraction absorbed PAR from year one of MODIS data. *Remote Sensing of Environment*, **83**, 214–231.

Nelder, J. A. and Mead, R. (1965). A simplex method for function minimization. *Computer Journal*, **7**, 308–313.

Noilhan, J. and Mahfouf, J.-F. (1996). The ISBA land surface parameterisation scheme. *Global Planetary Changes*, **13**, 145–159.

O'Neal, M. R., Frankenberger, J. R. and Ess, D. R. (2002). Use of CERES-maize to study effect of spatial precipitation variability on yield. *Agricultural Systems*, 73, 205–225.

Olioso, A., Chauki, H., Courault, D. and Wigneron, J. P. (1999). Estimation of evapotranspiration and photosynthesis by assimilation of remote sensing data into SVAT models. *Remote Sensing of Environment*, **68**, 341–356.

Olioso, A., Inoue, Y., Wigneron, J.-P., Ortega-Farias, O., Lecharpentier, P., Pardé, M., Calvet, J. C. and Inizan, O. (2001). Using a coupled crop-SVAT model to assess crop canopy processes from remote sensing data. In *IGARSS*, pp. 3006–3008. Sydney, Australia: IEEE Publications.

Olioso, A., Inoue, Y., Ortega-Farias, S., Demarty, J., Wigneron, J. P., Braud, I., Jacob, F., Lecharpentier, P., Ottlé, C., Calvet, J. C. and Brisson, N. (2005). Future directions for advanced evapotranspiration modeling: Assimilation of remote sensing data into crop simulation models and SVAT models. *Irrigation and Drainage Systems*, **19**, 377–412.

Oppelt, N. M. (2010). Use of remote sensing data to assist crop modeling. *Journal of Applied Remote Sensing*, **4**, 041896–041816.

Pauwels, V. R. N., Verhoest, N. E. C., De Lannoy, G. J. M., Guissard, V., Lucau, C. and Defourny, P. (2007). Optimization of a coupled hydrology-crop growth model through the assimilation of observed soil moisture and leaf area index values using an ensemble Kalman filter. *Water Resources Research*, **43**, W04421, doi: 10.1029/2006WR004942.

Plummer, S. E. (2000). Perspectives on combining ecological process models and remotely sensed data. *Ecological Modeling*, **129**, 169–186.

Prévot, L., Chauki, H., Troufleau, D., Weiss, M., Baret, F. and Brisson, N. (2003). Assimilating optical and radar data into STICS crop model for wheat. *Agronomie*, **23**, 297–303.

Press, W. H., Teukolsky, S. A., Vetterling, W. T. and Flannery, B. P. (1992). *Numerical Recipes in Fortran 77: The Art of Scientific Computing*. New York: Cambridge University Press.

Priestley, C. H. B. and Taylor, R. J. (1972). On the assessment of surface heat flux and evaporation using large-scale parameters. *Monthly Weather Review*, **100**, 81–92.

Reichle, R. H., Walker, J. P., Koster, R. D. and Houser, P. R. (2003). Extended versus ensemble Kalman filtering for land data assimilation. *Journal of Hydrometeorology*, **4**, 728–740.

Rodriguez, J. C., Garatuza, J., Chehbouni, A., Khabba, S., Boulet, G., Palacios, E., Lahrouni, A., Duchemin, B., Hadria, R. and Watts, C. (2004). Wheat yield estimation using remote sensing and the STICS model in the semiarid Yaqui valley, Mexico. *Agronomie*, **24**, 295–304.

Roebeling, R. A., Van Putten, E., Genovese, G. and Rosema, A. (2004). Application of Meteosat derived meteorological information for crop yield predictions in Europe. *International Journal of Remote Sensing*, **25**, 5389–5401.

Ruget, F., Brisson, N., Delécolle, R. and Faivre, R. (2002). Sensibility analysis of a crop simulation model, STICS in order to choose the main parameters to be estimated. *Agronomie*, **22**, 133–158.

Savitzky, A. and Golay, M. J. E. (1964). Smoothing and differentiation of data by simplified least squares procedures. *Analytical Chemistry*, **36**, 1627–1639.

Schmugge, T. J. and Kustas, W. P. (1999). Radiometry at infrared wavelengths for agricultural applications. *Agronomie*, **19**, 83–96.

Schneider, K. (2003). Assimilating remote sensing data into a land-surface process model. *International Journal of Remote Sensing*, **24**, 2959–2980.

Schuurmans, J. M., Troch, P. A., Veldhuizen, A. A., Bastiaanssen, W. G. M. and Bierkens, M. F. P. (2003). Assimilation of remotely sensed latent heat flux in a distributed hydrological model. *Advances in Water Resources*, **26**, 151–159.

Shen, S. H., Yang, S. B., Li, B. B., Tan, B. X., Li, Z. Y. and Thuy, L. T. (2009). A scheme for regional rice yield estimation using ENVISAT ASAR data. *Science in China, Series D, Earth Sciences*, **39**, 763–773.

Sinclair, T. R. and Seligman, N. A. G. (1996). Crop modeling: From infancy to maturity. *Agronomy Journal*, **88**, 698–704.

Singh, R., Goyal, R. C., Saha, S. K. and Chhikara, R. S. (1992). Use of satellite spectral data in crop yield estimation surveys. *International Journal of Remote Sensing*, **14**, 2583–2592.

Stol, W., Rouse, D. I., Van Kraalingen, D. W. G. and Klepper, O. (1992). *FSEOPT A FORTRAN Program for Calibration and Uncertainty Analysis of Simulation Models*, p. 24. Wageningen: CABO-DLO and Agricultural University.

Thenkabail, P. S., Ward, A. D., Lyon, J. G. and Van Deventer, P. (1992). Landsat thematic mapper indices for evaluating management and growth characteristics of soybeans and corn. *Transactions of the ASAE*, **35**, 1441–1448.

Timsina, J. and Humphreys, E. (2006). Performance of CERES-rice and CERES-wheat models in rice-wheat systems: A review. *Agricultural Systems*, **90**, 5–31.

Todoroff, P., De Robillard, F. and Laurent, J.-B. (2010). Interconnection of a crop growth model with remote sensing data to estimate the total available water content of soils. In *International Geoscience and Remote Sensing Symposium*. Honolulu, Hawaii, USA.

Tojo Soler, C. M., Sentelhas, P. C. and Hoogenboom, G. (2007). Application of the CSM-CERES-maize model for planting date evaluation and yield forecasting for maize grown off-season in a subtropical environment. *European Journal of Agronomy*, **27**, 165–177.

Tsuji, G. Y., Uehara, G. and Balas, S. (1994). *Decision Support System for Agrotechnology Transfer*, Version 3. Honolulu, Hawaii: University of Hawaii.

Tsuji, G. Y., Hoogenboom, G. and Thornton, P. K. (1998). *Understanding Options for Agricultural Production. System Approaches for Sustainable Agricultural Development*. Dordrecht, The Netherlands: Kluwer Academic Publishers.

Tubiello, F. N., Rosenzweig, C., Kimball, B. A., Pinter, P. J. J., Wall, G. W., Hunsaker, D. J., LaMorte, R. L. and Garcia, R. L. (1999). Testing CERES-wheat with free air carbon dioxide enrichment (FACE) experiment data: CO_2 and water interactions. *Agronomy Journal*, **91**, 247–255.

Vazifedoust, M., Van Dam, J. C., Bastiaanssen, W. G. M. and Reddes, R. A. (2009). Assimilation of satellite data into agrohydrological models to improve crop yield forecasts. *International Journal of Remote Sensing*, **30**, 2523–2545.

Verhoef, W. (1984). Light scattering by leaf layers with application to canopy reflectance modeling: The SAIL model. *Remote Sensing of Environment*, **16**, 125–141.

Wardlow, B. D. and Egbert, S. L. (2008). Large-area crop mapping using time-series MODIS 250 m NDVI data: An assessment for the US Central Great Plains. *Remote Sensing of Environment*, **112**, 1096–1116.

Wardlow, B. D. and Egbert, S. L. (2010). A comparison of MODIS 250-m EVI and NDVI data for crop mapping: A case study for southwest Kansas. *International Journal of Remote Sensing*, **31**, 805–830.

Weiss, M., Troufleau, D., Baret, F., Chauki, H., Prévot, L., Olioso, A., Bruguier, N. and Brisson, N. (2001). Coupling canopy functioning and radiative transfer models for remote sensing data assimilation. *Agricultural and Forest Meteorology*, **108**, 113–128.

Weissteiner, C. J. and Kühbauch, W. (2005). Regional yield forecasts of malting barley (Hordeum vulgare L.) by NOAA-AVHRR remote sensing data and ancillary data. *Journal of Agronomy and Crop Science*, **191**, 308–320.

Wigneron, J. P., Chanzy, A., Calvet, J. C., Olioso, A. and Kerr, Y. (2002). Modeling approaches to assimilating L-band passive microwave observations over land surfaces. *Journal of Geophysical Research*, **107**, doi:10.1029/2001JD000958.

Xie, Y., Kiniry, J. R., Nedbalek, V. and Rosenthal, W. D. (2001). Maize and sorghum simulations with CERES-maize, SORKAM, and ALMANAC under water-limiting conditions. *Agronomy Journal*, **93**, 1148–1155.

Zhang, X., Friedl, M. A., Schaaf, C. B., Strahler, A. H., Hodges, J., Gao, F., Reed, B. and Huete, A. (2003). Monitoring vegetation phenology using MODIS. *Remote Sensing of Environment*, **84**, 471–475.

Chapter 14

Simultaneous State-Parameter Estimation for Hydrologic Modeling Using Ensemble Kalman Filter

Xianhong Xie

College of Global Change and Earth System Science, Beijing Normal University, China

This chapter presents another merit of the Ensemble Kalman filter (EnKF) for state-parameter estimation of hydrologic models. EnKF provides a general framework to account for various uncertainties on one hand, and on the other, the hydrologic models hold large number of parameters and states to be identified and calibrated, which therefore limits the reliability of hydrologic modeling. With the framework of the EnKF, a state-augmentation technique is used to combine the states and parameters which are estimated simultaneously. This technique is demonstrated with a simple two-parameter rainfall-runoff model and a distributed hydrological model, Soil and Water Assessment Tool (SWAT). By assimilating runoff and other measurements, its sensitivities are analyzed with respect to the error specification, the initial realization, and the ensemble size. It is found that the state augmentation provides an efficient approach to obtain a set of acceptable model parameters and satisfactory hydrologic states. Sensitivity studies indicate the importance of consistent error specification and the less important setting for initial parameter distributions. We also caution that this technique may relevant to single-type parameter estimation rather than multi-type parameter problems.

14.1 Introduction

The hydrologic model is a critical important tool to characterize the nature hydrologic system, especially the distributed hydrological models that are capable of characterizing spatial patterns of catchments and useful for water resources management and environment remediation. But the

S. Liang, X. Li & X. Xie (eds.), *Land Surface Observation, Modeling and Data Assimilation*, 439–462.

modeling behavior is usually, even always, spoiled by errors resulting from the model structure, the input, and the model parameters. As a result, the uncertainty from parameter gains much more attention due to the fact that the parameter estimation can exploit a number of optimization methods and thereby reduce the parameter uncertainty. The parameter estimation procedure, termed as model calibration, aims to minimize the discrepancy between simulated and observed model outputs and is a mandatory element of a good modeling practice.

In the last few decades, the number of automatic calibration methods has surged and can be summarized into two types. One is based on an assumption which states that only one global optimal solution exists in the feasible domain of parameters and this solution can be detected by efficient searching algorithms. A typical algorithm of such types is the heuristic searching, encompassing the shuffled complex evolution (SCE-UA) and its variants (Chu *et al.*, 2010; Duan *et al.*, 1992, 1993; Vrugt *et al.*, 2003). The other type is based on an assumption that maintains a broad range of solutions exists and the solutions can be captured by an appropriate probability distribution. This type is greatly expressed in the generalized likelihood uncertainty estimation (GLUE) method (Beven and Binley, 1992; Beven and Freer, 2001). An advantage of GLUE is that it recognizes the equivalence or near-equivalence of different sets of parameters in the calibration of models (Beven and Binley, 1992). A comprehensive review on these methods can be referred to the Chap. 5 of this book and a book by Wang (2011). But a typical drawback of these methods is that they attribute all errors/uncertainties to the parameters.

In a separate research line, data assimilation methods show the strength that it accounts for various uncertainties in modeling. Typically, the sequential assimilation methods (e.g., the Ensemble Kalman filter (EnKF)) have been widely used. But the traditional data assimilation focuses on state estimation alone rather than on adequate consideration of the uncertainties associate with all sources in hydrologic modeling (Liu and Gupta, 2007). To realize the simultaneous state-parameter estimation, a dual state-parameter estimation strategy was developed based on the EnKF and has been demonstrated in conceptual rainfall-runoff models and ecosystem models (Moradkhani *et al.*, 2005). And then Xie and Zhang (2010) proposed a state-augmentation technique to estimate dynamic states and parameters simultaneously and thereby reduce the uncertainties of hydrologic prediction. This technique has been successfully demonstrated with a distributed hydrologic model.

The purpose of this chapter is to provide an extended illustration to the EnKF-based state augmentation technique for simultaneous state-parameter estimation. The theoretical framework is presented in Sec. 14.2. And then two cases are shown to examine the effectiveness of the EnKF-based state augmentation technique: in Sec. 14.3 focuses on a simple rainfall-runoff conceptual model and in Sec. 14.4 intends to apply the technique in a distributed hydrologic model. In Sec. 14.5, a discussion is given. Section 14.6 will close this chapter with summary and conclusions.

14.2 EnKF with State-Augmentation Technique

The EnKF methodology has been interpreted in several chapters in this book. Here we briefly expressed it within the framework of rainfall-runoff modeling. The EnKF operates sequentially by performing in turn a forecast step and a filter update step. In general, the forecast step for an ensemble member i between times t and $t-1$ can be written as

$$x_{i,t}^f = M(x_{i,t-1}^a, \lambda_{i,t-1}^a, u_{i,t}) + \omega_{i,t}, \quad \omega_{i,t} \sim N(0, Q_t) \tag{1}$$

$$u_{i,t} = u_t + \zeta_{i,t}, \quad \zeta_{i,t} \sim N(0, U_t) \tag{2}$$

where $x_{i,t}^f$ and $x_{i,t-1}^a$ are the forecast state vector at time t and the updated state vector at time $t-1$, respectively; $M(\cdot)$ stands for a nonlinear model operator; $\lambda_{i,t-1}^a$ is the model parameter vector; $u_{i,t}$ is the forcing input vector. The model error vector $\omega_{i,t}$ (or perturbation vector) and the noise term of the forcing input vector $\zeta_{i,t}$ are assumed to obey Gaussian distribution with zero means and covariances Q_t and U_t, respectively. Note that state vector x composes of the water storage state of the soil column (e.g., the soil water content) and the hydrologic response/output of the hydrologic system such as the runoff/streamflow and the evapotranspiration (ET).

As a conceptual extension, one might wish more generally to permit the system characteristics represented as "parameters" to vary slowly with time, in response to changes in the model state and/or system inputs (Liu and Gupta, 2007). To represent the slow evolution, a Gaussian random walk process is often used:

$$\lambda_{i,t}^f = \lambda_{i,t-1}^a + \varepsilon_{i,t}, \quad \varepsilon_{i,t} \sim N(0, E_t) \tag{3}$$

where $\lambda_{i,t}^f$ and $\lambda_{i,t-1}^a$ are the forecast parameter vector and the updated parameter vetor and ε is the random vectors associated with parameters.

A drawback of such parameter evolution is the overdispersion of parameter samples and loss of information between time points when the parameters are considered to be fixed (Moradkhani *et al.*, 2005). Alternatively, Kernel smoothing of parameter samples may be a remedy to this problem (Liu, 2000; Moradkhani *et al.*, 2005). But the Kernel smoothing method, employing a shrinkage factor that is prescribed experientially to restrict the sampling, could induce unstable estimation for the model parameters in the data assimilation process due to giving biased value of the factor. For these reasons, we make an extreme but simple and effective setting in which the parameters are not be forecasted, i.e., the covariance in Eq. (3) is $E_t \equiv 0$. Certainly, the application of the Gaussian random walk sampling or the Kernel smoothing sampling may be a good attempt.

To simultaneously estimate the dynamic states and model parameters, the state vector is augmented to include the parameter vector. Let $X^f_{i,t} = [x^f_{i,t}, \lambda^f_{i,t}]^T$, then the filter update step at time t can be expressed as

$$X^a_{i,t} = X^f_{i,t} + K_t(y_{i,t} - H(X^f_{i,t})) \tag{4}$$

where $y_{i,t}$ denotes the ith observation sample suitably generated by adding Gaussian noise to the actual observation, $H(\cdot)$ is the observation operator that maps the model states to the observation space. The difference of the two terms in the bracket of Eq. (4) is defined as a filter innovation. If the parameters are not assumed to vary in the forecasting step, then the $\lambda^f_{i,t}$ in Eq. (3) is expressed as $\lambda^f_{i,t} = \lambda^a_{i,t-1}$. The Kalman gain matrix K_t is given by

$$K_t = P_t H^T (H P_t H^T + R_t)^{-1} \tag{5}$$

where R_t is the observation error covariance that was used to quantify the observation uncertainties and to generate observation samples; P_t is the forecast error covariance that is calculated based on the extended state vector, $X^f_{i,t}$. It is not necessary to directly compute this forecast error covariance, and please refer to Evensen (2003) for detailed computation of the error covariance and the Kalman gain matrix.

Obviously, the correlation between the observed states (e.g., the runoff, soil water content) and the model parameters are expressed in the forecast error covariance P_{t+1}. Depending on this correlation, the unobserved states and the parameters will be updated sequentially when observations available. When two conditions are hold, i.e., the hydrologic model is linear and the modeling and observation errors is Gaussian, this state-augmentation technique will achieve optimal estimates for the states and parameters. As to a nonlinear and non-Gaussian problem, the optimal results may not render. In this case, however, a number of studies showed that the EnKF-based technique is still capable of achieving favorite state-parameter estimation (e.g., Chen and Zhang, 2006; Wang *et al.*, 2009; Xie and Zhang, 2010). So the condition of linear and Gaussian assumption can be relaxed for an operational hydrologic prediction system in which sub-optimal estimation for states and parameters is acceptable. That is why EnKF-based data assimilation method is still relevant for the nonlinear hydrologic problems.

14.3 Case Study for a Simple Rainfall-Runoff Model

We first present an illustrative case for the state-augmentation technique using the simple two-parameter reservoir model (TWOPAR) (Duan *et al.*, 1992; Sorooshian and Gupta, 1983), which bears the general characteristics of many conceptual hydrologic models, such as the Sacramento Soil Moisture Accounting Model (Burnash *et al.*, 1973). As shown in Fig. 14.1, this model holds two parameters, i.e., the maximum storage M (unit: length L) and the recession rate of baseflow K (unit: time inverse T^{-1}), and one dynamic state S_t (unit: length L) that represents the water storage condition of a catchment, so the total runoff can be expressed as $Q_t = f(S_t, M, K, P_t)$, where P_t (unit: length L) is the net precipitation.

A synthetic experiment is designed to evaluate the data assimilation behavior. Before this experiment, we perform a reference simulation. Let the parameters M and K take typical values, i.e., $M = 20.0$ and $K = 0.3$ (as shown in Duan *et al.*, 1993), and the initial state $S_0 = 10.0$, P_t from a precipitation series. By running the model with the reference setting, we obtain a series of total runoff Q_t, which will be viewed as observations

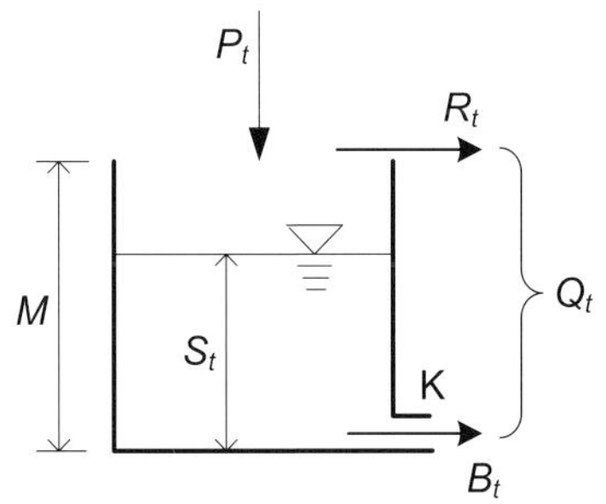

Fig. 14.1. The simple reservoir model TWOPAR. Adapted from Duan *et al.* (1992). Copyright 1992, John Wiley & Sons, Inc.

to be assimilated, adding noise if desired. The reason behind using this kind of synthetic experiment is that it will facilitate the evaluation of data assimilation due to the fact that the true parameter values are known.

The state vector is augmented to include the two parameters $X_t = [S_t, Q_t, M_t, K_t]^T$, and the observation operator is $H_t = [0, 1, 0, 0]^T$. The initial realizations of the parameters and the state are all prescribed with Gaussian distribution, i.e., $M_0 \sim N(30, 5^2)$, $K_0 \sim N(0.55, 0.2^2)$, $S_0 \sim N(20, 5^2)$, which are obviously biased to the reference values. The means of these distributions are referred to as prior estimates of the parameters and the state. To represent the observation and model uncertainties, moreover, we subjectively give them Gaussian noises, $N(0, 1.0^2)$, $N(0, 0.10^2)$, respectively. Based on these distributions, a hundred of realizations are generated to construct the ensemble.

Figure 14.2 shows the total runoff errors (estimates − observations) from the data assimilation and a simulation with the prior estimates of the parameters and the state. The data assimilation dramatically reduces the total runoff estimate errors, especially after the 130th step at which the surface runoff is activated due to the occurrence of large rainfall around the day. The generated surface runoff will make the parameter M be calibrated and consequently improve the runoff estimation. Figure 14.3 exhibits the two parameter estimations. The estimates of the parameter

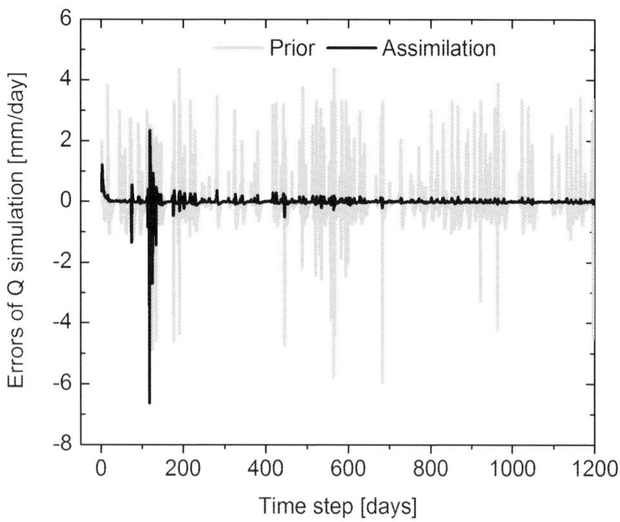

Fig. 14.2. Total runoff estimate errors from the data assimilation and the single model-run simulation given prior estimates of parameters.

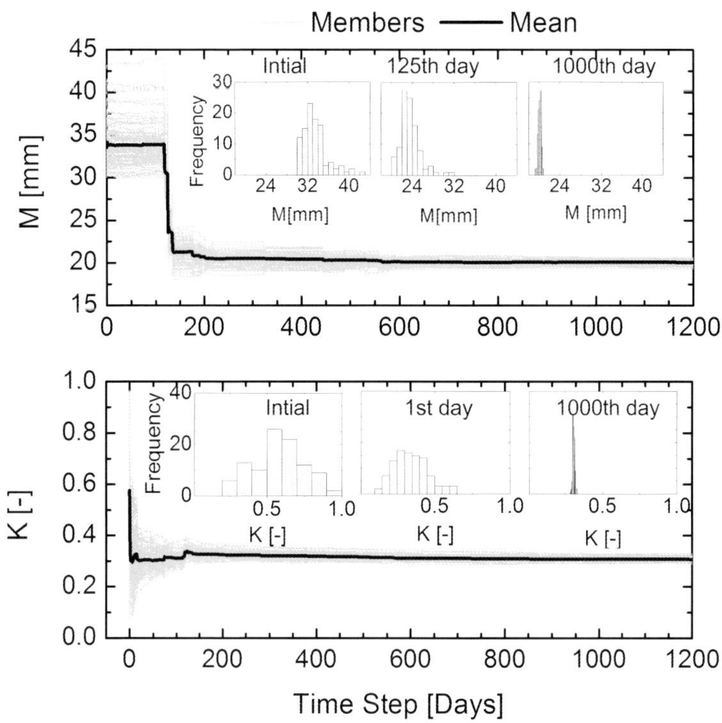

Fig. 14.3. Parameter estimation within the progress of data assimilation.

M keep constant somewhat before the surface runoff is activated, but they quickly approach to the true values around the 130th step. While the estimates of parameter K converge to the truth even after one step of updating.

For this simple rainfall-runoff model, we have also conducted many other experiments by using different reference parameter values and different combinations of modeling and observation errors and/or ensemble size. Most of the experiments provide acceptable estimates for the parameters and the state (the result not shown). So we conclude the success and robustness of the state augmentation technique for this problem. But we could not push this conclusion of reasoning too far, because the model used here is so simple that it is close to linearity between the dynamic state and parameters. When strong nonlinearity exists in a hydrologic model, this rough state augmentation technique may not perform well. To this point, a short discussion will be given in Sec. 14.5.

14.4 Application to a Distributed Hydrologic Model

14.4.1 *SWAT model*

The data assimilation algorithm is implemented on a popular distributed hydrological model, Soil and Water Assessment Tool (SWAT). Since it is physically-based and computationally efficient, uses readily available inputs, and enables users to study long-term impacts, this model has been widely used to predict impacts of land management practices on water, sediment, and agricultural chemical yields in large, complex watersheds with varying soils, land use, and management conditions over long periods of time (Di Luzio *et al.*, 2001; Gassman *et al.*, 2007; Neitsch *et al.*, 2001). For modeling purposes, a catchment is partitioned into a number of sub-catchments or sub-basins according to the property of Digital Elevation Model (DEM) data, and then input information for each sub-basin is grouped or organized into different hydrologic response units (HRUs) that are composed of unique land cover, soil, and management combinations.

The land phase of the hydrologic cycle is simulated at the HRU scale based on a water balance equation with a daily step:

$$SW_t = SW_0 + \sum_{i=1}^{t} (R_i - Q_{\text{surf},i} - ET_i - W_{\text{seep},i} - Q_{gw,i}) \tag{6}$$

where SW_t is the final soil water content (mm H_2O), SW_0 is the initial soil water content (mm H_2O), t is the time (days), R_i is the amount of precipitation on day i (mm H_2O), $Q_{\text{surf},i}$ is the amount of surface runoff on day i (mm H_2O), ET_i is the amount of evapotranspiration on day i (mm H_2O), $W_{\text{seep},i}$ is the amount of percolation and bypass flow exiting the soil profile bottom on day i (mm H_2O), and $Q_{gw,i}$ is the amount of return flow on day i (mm H_2O). In this equation, the term of $Q_{\text{surf},i}$ is the main component that determines streamflow in reaches and the soil moisture in soil profiles. It can be expressed with an empirical model, SCS runoff equation:

$$Q_{\text{surf},i} = \frac{(R_i - 0.2S_i)^2}{R_i + 0.8S_i} \tag{7}$$

where S_i is the retention storage of soil profiles (mm H_2O). It is empirically defined as:

$$S_i = \frac{25400}{CN_i} - 254 \tag{8}$$

where CN_i is the temporal curve number for the day. In order to capture the temporal-spatial variations of the retention storage (S_i) due to changes in soil water content, soil properties, land use, management and slope, the temporal curve number is implicitly expressed as:

$$CN_i = f(CN_2, SW_i) \tag{9}$$

where CN_2 is the static curve number corresponding to the average moisture condition, and it depends on the soil's permeability, land use, and antecedent soil water conditions (Neitsch *et al.*, 2001). Its value is typically between 30 and 100. It is a dominant parameter for the surface runoff generation in an HRU and consequently influences the streamflow process and the soil moisture states.

Although the SWAT model is capable of simulating other processes, such as the plant growth and the sediment movement, here we exclusively focus on the water movement in catchments. Eleven primary state variables are screened out to be included in the augmented state vector (Xie and Zhang, 2010). In these state variables, three model output, i.e., Runoff (R), Soil water content (SW), and Evapotranspiration (ET) are viewed as observable or prognostic states since they could be measured with specific instruments or techniques. Even there are a large number of static variables (parameters) needed to be identified for a simulation application, and each of them may make a different contribution. However, the SCS runoff curve number, CN_2, is the most sensitive parameter and its uncertainty could induce great effects on runoff modeling and other hydrological processes (Immerzeel and Droogers, 2008; Muleta and Nicklow, 2005; Rouhani *et al.*, 2007; Shen *et al.*, 2008). It is often needed to be estimated by calibration methods or data assimilation methods. In the SWAT model, it should be mentioned that the curve number is used to compute the maximum retention value and other two shape coefficients (Neitsch *et al.*, 2001). This computational process is implicitly expressed with Eq. (9). The three parameters will be continuously corrected with the CN_2 updated in the data assimilation process.

14.4.2 *Data assimilation procedure*

In this study, an experimental catchment is adopted for the Lake Fork Watershed in Northeast Texas, USA. Based on a digital elevation model, the catchment is partitioned into 20 sub-basins and subsequently 49 HRUs according to the land use and soil type information. As shown in

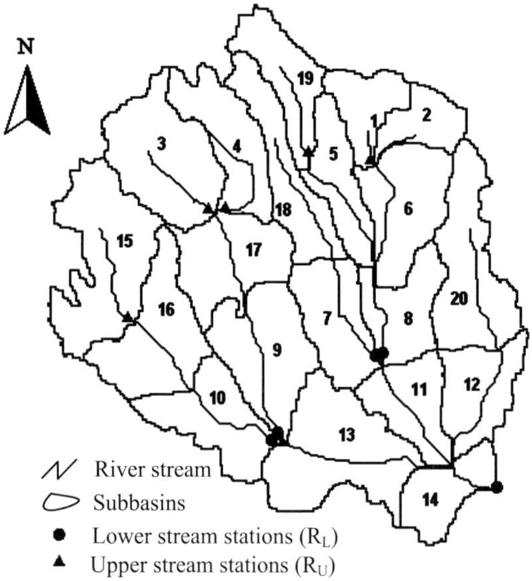

Fig. 14.4. Sub-basin distribution of the experimental catchment from Lake Fork Watershed in Northeast Texas, USA. Adapted from Xie and Zhang (2010). Copyright 2010, Elsevier.

Fig. 14.4, the serial number of sub-basins is not in a natural sequence because some outlets are added to get approximately uniform sub-basin areas.

For this experimental catchment, several synthetic experiments are designed to assess the capability and sensitivity of the data assimilation process. This is a general and effectual way to investigate data assimilation performance on predictive systems, since the true parameter values are known in the experiments. In these experiments, one reference field is randomly picked up from a predefined Gaussian distribution with given statistics. And then the hydrological model is integrated to obtain the reference state (or "true" state), referred to as the "control run" or reference modeling. At the measurement locations, the observations (e.g., the runoff) to be used in the assimilation system are drawn from the reference state. By merging the synthetic observations, the assimilation integrations are conducted based on a stochastic modeling platform.

This assimilation integration will experience three consecutive steps. The first is a warming-up process like any general hydrologic simulation. The second is the perturbation step, in which an ensemble of realizations is generated by perturbing the parameters and precipitation using predefined

Gaussian distribution. The aim of this step is to create reasonable distributions of hydrologic states to approximate the initial conditions and be beneficial to the subsequent data assimilation operations. And the third is the assimilation period the hydrological states are updated by assimilating observations drawn from the "control run". All the three steps are specified with different periods. It should be reminded that the data assimilation method used here focuses on the estimations of CN_2 and prognostic variables exclusively, and other parameters in the control run and assimilation run hold identical values.

The three prognostic variables, namely the runoff (R), evapotranspiration (ET), and soil water content (SW), are perturbed by additive white noises to represent the possible measurement and model errors. These noises are mutually uncorrelated in space and time with means equaling to zero (indicating unbiased estimations) and standard deviations scaling to the current values of variables. Likewise, the precipitations are perturbed, and the standard deviations also equal to the products of a scaling factor and the current precipitation values. Furthermore, two different scaling factors for precipitation perturbations could be used in the perturbation period and the assimilation period (period 3). Specifically, giving a larger scaling factor (here it is 0.30) can generate broad spreads of precipitation and hydrological response to account for uncertainty in the initial conditions when the prior knowledge is poor. In fact, the success of EnKF-based data assimilation method is quite dependent on the way of error prescription that should be consistent with the real uncertainties of the model and the measurement (Evensen, 2003; Pauwels and De Lannoy, 2006). Therefore, the errors prescribed with scaling factors have overestimated the real uncertainties of the interested hydrological system, and this inconsistency is favorable to examine the robustness of EnKF. All these prescriptions of errors are empirical to some degree but practical, and even a number of other ways may be also relevant.

14.4.3 *Result*

We first employ five runoff series, with respect to the sub-basins 7, 8, 9, 10, and 14, drawn from the reference simulation as observations to update the model states, since runoff is a common type of measurement in a catchment. In order to assess the capability of state augmentation technique, we also prescribe relatively small scaling factors for the standard deviations of the errors: 0.001 for the observed runoff and 0.01 for the precipitation and the

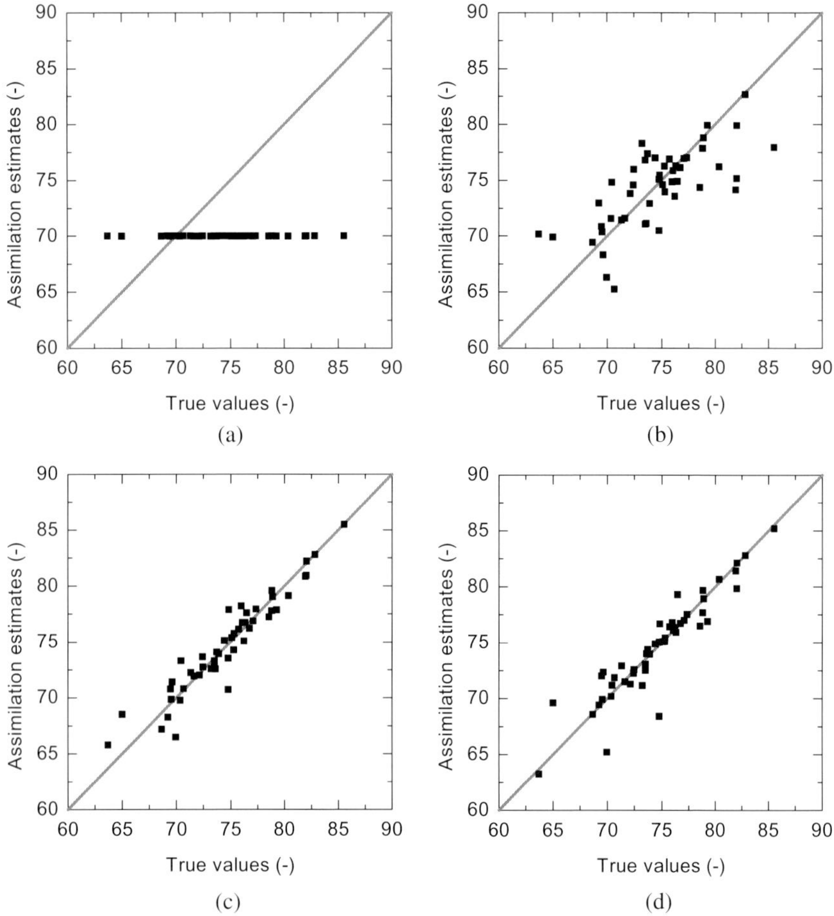

Fig. 14.5. Assimilation estimates of the CN_2s versus true (reference) values: (a) ensemble means of initial realizations; (b) ensemble means at the 60th assimilation step; (c) ensemble means at the 180th assimilation step, and (d) ensemble means at the 640th assimilation step (final results). Adapted from Xie and Zhang (2010). Copyright 2010, Elsevier.

three model output variables. The ensemble size is set as 200, which is affordable for a common personal computer.

The result of parameter estimation is shown in Fig. 14.5 as scatter-plots for the ensemble means of the CN_2s versus the reference field at the initial, the 60th, 180th, and 640th assimilation steps. It is apparent that the initial estimates of CN_2s are about 70 and do not have any resemblance of the reference field. But after a few times of data assimilation, the scatter points of the final result approach the 45° line, which indicates a good agreement between the ensemble means of CN_2s and the true values. Even just at the 180th assimilation step, the ensemble means capture the major features of the reference field, and there is a slight improvement for the CN_2 estimation in the rest of assimilation steps. It is of interest to note that some estimates are still discrepant to the true values, although the prognostic variables are matched very well. This implies the well-known non-uniqueness in the inverse problems. Specifically, the biased estimation values of CN_2s could provide acceptable hydrological responses, such as the runoff estimation. As typical examples taken from the seventh HRU in sub-basin 5 and the 28th HRU in sub-basin 10 (Fig. 14.6), the ensemble means of the CN_2s match the true values very well after 120 assimilation steps. In contrast, for the 31st HRU in the sub-basin 11, the value of CN_2 is underestimated, and the ensemble spreads to a broad range.

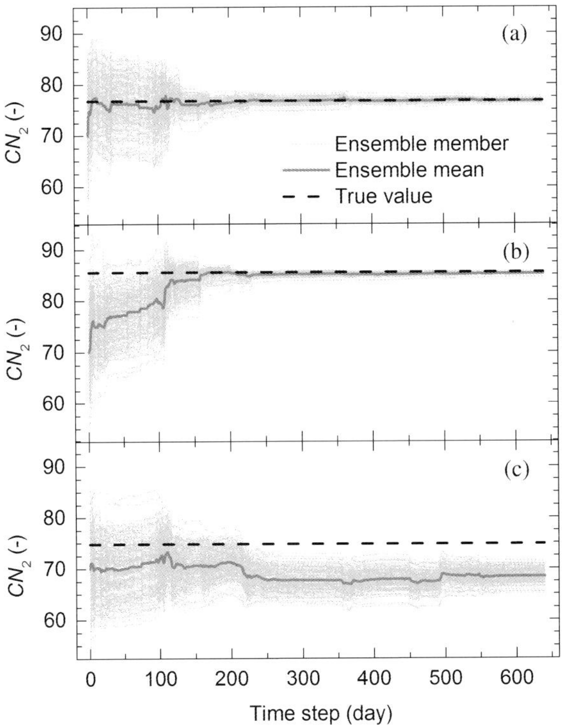

Fig. 14.6. Typical assimilation processes of the CN_2s: (a) for HRU 7, sub-basin 5; (b) for HRU 28, sub-basin 10; and (c) for HRU 31, sub-basin 11. Adapted from Xie and Zhang (2010). Copyright 2010, Elsevier.

Besides the success of parameter estimation, the data assimilation performs well for state variable estimation. A measure of root-mean-square errors (RMSEs) with respect to the true simulation is used to evaluate the performance of data assimilation. Moreover, an open-loop simulation is conducted using the same initial realizations of the CN_2s as well as the uncertain precipitation inputs but without assimilating the model with observations.

Figure 14.7 shows average time series of RMSEs of prognostic variables for the entire modeling domain. It is clear that the data assimilation reduces the RMSEs of the runoff and soil water content estimations. Especially in peak runoff occurring periods, the RMSEs are lower than that in open loop simulations. However, the assimilation does not give significant improvement for the evapotranspiration estimates as the two

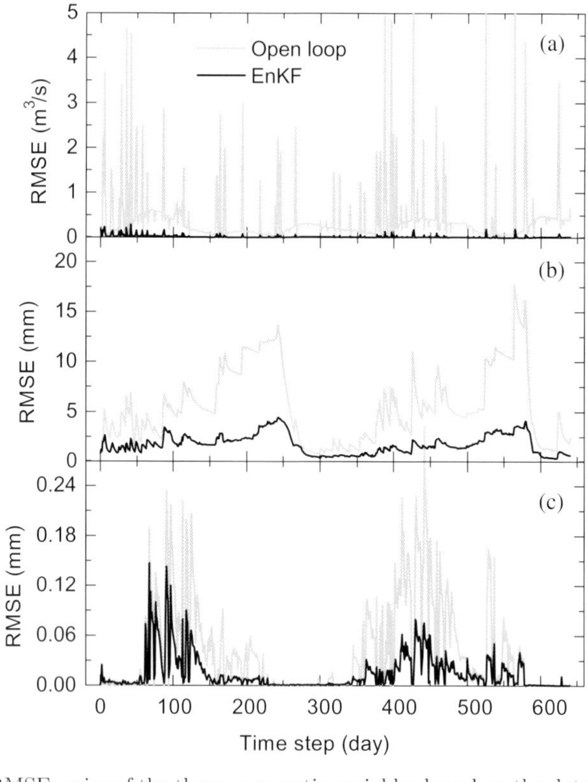

Fig. 14.7. RMSE series of the three prognostic variables based on the data assimilation and the open loop simulation: (a) stream runoff, (b) soil water content, and (c) evapotranspiration. Adapted from Xie and Zhang (2010). Copyright 2010, Elsevier.

RMSE series keep the same levels approximately. Evapotranspiration is a complex process containing the soil water evaporation, plant transpiration, and canopy free water evaporation, and these processes are dominated by many parameters and variables. In this study, however, we only focus on the curve number (CN_2) estimation that dominates the surface runoff generation and soil water balance (Eqs. (6)–(9)) and indirectly influences the ET processes. In order to diagnose the performance of ET estimation, other parameters related to these processes may be needed under this assimilation condition.

Although both parameter and state estimations succeed, a concern may arise that the location of runoff measurement and the number of observations can dramatically impact the data assimilation performance. To this comment, an illustrative analysis has been done and it suggested that the observations from lower stream stations, especially from the outlet, will facilitate the data assimilation (Xie and Zhang, 2010). This trait attributes to the fact that the runoff from lower streams contains the full information about the hydrologic states of the upper regions.

In addition to the runoff observations, many other types of measurements can also be combined to estimate the hydrological state. Especially the hydrologic remote sensing provide useful information about the land surface conditions at large scales, e.g., the radiance, surface soil moisture, and ET. So assimilating these remote sensing data into the hydrologic modeling is a good attempt, and they can serve as complementary information when the runoff observations are inadequate. To test this idea, two cases are designed: one only assimilates the runoff observations from the outlet of the catchment; the other assimilates multiple measurements, i.e., five-time-step (at the 1st, 31st, 61st, 91st, and 121st time steps) data of soil water content are added to the single runoff observation. The detailed setup can be found in Xie and Zhang (2010).

Figure 14.8 gives final results of the two cases for the CN_2s estimates versus the true values, indicating that their matches are acceptable despite the non-uniqueness and nonlinearity between parameters and hydrological responses. To the three prognostic variables, moreover, the case of combining the five-time data of soil water content provides the best estimates among the cases mentioned above after 100 assimilation steps, even though its estimation accuracy for CN_2 is not comparable to the case with five-runoff observations (Figs. 14.7 and 14.9). In fact, the soil water moisture condition plays an important role in controlling the surface runoff generation and the ET processes, as expressed in Eq. (6) through Eq. (9).

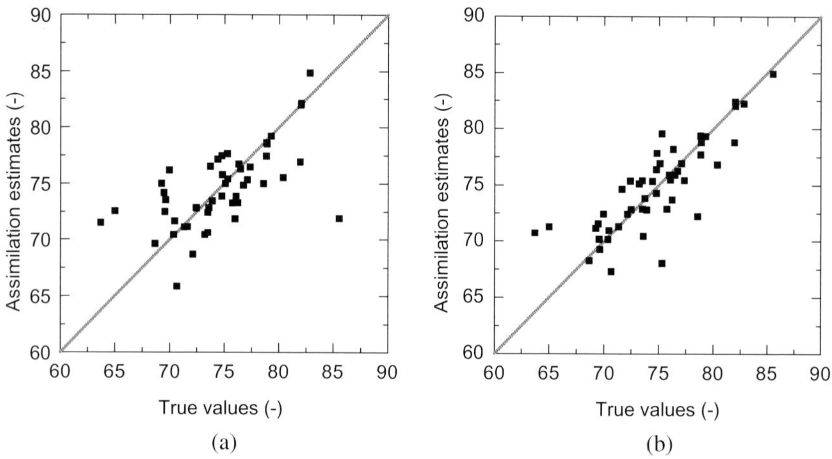

Fig. 14.8. Final results of the CN_2 estimates versus true (reference) values: (a) observations: runoff data from the outlet; (b) observations: runoff data from the outlet and the five-step soil water content (at the 1st, 31st, 61st, 91st, and 121st time steps). Adapted from Xie and Zhang (2010). Copyright 2010, Elsevier.

Therefore, the soil water content observation provides complementary data to the assimilation systems, even though it may not be available or utilized at every time step.

14.5 Discussion

EnKF is a Monte Carlo approximation based on a sequential Bayesian filtering process. Although the method has been successfully applied in a few cases for simultaneous state-parameter estimation, various aspects that will impact on accuracy and performance should be further examined before it is placed in operation.

First, the specification of model and observation errors is a dominant factor for the effectiveness of state-parameter estimation. A few methods have been developed to estimate modeling and observation error covariance information, such as the adaptive estimation methods (Crow and Reichle, 2008; Reichle *et al.*, 2008), the inflation covariance technique (Zheng, 2009), and the evolutionary algorithm-based method (Bai and Li, 2011). For comprehensive review, refer to the previous chapters by Crow (Chap. 7), Zheng *et al.* (Chap. 8), and Bai *et al.* (Chap. 9) in this book. We argue that most of them are far more practicable in the simultaneous state-parameter estimation, while the rough scheme mentioned above, using reasonable

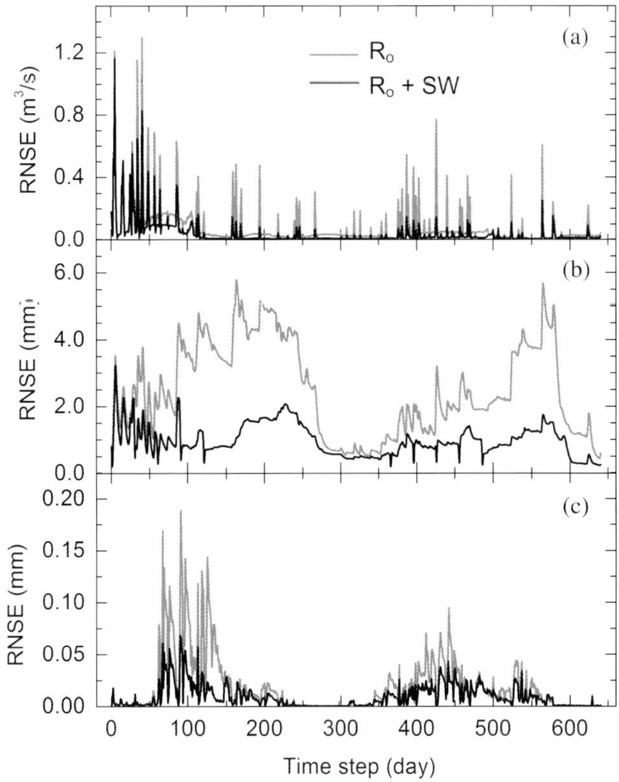

Fig. 14.9. RMSE series from assimilating the outlet runoff alone (R_0) and the combination: outlet runoff and five-time step soil water content ($R_0 + SW$): (a) stream runoff, (b) soil water content, and (c) evapotranspiration. Adapted from Xie and Zhang (2010). Copyright 2010, Elsevier.

scaling factors to approximate the model and observation errors, may be much more feasible for the assimilation proceeding. But three points should be noted: (1) biased specification of the error structure or magnitude will degrade the performance of data assimilation; (2) when the scaling factor scheme is used to represent the errors, it should be better to keep the magnitudes of the model and the observation errors at a consistent level. That is to say, both sources of errors should be overestimated at the same level. This point has been illustrated in Xie and Zhang (2010) and the chapter by Crow (Chap. 7). And (3), it would be better to overestimate rather than underestimate the magnitude in the case of poor knowledge about the errors (Clark *et al.*, 2008).

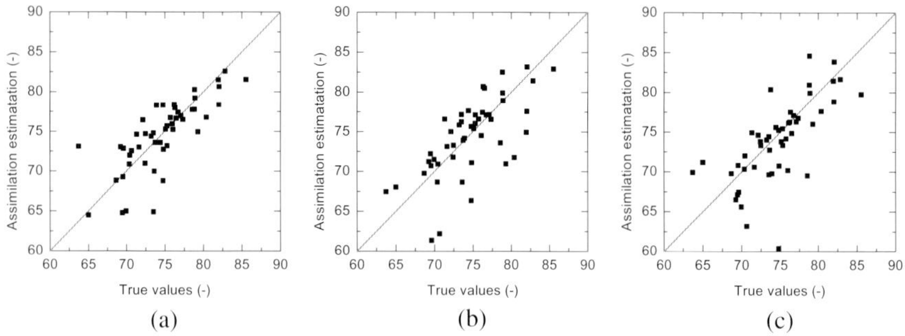

Fig. 14.10. Final results of the CN_2 estimates versus true (reference, $N(75, 5.0^2)$) values: (a) $N(70, 7.0^2)$, (b) $N(65, 7.0^2)$, and (c) $N(65, 5.5^2)$. Adapted from Xie and Zhang (2010).

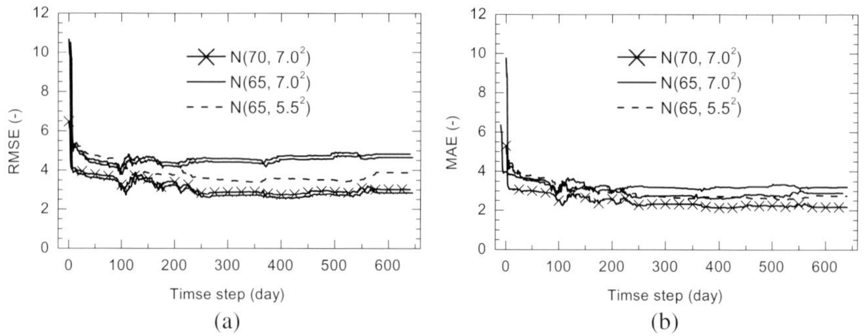

Fig. 14.11. (a) RMSEs and (b) MAEs of the CN_2 estimations under different initial guesses. Adapted from Xie and Zhang (2010).

Second, the initial sampling of model parameters is a necessary step for EnKF, but the setting of their distributions shows slight impact on the effectiveness of the data assimilation. To demonstrate this point, we conducted several experiments with different distribution for initial sampling of parameters based on the experiments in Sec. 14.4, and detailed information can be refereed to Xie and Zhang (2010). Figure 14.10 shows three cases ((a), (b), and (c)) for final estimates of parameters which match the true values with high correlation. Even using large bias to the true values (Case (b), $N(65, 7.0^2)$), the final estimates still capture the major trend of parameter distributions. Figure 14.11 exhibits their global estimation RMSEs and global mean absolute errors (MAEs), which indicate the close estimates of parameters. In addition, compared to Case (c) ($N(65, 5.5^2)$), Case (b), with larger initial biases, exhibits less RMSE and MAE within the beginning 90 assimilation steps, and even its final errors are not comparable to that of Case (c). So we roughly suggest that it is preferable to enlarge the magnitudes of the variances if the prior means are far away from the truths (i.e., with poor prior knowledge), especially for a short-period assimilation. This may be reasonable because wide range distributions of the initial realizations are beneficial for broadly covering true values and making each realization approach truths by assimilating observations.

Third, we shall pay attention to the ensemble size. Increasing the ensemble size enables the algorithm to propagate the error information more accurately, although it will pose heavy computational burden. So a balance should be made between the two ends, and it obviously depends on the problem of interest. But an empirical suggestion is that 200 realizations are good enough for watershed hydrologic modeling and geologic formations (Chen and Zhang, 2006; Xie and Zhang, 2010). Owing to the finite number of realizations used in EnKF, a doubt arises that the different ensemble of realizations may incur different performance of the data assimilation. To ease this doubt, we conducted an experiment using a number of different ensembles by sampling different ensembles. The result showed that there is a certain RMSE confidence interval for the random sampling, so it is a good application to try different random realizations and compare their results when we conduct data assimilation with the EnKF. Although this uncertainty exists, it should be emphasized that the EnKF-based state-augmentation technique is robust for the simultaneous state-parameter estimation, and the effect of random sampling is negligible. Generally, this simultaneous state-parameter estimation technique might be more efficient than traditional parameter calibration methods, e.g., the SCE-UA, since it employs the median ensemble size (say 200). More importantly, it holds a potential of assimilating relative short-term data (about 300 steps in the experiments shown in this chapter) to estimate the dynamic states and parameters as exhibited in Figs. 14.6 and 14.11. However, the traditional parameter calibration methods requires large amount of model evaluations (running the models), typically over 10^3 for hydrologic model calibration, and the data series should be long enough to get optimal estimates of parameters, e.g., approximate 8-year data for conceptual rainfall-runoff models (Yapo *et al.*, 1996). So these attractive advantages favor the application of the simultaneous state-parameter estimation in hydrologic modeling.

Despite these successes, the findings of this study should not be considered general. The simultaneous state-parameter estimation method mentioned in this chapter essentially is a rough-state augmentation scheme and it may not be feasible for multi-type parameter estimation, especially with high-dimensional parameter space because the rough-state augmentation technique still suffers from spurious or incorrect correlation between states and parameters, which will directly spoil the parameter estimation during data assimilation. This is basically due to biased model error quantification and large freedom for high-dimensional vectors of the augmented state. As the successful experiments exhibited in Sec. 14.4, there is a single type of parameters (i.e., the CN_2) that was selected to be estimated within the data assimilation framework. If we add another parameter to the extended state vector (e.g., the *esco* which is an ET-related parameter in SWAT, see Di Luzio *et al.* (2001)), the performance of the data assimilation may not as acceptable as that shown. The reader may recall the successful estimation for two types of parameters of a simple rainfall-runoff hydrologic model in Sec. 14.3, one highly relating to surface runoff (parameter, M) and the other determining the sub-surface runoff (parameter, K). Note that this model is so simple to be expressed or approximated with a linear function which will favor the data assimilation dramatically. A few studies presented acceptable state-parameter estimation by adding small random perturbation for the propagation of parameters (e.g., Wang *et al.*, 2009) or by using a Kernel smoothing of parameter samples (Liu, 2000; Moradkhani *et al.*, 2005). But these strategies are far more to address the spurious or incorrect correlation representation. Alternatively, Xie and Zhang (2013) proposed a partitioned update scheme in which the parameter set is partitioned into several types according to their sensitivities and each type of sensitive parameters is estimated in an individual loop by repeatedly assimilating observations. Here we intend not to introduce this scheme.

14.6 Conclusion

Data assimilation methods, especially the KF and its variants, have been widely used in hydrologic community, but they are mainly applied for dynamic state estimation, such as soil moisture. In this chapter, we extend the capacity of EnKF for the simultaneous state-parameter estimation by a state-augmentation technique. In this technique, the model parameters are appended to the state vector and are updated along with the dynamic states based on the covariance between the parameters and the dynamic states (or the hydrologic response, i.e., the runoff, ET). This technique is evaluated with a simple two-parameter hydrologic model and a distributed hydrologic model.

We conclude that the EnKF-based state augmentation technique is capable of estimating model parameters and dynamic states simultaneously. The performance of this technique is dominated by the model and observation errors quantification, but we provide an operational suggestion that consistent scale factors can be used to estimate the uncertainties of the model and observation errors. The initial setting for sampling parameter realizations is not an important factor but it is better to give large variances for parameters to obtain initial broad spread of the parameter distribution. Moreover, a balance with respect to the ensemble size should be achieved between the computational cost and the efficiency of the EnKF-based assimilation, a median size (say 200) is often sufficient for hydrologic problems. In addition, we shall caution that this rough-state augmentation technique is adept at single-type parameter estimation. For multi-type parameter, the dual state-parameter estimation (Moradkhani *et al.*, 2005) or the partitioned update scheme (Xie and Zhang, 2011) may be better alternatives.

Astute readers may notice that many indefinite words like "*may*" and "*should*" appear in this chapter. This choice of words is intentional because data assimilation is still an active area in hydrology and there is much work to be done before it is used in operation to characterize the hydrologic processes in different scales. First, the data assimilation methodology itself needs much improvement, and particular attention should be paid to the modeling and observation error estimation. Chapter 7 (by Crow), Chap. 8 (by Zheng *et al.*), and Chap. 9 (by Bai *et al.*) have been dedicated to this issue. Moreover, the strategy of assimilating various observation types (e.g., the remote sensing data and *in situ* measurements) is quite necessary to hydrologic system identification, parameter estimation, and state estimation, and ultimately to maximally reduce uncertainty in hydrologic predictions (Liu and Gupta, 2007). The improvement is motivated by the fact that data assimilation methodology is effective to favor the combination of the remote sensing technology and hydrologic modeling which is being sophisticatedly developed in the hydrologic community.

Acknowledgments

This work was supported by grants from the National Natural Science Foundation of China (No. 50911130509) and the Fundamental Research Funds for the Central Universities (No. 105562GK).

References

Bai, Y. and Li, X. (2011). Evolutionary algorithm-based error parameterization methods for data assimilation, *Monthly Weather Review*, **139**(8), 2668–2685.

Beven, K. and Binley, A. (1992). The future of distributed models: Model calibration and uncertainty prediction, *Hydrological Processes*, **6**(3), 279–298.

Beven, K. and Freer, J. (2001). Equifinality, data assimilation, and uncertainty estimation in mechanistic modelling of complex environmental systems using the GLUE methodology. *Journal of Hydrology*, **249**(1–4), 11–29.

Burnash, R., Ferral, R. and R. McGuire, R. (1973). A generalized streamflow simulation system: Conceptual modeling for digital computers, US Department of Commerce, National Weather Service and State of California, Department of Water Resources, Department of Water Resources, Sacramento, CA.

Chen, Y. and Zhang, D. (2006). Data assimilation for transient flow in geologic formations via ensemble Kalman filter. *Advances in Water Resources*, **29**(8), 1107–1122.

Chu, W., Gao, X. and Sorooshian, S. (2010). Improving the shuffled complex evolution scheme for optimization of complex nonlinear hydrological systems: Application to the calibration of the Sacramento soil-moisture accounting model. *Water Resources Research*, **46**(9), W09530, doi: 10.1209/2010WR00922.

Clark, M. P., Rupp, D. E., Woods, R. A. Zheng, X., Ibbitt, R. P., Slater, A. G., Schmidt, J. and Uddstrom, M. J. (2008). Hydrological data assimilation with the ensemble Kalman filter: Use of streamflow observations to update states in a distributed hydrological model. *Advances in Water Resources*, **31**(10), 1309–1324.

Crow, W. T. and Reichle, R. H. (2008). Comparison of adaptive filtering techniques for land surface data assimilation. *Water Resources Research*, **44**(8), W08423.

Di Luzio, M., Srinavasan, R. and Arnold, J. (eds.) (2001). Arcview interface for SWAT2000: User's guide. Blackland Research Center, Texas Agricultural Experiment Station, Temple, Texas.

Duan, Q., Sorooshian, S. and Gupta, V. (1992). Effective and efficient global optimization for conceptual rainfall-runoff models. *Water Resources Research*, **28**(4), 1015–1031.

Duan, Q., Gupta, V. and Sorooshian, S. (1993). Shuffled complex evolution approach for effective and efficient global minimization. *Journal of Optimization Theory and Applications*, **76**(3), 501–521.

Evensen, G. (2003). The ensemble Kalman filter: Theoretical formulation and practical implementation. *Ocean Dynamics*, **53**(4), 343–367.

Gassman, P., Reyes, M., Green, C. and Arnold, J. (2007). The soil and water assessment tool: Historical development, applications, and future research directions, *Transactions of the American Society of Agricultural and Biological Engineers*, **50**(4), 1211–1250.

Immerzeel, W. and Droogers, P. (2008). Calibration of a distributed hydrological model based on satellite evapotranspiration. *Journal of Hydrology*, **349**(3–4), 411–424.

Liu, F. (2000). Bayesian time series: Analysis methods using simulation-based computation Ph.D Thesis, Institutes of Statistics and Decision Science, Duke University.

Liu, Y. and Gupta, H. V. (2007). Uncertainty in hydrologic modeling: Toward an integrated data assimilation framework. *Water Resources Research*, **43**(7), W07401, doi: 10.1209/2008WR007401.

Moradkhani, H., Sorooshian, S., Gupta, H. V. and Houser, P. R. (2005). Dual state-parameter estimation of hydrological models using ensemble Kalman filter. *Advances in Water Resources*, **28**(2), 135–147.

Muleta, M. K. and Nicklow, J. W. (2005). Sensitivity and uncertainty analysis coupled with automatic calibration for a distributed watershed model. *Journal of Hydrology*, **306**(1–4), 127–145.

Neitsch, S., Arnold, J., Kiniry, J., Williams, J. and King, K. (2001). Soil and water assessment tool theoretical documentation, Version 2000. Grassland, Soil and Water Research Laboratory, Temple, Texas.

Pauwels, V. R. N. and De Lannoy, G. J. M. (2006). Improvement of modeled soil wetness conditions and turbulent fluxes through the assimilation of observed discharge. *Journal of Hydrometeorology*, **7**(3), 458–477.

Reichle, R. H., Crow, W. T.and Keppenne, C. L. (2008). An adaptive ensemble Kalman filter for soil moisture data assimilation. *Water Resources Research*, **44**(3), W03423.

Rouhani, H., Willems, P., Wyseure, G. and Feyen, J. (2007). Parameter estimation in semi-distributed hydrological catchment modelling using a multi-criteria objective function. *Hydrological Processes*, **21**(22), 2998–3008.

Shen, Z., Hong, Q., Yu, H. and Liu, R. (2008). Parameter uncertainty analysis of the non-point source pollution in the Daning River watershed of the three gorges reservoir region, China, *Science of the Total Environment*, **405**(1–3), 195–205.

Sorooshian, S., and Gupta, V. K. (1983). Automatic calibration of conceptual rainfall-runoff models: The question of parameter observability and uniqueness. *Water Resources Ressearch*, **19**(1), 260–268.

Vrugt, J. A., Gupta, H. V., Bouten, W. and Sorooshian, S. (2003). A shuffled complex evolution metropolis algorithm for optimization and uncertainty assessment of hydrologic model parameters. *Water Resources Research*, **39**(8), 1201, doi: 10.1209/2002WR001642.

Wang, D., Chen, Y. and Cai, X. (2009). State and parameter estimation of hydrologic models using the constrained ensemble Kalman filter. *Water Resources Research*, **45**(11), W11416.

Wang, S. (2011). *Studies on Parameter Estimation Methods for Hydrological Model and Associated Uncertainties*: China, Yellow River Water Conservancy Publisher.

Xie, X. and Zhang, D. (2010). Data assimilation for distributed hydrological catchment modeling via ensemble Kalman filter. *Advances in Water Resources*, **33**(6), 678–690.

Xie, X. and Zhang, D. (2013). A partitioned update scheme for state-parameter estimation of distributed hydrologic models based on the ensemble Kalman filte. Submitted to *Water Resources Research*.

Yapo, P. O., Gupta, H. V. and Sorooshian, S. (1996). Automatic calibration of conceptual rainfall-runoff models: Sensitivity to calibration data. *Journal of Hydrology*, **181**(1–4), 23–48.

Zheng, X. G. (2009). An adaptive estimation of forecast error covariance parameters for Kalman filtering data assimilation. *Advances in Atmospheric Sciences*, **26**(1), 154–160.

Index